ZEPPELIN BLITZ

Zeppelin caught in the searchlights during a raid on London, 1915.

ZEPPELIN BLITZ

THE GERMAN AIR RAIDS ON GREAT BRITAIN DURING THE FIRST WORLD WAR

NEIL R. STOREY

The
History
Press

Zeppelin crossing the North Sea to bomb Britain, with the cruiser *Ostfriedland* in the foreground.

First published 2015

The History Press
The Mill, Brimscombe Port
Stroud, Gloucestershire, GL5 2QG
www.thehistorypress.co.uk

British Library Cataloguing in Publication Data.
A catalogue record for this book is available from the British Library.

ISBN 978 0 7509 5625 3

Typesetting and origination by The History Press
Printed in Malta by Melita Press

CONTENTS

AUTHOR'S FOREWORD

… Came the Thing they were sending us, the Thing that was to bring so much struggle and calamity and death to the earth. I never dreamed of it then as I watched; no one on earth dreamed of that unerring missile.

H.G. Wells, *War of the Worlds* (1897)

Since the first flight of Count Ferdinand von Zeppelin's LZ-1 over Lake Constance (Bodensee) on the Rhine on 2 July 1900, the vast size (128m long) of these lighter-than-air craft prompted feelings of awe and, as the reality of their military potential was realised, they were increasingly regarded as an ominous presence in the sky, especially by people in the European countries surrounding the German Empire.

In a world which had witnessed unthinkable progress in industry and engineering throughout the nineteenth century, there were those who feared the technology of military development had gone too far. Authors contemplated the machinery of war with fascination, the most enduring of these being H.G. Wells' *The War of the Worlds*, in which an invasion and brutal occupation by extra-terrestrials was unstoppable by even the most modern of our weaponry. Add to this the tenor of over sixty books and numerous magazine articles describing invasions of Great Britain by foreign powers published between 1871 and 1914. The seminal *Battle of Dorking* by George Tomkyns Chesney, originally published as a story in *Blackwood's Magazine* in

1871, tells of a fictional German attack and landing on the south coast of England that had been made possible after the Royal Navy had been distracted in colonial patrols, and the army by an insurrection in Ireland – a situation that was to have resonances in the Zeppelin air raid and bombardment of Lowestoft on the evening of 24/25 April 1916.

The distrust and perception of anti-British feeling and the sinister machinations of the German race continued to feature regularly in the popular press, and books too, such as *The Riddle of the Sands* (1903), in which Erskine Childers weaves a tale of two young amateur sailors who battle the secret forces of mighty Germany. Their navigational skills prove as important as their powers of deduction in uncovering the sinister plot that looms over the international community.

Anglo-German relations were strained further after the launch of the revolutionary HMS *Dreadnought* – the first big-gun, turbine-driven, iron-clad battleship – in 1906, when the Germans started building their own warships, akin to and rivalling the pride of the British fleet. To many British observers this meant nothing less than a challenge to the maritime supremacy of Great Britain. This fire was further kindled by Anglo-French journalist and writer William le Queux, who produced such best-selling titles as *The Invasion of 1910* (1906) and *Spies of the Kaiser* (1909). He made no secret that his books were fiction, but claimed that they were based on his own secret knowledge and the insight he gained from his connections within the European intelligence community.

In the spring of 1909, newspapers had columns of reportage about the advances, and planned advances, of the parameters of manned flight. In May 1909, John Moore-Brabazon became the first resident Englishman to make an officially recognised aeroplane flight in England. The previous year, Alfred Harmsworth (Baron Northcliffe), proprietor of the *Daily Mail* newspaper, had put up a reward of £1,000 for the first flight across the English Channel, a feat which was achieved by Louis Bleriot on 25 July 1909. So, no aircraft had crossed the English Channel before July 1909, but after reports of successful distance trials of Zeppelins in Germany, in the climate of anti-German fears many considered that the purpose of these monstrous creations could only be a sinister one and worried how long it would be before they appeared over Britain.

Curiously, in March 1909, Police Constable James Kettle was walking his beat on Cromwell Road in Peterborough, Cambridgeshire, when, upon hearing the 'steady buzz of a high-powered engine' was caused to look up, where he saw a 'bright light attached to a long oblong body outlined against the stars as it crossed the sky at high speed.' His report was met with interest from national newspapers, but no doubt fearing the panic such accounts could create, a senior police officer was sent to front the explanations. Kettle, it was claimed, had simply been 'mistaken'; the whirring he had heard was attributed to the noise from the nearby co-operative bakery, and the object in the sky was explained as a Chinese lantern attached to 'a very fine kite flying over the neighbourhood of Cobden.' The speed Kettle had mentioned was just brushed off as 'a little poetic touch for the benefit of you interviewers.'

However, more sightings of these mystery airships began to occur across the country, and by May 1909 newspapers across Britain were full of accounts of unexplained airships which had been seen and heard traversing the night sky over Great Britain, at locations as far apart as Wales and Suffolk. In one notable instance, on Friday 7 May 'a long sausage-shaped dirigible balloon' was spotted over New Holland Gap, 1½ miles from Clacton in Essex. On a spot over which the airship had passed was found what was described as 'a stout ovoid dark grey rubber bag, between 2ft and 3ft in length, enclosed in a network mesh, with a stout steel rod passing through the centre of it and projecting about 1ft from each end. One end of the rod is capped with a steel disc resembling a miniature railway-waggon buffer.' It was thought that the object was some sort of 'fender' designed to break the contact of a descending aerial machine with the earth – the bag was stamped 'Muller Fabrik Bremen.'

On 14 May, the *Daily Express*'s Berlin correspondent had reported:

… It is admitted by German experts that the mysterious airship which has been seen hovering over the eastern coast of England may be a German airship. England possesses no such airship, and no French airship has hitherto sailed so far as the distance from Calais to Peterborough. On the other hand, the performance of several German airships, including the *Gross* airship, which has made one voyage of thirteen hours, would render it possible for them to reach the English coast. At the same time it is improbable that the German airship seen above England ascended from German soil. An aerial voyage to the English coast would still be a dangerous and formidable undertaking even for the newest airships …

However, by this date sightings of mystery airships had been reported from Ely, St Neots, Wisbech, Peakirk, Orton and Wingland in Cambridgeshire; Ipswich, Saxmundham, Bradfield St George and Woolpit in Suffolk; Southend-on-Sea in Essex, and even Sandringham, where it was said the mystery airship had been spotted by royal servants. The witnesses generally agreed that the mystery airship was cigar-shaped, at least 100ft in length, the engines made a 'throbbing' noise and that it appeared to perform its manoeuvres with ease. Some also claimed to have seen 'the glare of its searchlight.'

On Monday 17 May, the subject of the mystery airships was brought up in a debate in the House of Commons. Sir Arthur Fell, MP for Great Yarmouth, had asked Richard Burdon Haldane, the Secretary of State for War, if he could give the numbers of dirigibles, either constructed or in the course of construction, by Germany. Mr Haldane replied that seven dirigible airships had been built, and another five were under construction, with more than £100,000 being earmarked specifically for the craft in 1908. Mr Horatio Myer (MP for Lambeth North) followed up by asking Haldane: 'Will the honourable gentleman, in any report he may circulate, tell us about a certain dirigible supposed to be hovering about our coast?' This question was greeted with laughter, and no reply was received.

And there the sightings of mystery airships over Britain ended as abruptly as they had begun. Most of the local and the serious national press appeared to have tired of the stories, and even spurned them. The *Yarmouth Mercury* commented that, since the first 'flittings of a mysterious airship over the eastern Counties at night time … the halfpenny London press, hungry for sensation, have worked up a very fair scare.'

This climate of fear was revisited in October that same year by the twenty-minute film *Der Luftkrieg der Zukunft*, also entitled *The Battle of the Clouds*, directed by Walter R. Booth, in which a squadron of airships conduct a bombing raid upon Britain, firing a town with incendiary bombs, wreak havoc in the countryside and bring down an aeroplane which had tried to shoot one of the airships down, and are only foiled by the launch of a missile that had only recently been perfected by a jobbing inventor. *The Battle of the Clouds* is now regarded by many to be the very first film of the science fiction genre.

Many of the questions raised by these mystery airship sightings still remain, over 100 years later, and those from 1909, before the Channel had been officially crossed by any flying machine, are by far the most enigmatic. No evidence has yet been uncovered, in the German archives or from British Intelligence, to prove that covert missions had been undertaken by Zeppelins over Britain under the cloak of darkness in 1909. So, what did the people see? It is clear that far more people saw the airships and lights than just those who gave their names and stories to the newspapers. Could all these witnesses really have been mistaken, or deluded? Perhaps Britain had been gripped by a mass panic or, tantalisingly, was there more to the mystery 'scare ships' seen over Britain in those spring skies of 1909?

Whether Zeppelins made their appearance over Britain before this time or not, fears of offensive attacks by aircraft had already been voiced in government circles. As a result, in 1908 Prime Minister Herbert Asquith had approved the formation of an Advisory Committee for Aeronautics and an Aerial Sub-Committee of the Committee of Imperial Defence to investigate the matter. Both committees were composed of politicians and officers selected from the navy and the army.

Then there was the Aerial Navigation Act 1911, passed when Winston Churchill was Home Secretary, which allowed prohibition orders to be issued over certain areas for certain events or occasions. For example, one of the first orders issued under the Act prohibited the navigation of aircraft of every description over any place within 4 miles of Norwich during King George V's visit to the city for the annual Norfolk Agricultural Association Show there on 28 June 1911. The problem was the Act had no teeth; admittedly the punishment for such an infringement was a hefty £200 fine or imprisonment for six months (or both) for aviators caught flying over a prohibited area, but it did not outline guidelines for a military reaction to such a situation. Such guidelines only came in January 1913, after a new flurry of sightings of 'scare ships' across Britain – in Yorkshire, Norfolk, Kent, and Wales, among other places – in 1912, when the Aerial Navigation Bill, to amend the 1911 Act, was introduced to the House of Commons by Colonel John Seely, Secretary for War. This gave the Secretary of State powers to prohibit aircraft from flying over certain areas, which could include the whole of the coastline and the 'territorial waters adjacent thereto' (i.e. within three miles of the coast) and, significantly, if an aircraft flew over a proscribed area or failed to comply with the landing conditions a signal was to be given by 'the officer designated for the purpose'. If the aircraft still failed to comply it would 'be lawful for the officer to fire at or into such aircraft and to use any and every other means necessary to compel compliance'.

By 1913 Churchill had been appointed First Lord of the Admiralty and was only too aware of the threat that aerial warfare could present in any future war. He had already created an Air Department within the Admiralty in 1912 and had appointed Captain Murray Sueter as director to oversee the creation of a new branch – The Royal Naval Air Service (RNAS). Churchill actively involved himself in the debates on how to counter the Zeppelin menace, producing the Admiralty report *Aerial Defence*, published in January 1913. He believed that our aeroplanes could potentially deal with Zeppelins in the air if we had enough of them; that rifle fire from a battalion of troops could be effective in keeping an offensive airship high enough that it would have difficulty aiming its bombs; and he raised questions about how fast British

Howitzers could be deployed in an anti-aircraft role. On the outbreak of war Churchill proposed that Zeppelins should be attacked on the ground when they were vulnerable and suggested that a pre-emptive strike on the Zeppelin sheds would reduce or remove the possibility of air raids on Britain.

Features appeared during the opening weeks of the war in newspapers and magazines with such alarmist titles as 'The Zeppelin Menace' or 'How Zeppelins might Threaten Great Britain and Ireland', which included illustrations of maps with overlays showing the range of Zeppelins.

Britain suffered the first direct attacks upon its soil in modern warfare before the end of 1914. Bombardments by scouting groups of the Imperial German Navy were carried out first on Great Yarmouth and Gorleston, on 3 November 1914 – the shells fell short of the shore there, but the bombardments of Scarborough, Hartlepool and Whitby, on 16 December 1914, resulted in 137 fatalities and left 592 injured, most of whom were civilians. A total of three bombs had also been dropped by lone German aeroplanes; the first falling on Dover on 24 December, the next two on Cliffe, in Kent, on Christmas Day.

The media hype was growing, and hit its apogee when the first air raids were actually carried out by these long shadowy monsters, on the night of 19 January 1915. Zeppelins had been well established in the minds of the British populace as a terror weapon that rained death from the skies. In the wake of the raid, a barrage of press coverage was unleashed, proclaiming the attack an outrage and branding the Zeppelins as aerial 'baby-killers' (although no babies were actually killed in England during the first raid). Further scaremongering, and even fear of collaborators guiding the Zeppelins from the ground with lights and motor cars, ensued.

Fears of panic, and the effect of these and any further raids on morale, led the government to take stringent measures to minimise or censor the reportage of most future air raids. Formal action was taken after the Harwich Garrison Commander, Brigadier General C. Reginald Buckle, wrote a detailed formal complaint to his superior, the Lieutenant General Sir Charles Wollcombe, Commander-in-Chief, Eastern Command on 30 April 1915. After suffering an air raid the press had reported that his military installation had been attacked and that the Germans had intercepted the information, he added. 'I deem the matter one of urgency and as no useful purpose is served by the publication of these facts, and possibly very much damage may be done thereby, I suggest that steps should be taken to terminate it forthwith.'[1]

General Staff Colonel Arthur Warre Elles forwarded the letter to Horse Guards for Commander-in-Chief, Eastern Command, adding 'The general public is really not interested in exact details of time of attack and route of the hostile craft. The local people are in possession of the information at first-hand, and do not need to be informed by the press.'[2]

Military Intelligence had already considered the question and concluded that 'press reports are no real help to the enemy'[3] but the characterful General William Sefton Branker, Director of Military Aeronautics, stated that although he did not consider the press reportage much help to enemy intelligence:

> I see no harm in preventing carefully worked out itineraries appearing in the press; their suppression will give the Iron Cross hunters of substantiating their claims of having bombarded a fortress and obtaining their coveted distinction; but any attempt at complete suppression will surely give rise to all sorts of disquieting rumours and demands for the truth.[4]

As a result, a 'D-Notice' was issued on 1 June 1915.

The D-Notice scheme, introduced in 1912, was run as a voluntary system by a joint committee headed by an Assistant Secretary of the War Office and a representative of the Press Association whereby the request was only advisory and as such was not legally enforceable. News editors could choose not to abide by a D-Notice, but it was very rare such a request from such an authority was not complied with and if it was breached the displeasure of the War Office would be sent in letter form and an explanation requested from the editor. The wording was simple:

> In the public interest and to prevent the publication of information useful to the enemy, it has been decided nothing may appear in the Press in regard to raids by enemy aircraft except the official statements issued by the Government. Newspapers which advertise Insurance against air risks should take care to avoid publication of addresses of claimants or other details which would indirectly afford information of places attacked. The publication of photographs illustrating damage done by enemy bombs is prohibited.

As a result, exact locations were not named in the British newspapers; neither were casualties. Reports from the coroner's courts and funeral notices that would usually be published as a matter of course in newspapers were prohibited in the case of victims of air raids. As a consequence the details of many air raids simply did not appear in newspapers and have not appeared in books on the subject, and many casualties have remained numbers rather than names.

Despite the restricted news coverage, the feelings of hatred towards those carrying out the raids by those who fell victim to their bombs or suffered the dread a bomb might just be dropped on them as the Zeppelins passed over, continued unabated. The problem was that to those who witnessed British anti-aircraft defences firing shells and pom-pom rounds from the ground, or the efforts of aircraft equipped with machine guns and bombs, they appeared so tiny and so impotent in comparison to the huge Zeppelins that they were only too aware that we were, initially, unable to defeat them. For many, the echoes of H.G. Wells' *War of the Worlds* were becoming only too real, and people gathered around the potentially lethal unexploded bombs dropped by Zeppelins, or the craters left by those that exploded, and regarded them with such curiosity that they may as well have fallen from Mars.

Many of those who lived in the East End of London through both the bombings by Zeppelins in the First World War and the Blitz during the Second World War would recall how the Zeppelin raids were far more frightening than the raids of the Luftwaffe. Whereas the German raiders of the Second World War would drop many more bombs, they would keep on flying over, while the Zeppelins would simply hang in the air dropping bomb after bomb with impunity.

And so the situation remained; especially as Britain's military did not have a comparable terror weapon nor an effective means of retaliation until Brock, Pomeroy (explosive bullets) and Buckingham (phosphorous incendiary bullets) ammunitions were developed and used to form a lethal combination against Zeppelins. Indeed, it was with the combination of Brock and Pomeroy, used by RFC Lieutenant William Leefe Robinson VC to bring down SL-11 on 3 September 1916, that the offensive was finally turned against the Zeppelin raiders.

From 1915–18, Imperial German Zeppelins and airships made fifty-one bombing raids over the English, killing 557 and injuring 1,358. More than 5,000 bombs were dropped on towns across Britain, causing £1.5 million in damage.

Both Captain Joseph Morris in his book *The German Air Raids on Great Britain 1914–1918* (1925) and H.A. Jones in his official history *War in the Air: The story of the RAF in the Great War* (Volume III, 1931) drew information from the various *Reports of the Intelligence Section General Headquarters: Air Raids*, but it is clear there were constraints imposed upon them about what details they could or could not include because the reports were classified 'Secret' and remained so until their formal release in 1966. Some aviation historians have subsequently drawn on parts of the published reports in relation to particular raids, but few have looked at the original files in detail and many are unaware of their existence at all; as a result many fascinating aspects and details of the story of the Zeppelin air raids have remained untold to a wider audience, until now.

The reports were compiled from information obtained through wireless interception stations, intelligence sources, and from direct observation by anti-aircraft units, military observers and police reports, supplemented by the check afforded by the dropping of bombs. They were printed without any attribution to their author, but the man under whose direction they were collated and who wrote them for their limited audience during the First World War was Lieutenant Colonel Hermann Gaston de Watteville (1875–1963). Best known to military historians for his post-war biographies of Lord Roberts (1938), Lord Kitchener (1939) and 'The British Soldier' (1954), he was also editor of the journal of the Royal United Service Institution from 1924–35 and produced an authoritative account of the campaign in Waziristan 1919–20 (1925). It is less well known that he made contributions to a host of military publications throughout his long military career, often signed off simply as 'H.G. DE W.' and that he is acknowledged for his contributions to the 1911 *Encyclopædia Britannica*, which he made while he was an instructor at the Staff College, Camberley. He was also a good friend of John Buchan, author of the classic spy thriller *The Thirty-Nine Steps* (1915), from the time they met at Oxford, and both of them served in military intelligence during the First World War.

De Watteville's work within military intelligence remained, for the most part, secret. His obituary in *The Times* stated he served in the Royal Artillery from 1900 to 1924 and during the war was Mentioned in Despatches and made a CBE. It only mentioned that he served in the Political Intelligence Department of the Foreign Office during the Second World War. It took his friend Captain Cyril Falls, the author of a number of the military operations official campaign histories and numerous books on military history, to set some of the record straight in a tribute published a few days later, in which he quotes Brigadier General Sir James Edmonds as stating that de Watteville's intelligence work at the War Office before the First World War was 'outstanding'. During the war de Watteville had a reputation for speaking and writing French and German 'like a native' and was awarded the Legion d'Honneur by the French. Falls summed up de Watteville:

> As a talker he was one of an old school, witty, biting when a snap was deserved, above all polished and precise – in fact a delightful companion. Finally, he was the best critic of military history I have known. He read

before they went to press all my official volumes and most others on the First World War and never made a criticism that was unwanted.

So, one may ask, were the *Reports of the Intelligence Section General Headquarters: Air Raids* accurate? When Jones was researching for his *War in the Air* immediately after the First World War he visited Berlin, and through the courtesy of the president of the marine archive he studied the original reports and had conversations with wartime Zeppelin commanders. He also took the opportunity to compare notes with Reichsarchiv officials and found that the German airship reports confirm in general, but not in detail, the British air raid reports. He commented:

> The Zeppelin commanders had to face many difficulties. They set out on their long journeys with no reliable knowledge of the weather conditions over the British Isles. Slight miscalculations of the changes of wind strength and direction led to errors of navigation which are reflected in the places named as bombing targets in the commanders' reports. Nor was there much opportunity to check navigation by direct observation. It is clear that the airship crews were continuously baffled by the darkening of English cities. The German directional wireless stations were not of much help. The angle subtended by the bearings obtained when an airship made a wireless call for her position was, of necessity, narrow, and the resultant calculation was at best approximate, and, sometimes, definitely inaccurate.

The difficulties in navigation outlined by Jones also concur with those discussed by Admiral Reinhard Scheer in his chapter on airship attacks in *Germany's High Sea Fleet in the World War* (1920).

Zeppelin Blitz is the first published book to draw extensively on the *Reports of the Intelligence Section General Headquarters: Air Raids*. Much of the material is transcribed directly from the original texts, including de Watteville's pithy comments on the actions of the Zeppelin raiders and their commanders. The accounts drawn from the reports are combined with information recorded in Metropolitan, Borough and County police reports and a host of contemporary sources from my extensive research over the last twenty-five years in both public and private archives. Recorded are the flight paths of the raiders, where the bombs fell, the names of numerous casualties and the locations of anti-aircraft guns. There are also rare and previously unpublished photographs of bomb damage, Zeppelin crews and crash sites, and first-hand combat reports by both Royal Flying Corps (RFC) and RNAS pilots that vividly recount the engagements with Zeppelins, all of which combine to provide an authentic and revealing insight into the Zeppelin raids of the First World War like never before.

Neil R. Storey

Notes

1. TNA AIR 1-604-16-15-242 Air Raids on England: Miscellaneous p.4b
2. Ibid p.4a
3. Ibid Minute Sheet 3
4. Ibid Minute Sheet 2

Charles Hunt, chief constable of King's Lynn, examines an unexploded HE bomb dropped on King's Lynn during the night of 19 January 1915.

INTRODUCTION

The First Zeppelin Air Raid – 19 January 1915

While Britain's navy still ruled the waves, the coastal class airships, or 'blimps', and aircraft of the RFC and RNAS were fearfully inadequate in comparison to the German Zeppelins which were, in effect, the long-range heavy bombers of their day. Zeppelins could cross the North Sea from their bases in Germany and, flying far beyond the range of battleship shells or field artillery, they could literally take the fight to the enemy.

General von Falkenhayn, Chief of the Imperial German General Staff, was keen to commence combined operations against Britain, but his plans and airships failed to impress the Imperial German Navy chiefs – they had their own plans.

Admiral Hugo von Pohl, Chief of Naval Staff, sought an audience with the Kaiser to obtain sanction to conduct air raids on Britain. The Kaiser was initially against any such action but, after conference, a compromise of allowing bombs to be dropped only on specifically targeted military installations and docks was agreed.

During December and early January, a few airship reconnaissances pushed close to the English coast, and fears did grow of aerial bombing attacks being carried out by Zeppelins. These fears led to what may arguably be the first civilian casualty caused by Zeppelins (or at least the fear of their attacks).

On Friday 15 January 1915, Mrs Alice Mary Cubitt (50), the wife of Gorleston bank manager Henry Cubitt, was helping her neighbour, Dr David Walter, in the construction of a 'dugout' place of safety against air raids and bombardment, in his garden at 2 Park Road, Gorleston Cliffs, when the structure suddenly collapsed (probably due to the sodden soil caused by continued rainfall). This buried the pair under a large quantity of struts, boards and earth. Dr Walter managed to extricate himself, and as he did so, he called to his servants to summon the assistance of some of the military sappers at work on the cliffs, who came running with spades to release Mrs Cubitt, but she was found to be dead. Medical examination revealed that she had a weak heart and the shock of the collapse had killed her.

Four days later, on the morning of 19 January 1915, Zeppelins L-3, under the command of Kapitänleutnant Johann Fritz, and L-4, under Kapitänleutnant Count Magnus von Platen-Hallermund, set out from their base at Fuhlsbüttel with orders to bomb key installations along the River Humber. Both Zeppelins were carrying crews of sixteen men, eight 110lb explosive bombs, ten or eleven 25lb incendiary bombs and enough fuel for thirty hours' flying.

The two Zeppelins were to be led across the North Sea by L-6, which flew out of Nordholz carrying no lesser man than Peter Strasser, Chief of the German

Admiral Hugo von Pohl, Chief of Imperial German Naval Staff.

Naval Airship Division. L-6 had been given the 'prize' target of the Thames estuary. The mission of L-6 was, however, cut short when the crankshaft of the port engine broke while she was north-east of the Dutch island of Terschelling. It was still 90 miles to the English coast and, fearing that the weight of ice that might collect on the envelope of the Zeppelin would prove too much for her to carry with only two engines, Strasser recorded that he: '… decided in agreement with the commander of L-6, but with a heavy heart, to turn back.'

When reading what follows next, it is worth keeping in mind the words of Horst von Buttlar-Brandenfels, commander of the ill-fated L-6, and Zeppelin commander on numerous later raids, who recorded the problems of navigation during these early operations in *Zeppelins over England*:

Fregattenkapitän Peter Strasser, Chief of the Imperial German Naval Airship Division.

In those days we always flew as far as possible in sight of land in order to be able to determine our ship's position the more accurately and, above all, to have some means of checking the speed at which we were going. Our wind measurements, which we received before ascending, were extremely inadequate, and did not give us anything like a true idea of the conditions, more particularly as all observations regarding the west – and these were the most important and most valuable from our point of view – were altogether lacking. Whether the wind was increasing or abating could be determined only by ascertaining the position of the ship at various intervals, and these positions, as I have already pointed out, were fixed by keeping observation on the land. Later on, all this was entirely changed and we navigated our ships in accordance with wireless information.

Flying in sight of the Dutch coast had, of course, this disadvantage, that the ships were reported to England the moment they steered a westward course but, for the reasons above, we had to put up with this drawback. Moreover, in any case it was probably that English submarines, which were constantly busy in the North Sea laying mine barrages, reported the approach of the airships by wireless. Whether … we should succeed in reaching England or not was entirely dependent on whether our engines would be able to hold out.

L-3 and L-4 Zeppelins approached the Norfolk coast in company, passed over the *Haisborough* lightship and were first spotted from land at 7.40 p.m., flying low. Each 'carried a light' and were described by an observer at Ingham, 1½ miles from the sea, as 'like two bright stars moving, apparently 30 yards apart.' Before coming overland at 7.55 p.m., they separated. L-3, making landfall over Haisborough, turned south-east, and L-4 made for Bacton.

L-3 was spotted by a patrol of 6th Battalion, Norfolk Regiment (Cyclists) TF, as it passed over Eccles Gap. It then steered over Lessingham, Ingham (8.05 p.m.) and Martham (8.15 p.m.), where Kapitänleutnant Fritz attempted to find out where he was by dropping a flare. Mr Frank Gray, the landlord of the Royal Oak, happened to be looking out and saw the flare, which he described as looking like 'a big star' and was astonished at how the phenomenon 'stood out of the starless inkiness of the rest of the sky.' The casing from this flare was picked up in a lane near the hall at Martham, and was described in the *Yarmouth Mercury* as being 'about 2ft long and 3½in diameter. At one end, which was solid except for a sort of turn screw in the centre, were the figures 6, 12, 15, 18 at even distances round the edge and there were various German words on the tube itself.'

Shortly after 8.15 p.m. L-3 dropped its first bomb, an incendiary, on farmer George Humphrey's water-logged paddocks near St Michael's Church at Little Ormesby, leaving a small crater about 1½ft wide. Heading seaward to skirt the coastline, it was spotted south of Caister (8.22 p.m.) out to sea.

Zeppelin L-3, which, under the command of Kapitänleutnant Johann Fritz, dropped the first bombs during the first Zepplein air raid on Britain, 19 January 1915.

Turning almost immediately towards Great Yarmouth and spotting the town, L-3 dropped a parachute flare to illuminate its target. The people below believed they were being swept by a searchlight, and another small detachment of soldiers from the 6th Battalion, Norfolk Regiment (Cyclists), who were on coastal defence duty, opened up with rifle fire. Crossing the town from north to south, L-3 dropped its second bomb, another incendiary, at 8.25 p.m., which landed on the back lawn of Mr Norwood Suffling's house at 6 Albemarle Road, overlooking the Wellesley Recreation Ground. This bomb 'burst with a loud report' but did little damage, apart from gouging a 2ft crater and splashing mud up the house.

The third bomb, the first of the explosive bombs to be dropped, described in the *Yarmouth Mercury* as a 'diabolical thing', fell at the back of 78 Crown Road, narrowly missing one of its elderly occupants, Mrs Osborne, who, at the moment of impact, was crossing the small back yard to the back door. Still shaking as she spoke to the reporter, she said of the sound: 'It was like a big gun … If I had gone just a step or two further I must have been killed by it.'

On the morning after the raid, the bomb was dug out of the small crater it had made in the pavement by Norfolk National Reservists and taken to the York Road Drill Hall, where the detonation mechanism (including a small air propeller operating on a fuse) was removed. The defused bomb became the object of much interest and curiosity to the large number of people who came to see it through the day.

The fourth bomb, an incendiary, fell a few yards further west, failing to detonate, and it buried itself harmlessly against the gate post of Mr W.F. Miller's livery stables behind Crown Road.

The people of the St Peter's Plain and Drake's Buildings area of the town were not to be so lucky. Here landed the fifth bomb, a devastating high explosive (HE). Some of the windows of St Peter's Church and parsonage were blasted in, and the front of St Peter's Villas, the home of fish worker Mr E. Ellis, was brought down by the explosion. Luckily he was in the kitchen; the back door was blown off its hinges and fell on top of him, as did the kitchen window and sundry other wreckage, but he only suffered cuts from flying glass and debris, and he was thankful – only minutes before he had been in the room that took the full blast. He did suffer wounds severe enough to receive hospital treatment – a gash to his knee caused by the falling glass penetrated deep and caused him a lot of pain. He appears in several photographs, standing indignantly in front of his house with his head bandaged. Luckily, his wife and family were away in Cornwall, where Mr Ellis was soon to join them for mackerel fishing.

Opposite the St Peter's Villas were the premises of J.E. Pestell, builder and undertaker, which received the full force of the explosion and suffered such

The third bomb dropped by L-3, at the back of 78 Crown Road, Great Yarmouth. Taken on the morning after the raid, shortly after it had been dug out of the small crater it had made in the pavement by soldiers of the Norfolk National Reserve.

Injured but undaunted, Mr E. Ellis stands in front of the wrecked frontage of his home, St Peter's Villa, Great Yarmouth.

The bomb damage to St Peter's Plain viewed from Lancaster Road, Great Yarmouth.
Miss Martha Taylor was killed by Pestell's corner office window, near where the policeman and soldiers are standing to the right of the photograph.

extensive damage that they had to be demolished. Pestell also lived on the premises with his young family, and by some miracle they were all unhurt.

However, the first casualties from the first Zeppelin air raid were struck down close by. Martha Taylor, a 72-year-old spinster who lived at 2 Drakes Buildings with her twin sister, Jane Eliza, was returning from a trip to the grocer's shop on Victoria Road when the bomb landed on St Peter's Plain. Her body was discovered by Private Alexander Brown of the National Reserve who was on his way back from the Hippodrome when, turning the corner of St Peter's Plain, he stumbled over a 'bundle' outside Mr Pestell's corner office window. On closer examination he discovered it was the body of Miss Taylor, whose clothes had been blown to rags. He ran to the Drill Hall, informed Corporal Henry Hickling and they took a stretcher to where she lay. It was only when she was moved that it was revealed she had been badly injured in the lower part of her body, and part of her arm had been blown clean off and lay nearby.

Dr R.H. Shaw examined Miss Taylor's body at the hospital, and explained the extent of the injuries she had suffered at her inquest: 'The left side of the body was torn open from the hip to the shoulder and the organs dislodged. The right shoulder joint and bone and the right knee joint were broken. The ankle was also injured and the greater part of the left arm missing. Death was instantaneous.'

A second body, that of shoemaker Samuel Alfred Smith (53) was next reported. He had evidently been standing at the end of the passage in which the door of

Miss Martha Taylor, killed in the Zeppelin raid on Great Yarmouth, 19 January 1915.

his shop was situated, to watch the passing raider. The passage, located opposite St Peter's Villas, had a large double gate at the end, one half open and the other half closed. The bomb fell 30ft away from Mr Smith. Several fragments of the bomb blast had been hurled in his direction, the gate was badly peppered, and tragically Mr Smith was caught by some of the shrapnel, part of his head was torn away and his left thigh badly lacerated. Found lying in a pool of blood, it was clear he had stood no chance.

Other people close by were knocked off their feet by the blast, among them Mr W.J. Sayers and his 11-year-old son, Louis. They were just yards away from St Peter's Villa when the bomb went off. Mr Sayers got a 'rather nasty shaking', but his little boy received a flying fragment of glass in the shoulder. Mr Sayers said: 'We went down like a pair of shot rabbits … I feel I must thank God that we are still alive. Less than thirty seconds before the bomb fell we had hurried over the very pavement it pulverised. There would not have been even enough of us for an inquest had we been slower.'

William Storey and his family had recently occupied and furnished 17 St Peter's Plain; he had been in the kitchen with his wife and their two babies, one aged 2, the other 9 months, along with his sister and a female family friend. None of them heard anything until the explosion. Mr Storey explained: 'The gas went out, glass and doors flew in every direction. The women screamed but when we got a light I was relieved to find no one was hurt but we were all unrecognisable because of soot and dust.' They had a narrow escape; a large bomb fragment had 'carried the front window away', tore a hole through the stairway door and penetrated nearly 2ft into the solid bricks beyond.

The Reverend J. McCarthy, Minister of St Peter's Church, had just finished an intercession service and was talking with a parishioner when suddenly there was a crash, and the windows of the north side of the church fell into the aisle and the vestry door was blown open, bending a heavy lock back at right angles. He fled with his family to take refuge in the cellar. When they emerged they discovered that every window had been broken in the parsonage.

Mr Frank Burton, Clerk to the Guardians, had a very narrow escape from injury. He had been returning home from a meeting of the Guardians, and was passing Mays' butcher's shop on the corner of Victoria Road when the bomb burst on St Peter's Plain. He was protected by the New Royal Standard pub which received the force of the blast, losing its windows and receiving several flying fragments that damaged the walls – any one of which could have proved fatal if they had hit someone in their path.

One notable incident was the case of Private Poulter, a Territorial Army soldier from the Essex Regiment, who was leaving the lavatory near St Peter's Church when the bomb went off and was wounded by shrapnel in the chest. It was Dr Leonard Ley who had the distinction of being the first doctor to operate on

The wrecked premises of J.E. Pestell, builder and undertaker, on St Peter's Plain, Great Yarmouth. Samuel Smith was killed at the end of the passage to his shops near the large double gates.

Cobbler Mr Samuel Smith, killed in the Zeppelin raid on Great Yarmouth, 19 January 1915.

Some of the damaged windows of St Peter's Church, Great Yarmouth.

an air raid victim. After successfully removing the piece of shrapnel, he kept it and had it mounted as a tiepin, which he wore with great pride for years afterwards.

Harry Tunbridge, the manager of Britannia Pier, was a section leader in the 1st Yarmouth Voluntary Aid Detachment, British Red Cross Society, and was drilling with the other members of the detachment in the basement of Boots' stores on King Street when the caretaker rushed down and said, 'Bombs are being dropped in the town!' The detachment divided themselves into three sections, one going down The Drive, another direct to the hospital and the others scoured the area making enquiries as to anyone who had been injured.

Mr Richards, the detachment's pharmacist, accompanied Mr Tunbridge. They obtained a lamp and rapidly received a report of a man being injured on St Peter's Plain. Rushing to the spot, they found the body of Samuel Smith in the opening next to his workshop. Pharmacist T. Richards saw that Smith was past any treatment and, when a group of soldiers with a stretcher arrived with Inspector Crisp soon after, they removed the body to the mortuary. Members of the detachment then accompanied the police in their search for any wounded or unconscious people, and remained on duty through the night at Lady Crossley's Hospital, where Messrs Arnold's large covered motor car was in readiness for emergencies. The detachment was finally dismissed at 6 a.m. on Wednesday.

The sixth and seventh bombs fell almost simultaneously, the first crashing through the roof of a stable abutting Garden Lane, near South Quay, owned by butcher William Mays. Failing to detonate, the bomb was found resting on a truss of hay beside a pony the following morning – bomb and pony both intact! This bomb was also recovered by National Reservists, and was removed out to sea, sunk in 12 fathoms of water and exploded by a time fuse. This caused a great disturbance in the water and killed one fish – a 20lb cod – which showed its white belly on the surface and was brought ashore for a meal.

The eighth bomb fell with a 'huge, fiery flame', and landed opposite Messrs Woodgers' shop, near the First & Last Tavern on Southgate Road, near the Fish Wharf. There were no more casualties here, but a number had narrow escapes. The damage was confined to a number of broken windows, lots of spattering of 'some grey substance on the walls of the houses' (probably the accelerant from the incendiary), and a granite paving stone was fractured by the impact. Fragments from this bomb were soon on display in the pub.

The Zeppelin then appears to have passed along the edge of the river, dropping its ninth bomb which fell between two vessels, the drifter *Mishe Nahma*, undergoing repairs, and the pilot boat *Patrol*. It struck the river side of the dock gates of Beeching's South Dock, where it smashed through two planks causing it to flood on the tide. The bomb failed to detonate, and bounced off the stone quay of Trinity Wharf, narrowly missed a sentry from the National Reserve and the base of a crane turntable, before it fell harmlessly into the river.

The Zeppelin bomb that crashed through the roof of butcher William May's stable, which abutted Garden Lane, near South Quay, Great Yarmouth. Failing to detonate, the bomb was found resting on the hay beside the pony the following morning.

The tenth bomb fell into the 'swill' ground at the back of the Fish Wharf, blasted the water tower and made a large hole in the ground, fractured a water main and blew a nearby electric light standard to smithereens. Almost the entire glass roof of the wharf was smashed, and the fish sales offices badly damaged (it was estimated that £500–£600 of damage was caused to the Fish Wharf).

Several enormous chunks of the building's foundation had simply been blasted away and the refreshment rooms opposite had every window smashed – both front and back. Most of the family and staff were out, and it was miraculous that Miss Steel, who was in the building playing the piano, escaped injury but was severely shaken. Two small children also had a narrow escape; they were in bed and uninjured after the blast but their covers were smothered in broken glass fragments. Miraculously, the only casualty here was Captain Smith, the Fish Wharf master, who suffered a cut to his hand from flying glass.

The eleventh bomb was another high explosive (HE) that fell by the river blowing a hole in the quarter, 'started' the timbers and blew the rigging wire 'out like cotton' of Mr Harry Eastick's steam drifter *Piscatorial*. It was to be his second casualty of the war, after he had recently lost his drifter *Copious*, which had sunk taking nine lives with it when it struck a mine shortly after the bombardment in November 1914. Debris and bomb-casing from the blast also damaged the grain store of Messrs Combe and Co., maltsters on Malthouse Lane, where a chunk of ragged steel casing was recovered measuring 7½in by 1½in, and a number of windows were smashed on the south town side.

The last bomb, the twelfth to fall during the ten-minute attack, landed at the back of the racecourse grandstand on South Denes, a short distance from the Auxiliary War Hospital. Leaving the largest crater of the attack, it blew down the paddock palisading, destroyed a number of fish baskets and killed a dog.

L-3 left a trail of bombs behind her and droned back off across the sea, following the coast to Runton where she turned seaward and headed back to base at 10 p.m.

Bomb damage to the swill ground at the back of the Fish Wharf, Great Yarmouth.

Crew indicating the hole torn in the quarter of Mr Harry Eastick's steam drifter, *Piscatorial*, by Zeppelin bomb shrapnel.

L-4 followed the coast from Bacton to Cromer from the east, shortly before 8.30 p.m. The lights of the town had been shaded for a number of weeks, but when the Zeppelin arrived the local military had received warning of 'hostile aircraft' over Great Yarmouth and told all shopkeepers to turn out all their lights, so the town was in darkness.

Local people, not considering the danger, ran onto the streets and looked up at the raider. Most folks who saw it over Cromer agreed that it was so low it almost caught on the pinnacles of the church tower. As the Zeppelin passed over the Great Eastern Railway (GER) station it appeared to lose its bearings, missed the main street and struck right across to the electric light station, where it appeared to encircle the tall chimney and then left in the direction of Sheringham. It was spotted circling round to the south-east, between Weybourne and Sheringham, at 8.35 p.m.

Travelling in an easterly direction over Sheringham, von Platen brought L-4 down to 800ft and dropped a flare. Mr Stanley Simons, who lived on Augusta Street, was watching the Zeppelin and saw it drop the flare, which he described as 'a small object which burst with a slight noise and a bright glare.'

L-4 then dropped an incendiary bomb which fell on Whitehall Yard, Wyndham Street, where it entered the roof of one of the houses and passed through a bedroom down to the ground floor, dropping near the fireplace. The house was occupied by local bricklayer Robert Smith, his wife and daughter, May (14), who was fortunate to escape with only a few scratches. *The Cromer and North Norfolk Post* reported:

> The chair in which she was sitting was slightly damaged. The family were naturally terrified. Our representative who visited this house found the bedroom in great disorder. Hardly a pane of glass was left in the windows, a large hole was made in the roof and the boards were torn up. Considerable damage was also done downstairs to the floor and furniture. The scene was one of utter disorder. The landlord, Mr Jordan, is not insured.
>
> Another little girl, a companion of May Smith who was sitting in the same room, received an injury to her wrist and her hair was singed. 'I never had such a fright in my life. I shall never forget it to my dying day,' remarked Mrs Smith, 'I never want to go through it again.' The bomb was about 4½ inches in diameter. The shell itself must have weighed about twelve pounds. If it had exploded probably the whole square of houses would have been wrecked and lives lost. Mr Smith, the occupier of the house, said 'It all came so suddenly. The bomb fell near the fireplace and it is a wonder how we escaped serious injury. My impression is that it was a fire-ball and that the object of the raiders was to burn the houses.'

Many Sheringham residents ran out onto the streets to see what was going on, and a missionary meeting being held in the Church Room was stopped and abandoned as several members of the audience rushed out onto the street to watch the murderous machine make off.

Special and emergency constables went on duty at once, and proceeded to give instructions as to lights and generally assist the police in dealing with the situation. The night, although a little hazy, was 'decently fair' and there was no wind. Mr R.C. West, a coal merchant who was on duty as a special constable with Mr Gooch at the Gas Works, near the Sheringham Hotel, saw a huge body floating in the air and apparently coming in the direction of the lifeboat sheds.

L-4 then dropped a second incendiary on a building plot between the back of The Avenue (near Mr Lee's house) and Priory Road, leaving a small crater. A man who was walking along Beeston Road said that the bomb nearly struck him. He had an anxious time, as for some yards he was endeavouring to get out of the way, but was firmly of the belief that the Zeppelin was so low down if he had had a gun he could have shot it.

Inspector Carter was on duty on the Cromer Road; he witnessed the dropping of the bomb upon The Avenue and rushed to the police station to telephone Norwich and other places. Newspaper reports stated that the engine of the Zeppelin made 'a terrific noise' as it passed low over Sheringham, eyewitnesses believed that it nearly touched the Roman Catholic church in Cromer Road, and it also passed close to St Peter's Church and the Grand Hotel. The airship, which had a tremendously long body, was described by locals as looking 'like a gigantic sausage.'

The first bomb dropped by Zeppelin L-4 on 19 January 1915 fell upon the home of Robert Smith and his family at Whitehall Yard off Windham Street, Sheringham. Fortunately it did not explode.

With its first two bombs gone, L-4 then flew over the golf links, and was lost to sight, travelling in a north-west direction off the coast towards Blakeney and Wells.

L-4 followed the coast and came overland again at Thornham at 9.50 p.m., dropping its third bomb on the green. It headed eastward to Brancaster Staithe, before returning to Brancaster village, where a fourth bomb, another incendiary, was dropped near the church, about 50 yards from Dormy House and approximately 150 yards from the local auxiliary war hospital.

Then, after passing over Holme-next-the-Sea, the Zeppelin proceeded to Hunstanton where its fifth bomb landed at 10.15 p.m., but did not explode, near the centre of a field on the High Road leading from Old Hunstanton to New Hunstanton. It was suggested in the press at the time that the Zeppelin may well have been drawn by the beam of the lighthouse, whereas the *Report of the Intelligence Section GHQ GB on the Airship Raids from Jan to Jun 1915* states: 'at 10.15 an HE bomb aimed at the wireless station dropped in a field about 300 yards away. After circling the town, which was in total darkness, it went out to sea twice, but returned each time and then made off along the coast to Heacham …'

L-4 droned over Heacham at about 10.40 p.m. and a number of residents came onto the roads or craned their heads out of bedroom windows to look at the Zeppelin as it passed overhead. Bombs six and seven were dropped here; one HE bomb fell near Mrs Pattrick's cottage in Lord's Lane. After clipping the edge of a windowsill and damaging some of the bricks in the wall, it smashed part of the roof of the adjoining wash house and fell into a rainwater tub, promptly blowing it to pieces – a narrow miss indeed. The second bomb did not explode, and was only discovered a couple of weeks later by a lad named Dix, who had been walking across Mr Brasnett's field between the council school and the chalk pit.

The 1/1st Lincolnshire Yeomanry, who were based in Heacham at the time, dug the bomb from the ground and removed it to the lawn of 'Homemead', where the officers were staying. A sentry was posted and, with the local policeman in attendance, the local populace soon gathered to look at the bomb. On the Sunday, an officer drove all the way up from Woolwich Arsenal to collect the bomb and remove it for further examination, but before leaving he did confirm it was a 100lb bomb, and that if it had gone off it would have damaged anything within a 100-yard radius.

Zeppelin L-4 continued its flight south, reaching Snettisham at about 10.45 p.m. The Reverend Ilsley W. Charlton, Vicar of Snettisham, wrote his account of what happened next, published in the *News and County Press*:

Supposing that the distant noise was the hum of an ordinary aeroplane, and that some lights would be visible, my wife and I and a lady friend were walking about in the garden, trying to penetrate the darkness and discover the aircraft. The drone of the engine was so much louder than usual that we were quite prepared to descry at length, exactly overhead, the outline of a Zeppelin hovering over the church and Vicarage at a great height, appearing at the distance, to be only about fifteen or twenty yards long.

No sooner had we identified it as probably a German airship, that suddenly all doubt was dispelled by a long, loud hissing sound; a confused streak of light; and a tremendous crash. The next moment was made up of apprehension, relief and mutual enquiries, and then all was dark and still, as the sound of the retiring Zeppelin speedily died away.

The Zeppelin had circled the village and was so low that the lights revealed by the opening of the trapdoor for the release of the bomb were clearly seen. L-4 then dropped its eighth bomb, an HE, which landed about 4 yards from the Sedgeford Road in Mr Coleridge's meadows, causing an explosion that was felt across the village. The houses in the immediate vicinity suffered a number of broken windows, but the worst of the blast was received by St Mary's Church. Reverend Charlton concluded: 'That there was no loss of life, and that the church (with the exception of twenty-two windows) escaped damage, we owe, humanely speaking to the fact that the bomb fell on a soft, rain soaked meadow, with a hedge and a wall between it and the church.' After the war the windows were repaired, and the east window replaced with a stained glass, 'as a thank offering for preservation and in memory of the men of this parish who fell in the Great War'.

The Zeppelin then passed over Ingoldisthorpe, Dersingham, Sandringham (Wolferton), Babingley, Castle Rising, South Wootton and Gaywood, but dropped no bombs. However, the *News and Country Press* were keen to point out the 'peculiar gusto' and 'special pleasure' of the reportage in the *Hamburger Nachrichten*, which claimed:

On the way to King's Lynn, Sandringham, the present residence of King George was not overlooked. Bombs fell in the neighbourhood of Sandringham and a loud crash notified the King of England that the Germans were not far off … Our Zeppelins have shown that they could find the hidden royal residence. In any case, they did not intend to hit it, and only gave audible notification of their presence in the immediate neighbourhood.

As he approached King's Lynn, von Platen saw many lights and was actually (wrongly) convinced that he was north of the Humber when he spotted what he thought to be 'a big city' and, claiming he had been fired upon by both 'heavy

Boarding up the shattered windows of St Mary's Church after the blast damage from the Zeppelin bomb that exploded in Mr Coleridge's meadow near the Sedgeford road, Snettisham.

artillery and infantry fire', he proceeded to bomb King's Lynn. He steered L-4 towards the town from the Gaywood district, and appeared to take the railway lines as his guide.

King's Lynn's chief constable, Charles Hunt, had received an unofficial report of a Zeppelin raider dropping bombs on Yarmouth and Sheringham. In his report of 5 February 1915 he stated that upon hearing the explosion:

> I immediately communicated with the Electrical Engineer of this Borough and asked him to put the street lights out as soon as possible. He stated that his men had started putting them out and he would put further men on and get them out as soon as he could … I at once communicated with Major Astley who is in charge of the National Guard in this town, also the Officer Commanding the Worcestershire Yeomanry who are billeted here … About 10.45 p.m. when I was trying to get through to Dersingham the Superintendent there rang me up and stated that a Zeppelin had passed over Dersingham and had dropped bombs in that neighbourhood. Before a message was complete I heard bombs being dropped close to this Borough. Immediately on hearing these explosions the Electrical Engineer put out all lights by switching off at the main, not

only putting out lights in the streets but also in private residences as well. The aircraft was soon over our building and several bomb explosions were heard almost immediately.

On the commencement of the bombardment, the members of the fire brigade assembled at the station to be ready if needed. Mr G.E. Kendrick and the engineer were the first to arrive, and they made the appliances ready. The Intelligence Report commented, however: 'Lynn seems to have had a considerable amount of light showing and probably presented a very clear target.'

The airship picked up the railway at Gaywood and passed over the station at King's Lynn. In all, eight bombs (seven HE and one incendiary) were dropped on the town.

The experiences of Inspector R. Woodbine at King's Lynn Station were recorded in a letter published in the *Lynn News*:

I had arrived in our junction signal box just before the first bomb fell. The signalman had told me he noticed a distant report which had shaken his windows and looking out of his box windows I heard a noise resembling an aeroplane. I remarked, as the noise grew in volume, that is no Britisher and told him to put his lights down as the visitor was evidently coming for us. He also told Exton's Road signalman to do the same. The noise seemed in a direct line with the signal box coming from the direction of the Ship Inn on Gaywood Road. I could distinctly see the movement of the propeller but not the body of the machine owing to it being straight in front of me. I called the signalman to have a look but before he reached the window the first crash came. You may imagine, if you can, our feelings. I cannot describe them and never wish to experience such another fright. I remained in the signal box until three bombs had fallen; they appeared to fall all over the station and I thought from the sound that the royal carriage shed and the station were struck by the second and third bombs. I was anxious to get back to the station to see what was done but found no damage to either building of course.

The first bomb had landed on a field adjoining the railway, at the junction of the Hunstanton and Norwich lines near Tennyson Avenue; a number of houses in the area had their windows smashed by the blast. This was followed, soon after, by a second bomb which exploded 'with a grey-blue flash' on allotments that ran along the Walks side of the railway at the Tennyson Road end, making a crater some 16ft across and 7ft deep and blowing in the windows of the carriages standing close to nearby railway sheds.

After droning its way menacingly over St John's Church and St James's Park, the carnage began when the third bomb was dropped on Bentinck Street – a typical

Victorian street, lined with poor-quality terraced housing built for working folk. It was also one of the most densely populated parts of the town, and it was nothing short of a miracle that the casualties were as few thery were. The home of fitter's mate John Goate and his family at 12 Bentinck Street received a direct hit; 14-year-old Percy was killed outright, his father and mother were crushed and wounded and his 4-year-old sister Ethel was stunned. Mrs Goate's testimony at the inquest revealed what happened on that fateful night:

> We were all upstairs in bed, me and my husband, and the baby and Percy, when I heard a buzzing noise. My husband put out the lamp and I saw a bomb drop through the skylight and strike the pillow where Percy was lying. I tried to wake him, but he was dead. Then the house fell in. I don't remember any more.

PC John Fisher also gave evidence that, just after the bombardment, a message was received at the police station stating that an ambulance was required at Bentinck Street. He and other constables rushed there with two ambulances. On arrival they were informed that young Percy Goate had been pulled out of the ruins, but he was dead and his body had been taken to a house in Clough Lane. When one of the ambulances returned after their first run to the hospital, the boy's body was removed to the mortuary.

The wreckage of the homes of the Goate and Fayers families, where Percy Goate and Alice Gazley were killed.

Mrs Alice Maud Gazley, killed in the Zeppelin air raid on King's Lynn, 19 January 1915.

The other King's Lynn fatality was Mrs Alice Maud Gazley, a widow at just 26 – her husband, Percy, had been a porter at South Lynn Station, but as a Reservist had been called up and sent to France with the BEF (British Expeditionary Force) and had been killed in action on 27 October 1914 while serving with the 3rd Battalion, Rifle Brigade. Mrs Gazley lived at Rose Cottage on Bentinck Street, but she had gone to the Fayers family, about four doors away, for supper. Giving her account of the events on that fateful night at the inquest, Mrs Fayers recalled they had just finished their meal together when Mrs Gazley remarked, 'There's a dreadful noise!' Shortly after there was 'a terrible bang', which frightened both of them. Mrs Gazley said, 'Oh, good God, what is it?' and made to rush out into the street.

Tragically, the Fayers were neighbours of the Goate's and their house collapsed moments after the direct hit upon their neighbours' house. The Fayers all sustained minor injuries. Shortly after the raid, Mr Henry Rowe (Mrs Alice Gazley's father) went to check on his daughter at Rose Cottage. He found the windows shattered but there was so sign of her. He spent the rest of the night searching for her.

At about 6.30 a.m. Mr Rowe went to the police office to let them know that he had not found his daughter and stated, 'I think she is under the ruins of Mr Fayers' house.' Police Sergeant Beaumont asked Rowe to wait until daylight then, between the hours of 7 and 8 a.m., Sergeant Beaumont, Mr Rowe and others began to search the wreckage of Mr Fayers' house. It was there they found the body of Mrs Gazley, which was then recovered from the wreckage and removed to the mortuary by police ambulance.

A stable close to the two wrecked houses on Bentinck Street contained a horse belonging to Mr Cork the baker, of Blackfriars Street. Incredibly, despite the stable being badly damaged the horse was released safely and unscathed the day after the raid.

At the inquest, Dr G.R. Chadwick stated he had examined both victims of the air raid at the mortuary. He found that Percy Goate had suffered wounds on his face and one on his chest. Mrs Gazley suffered bruising to her face and abrasions on the front of the right thigh. Dr Chadwick was very much of the opinion that the injuries suffered by both victims had not been sufficient to be life threatening and that both had died as a result of shock. Their death certificates both recorded their cause of deaths, as suggested by Dr H.C. Allinson, the deputy Borough Coroner, as 'from the effects of the acts of the King's Enemies', although the foreman of the inquest jury was compelled to state that some of the jury felt it should be recorded as murder.

The fourth bomb exploded at the junction of Albert Avenue and East Street, at the back of the property owned by vet and blacksmith Mr T.H. Walden, causing extensive damage to the terraced houses in the area. Several people were trapped and required assistance to get out of the rubble of their former homes.

The fifth bomb, an incendiary, fell on 63 Cresswell Street, the home of Mr J.C. Savage and his family. They all had a lucky escape; the bomb fell through the roof, crashed through the floor in the back bedroom, through a tin box and into a basket of linen in the kitchen. It had caused a fire to the bedroom on the way through. The fire brigade were summoned but, by the time they arrived, it had been extinguished by neighbours.

The sixth bomb fell on Mr Wyatt's allotment, at the end of Sir Lewis Street near Cresswell Street. Causing a crater 15ft across, it wrecked fences, trees and shattered windows, but luckily no one was hurt.

The seventh device buried itself in a garden at the back of a house in the occupation of Mr Kerner Greenwood near the docks, also causing minimal damage.

The eighth hit the power station for the King's Lynn Docks & Railway Co., causing extensive damage to the engine house, destroying its two boilers and the hydraulic gear that operated the Alexandra Dock gates, and caused considerable damage to surrounding buildings.

Mr Fayers and neighbours look at a part of the Zeppelin bomb from the wreckage of his home, as a sentry from the Worcestershire Yeomanry looks on.

Just as in Yarmouth, the Zeppelin had been over King's Lynn for about ten minutes before it steered off into the darkness again. Many were treated for shock as a result of the raid on King's Lynn, a total of thirteen people received treatment in the West Norfolk & Lynn Hospital, and many others suffered minor injuries caused by the flying debris and were treated at home. A list of those who required hospital treatment was published in the *Lynn Advertiser*:

Zeppelin bomb damage on East Street, King's Lynn, after the Zeppelin raid of
19 January 1915.

G.W. Clarke – cut lips
Mr Goate – cut face and swollen ankle
Mrs Goate – leg damaged
Ethel Goate, aged 4 years – stunned
Mr Fayers – cut on head
Mrs Fayers – cut on face
G. Hanson – back of hand cut by glass
R. Howard – face cut in two places
G. Parlett – forehead cut and head wound
D. Skipper – face and head cut
Mrs Skipper – injured leg
R. Wykes – cut head
W. Anderson – wrist lacerated

L-4 droned over Grimston Road Station at 11.15 p.m., then turned south-east over Gayton and Westacre, then east again, passing north of East Dereham at 11.35 p.m., flying over Mousehold Heath near Norwich at 11.50, then over Acle at about midnight and finally passed out to sea north of Great Yarmouth before 12.30 a.m.

Aftermath

Francis Perrott, reporter on the *Manchester Guardian* (until his death in 1926 regarded as one of the finest reporters of his day), visited Great Yarmouth the day after the air raid occurred and recorded his impressions of what he saw in the town:

Yarmouth has taken the visitation with remarkable calm and cheerfulness. I found people even pleasantly excited by what has happened and willing to make the most of it as a winter's tale. When first the threatening hum of the propellers was heard in the starless sky, people seem to have obeyed the powerful human instinct of curiosity and ran out into the street to see what was going on but in a short time they remembered the official instructions and made for cover. The authorities turned off all the lights and Yarmouth spent the rest of the evening in dead darkness. Towards midnight the throbbing of engines was heard again over the town but if this really was the noise of Zeppelins returning home no one saw them. [This was L-4 taking its circuitous route across the county after bombing King's Lynn.]

I could find no one in Yarmouth who actually saw the aircraft, although there are vague stories of 'long black shapes' and things in the sky 'like a big black cigar.' The best evidence on the point of whether they were airships or aeroplanes is the size of the bombs dropped. Two unexploded bombs were on view in the Drill Hall this afternoon, where an interested crowd inspected them. They are bulky pear-shaped things two feet long, forty inches round the base and weigh about sixty pounds. Yarmouth is convinced that only a Zeppelin could throw bombs of that size about wholesale.

The first sign of interest seen in the streets was a group of sightseers round a shop window where bits of bomb were shown as relics. Down on the South Quay the publican of the 'First and Last' tavern showed me a handful of fragments picked up in his bar just after a bomb had burst outside his doorstep. 'There was some nasty, sticky, yellow stuff inside it,' he said. I came nearest the reality of the raid in the little open space behind St Peter's Church where people were collected staring hard at a mess of ravaged houses, broken windows and littered roadway. Here the bomb fell which killed old Miss Taylor and Samuel Smith the cobbler. It dropped on the pavement across from the St Peter's church and outside a modest villa. It stripped half the wall of the villa and made mincemeat of the content. The man inside escaped with a cut head. I found him hovering about the hole in the road as if he could not leave it twenty hours later … The dwellers in the houses round the church that had been torn open to daylight, like doll's houses with the front opened, were standing about looking as if they felt the importance conferred by the calamity. They gladly took you over what was left of their homes and related marvellous escapes. A queer sight was the furniture of the villa piled in the street, with a forlorn doll holding out its arms in horror at the ruin wrought by the bomb.

From St Peter's church the raiders steered along the quay. Following in the track, I found the deep bruise in the roadway, where the Germans tried to destroy the tavern, only to give the innkeeper a grievance because the War Office had claimed his cherished bits of shell and also the ruin caused on the Fish Quay. Here the bomb fell in a heap of empty basket nets close to the salt water tank. It burrowed a hole in which you could stand up to your knees, knocked a warehouse about and damaged an eating house. On the wall of a wrecked office the only thing left untouched was the barometer at 'set fair.'

The last call of the Germans before quitting Yarmouth was the race course, where the grand stand was splintered. In some parts of the town men were filling carts with tons of broken glass. The fishermen on the quay were chiefly interested in a steam drifter which was riddled with shell fragments. 'Looks as if she'd been taking on the German Navy all by herself' said one.

The relics of the air raid also drew considerable interest in King's Lynn, as the *Lynn News* reported:

The bomb shell in the possession of the Lynn police, with portions of other bombs discovered in the borough, was placed in the Stone Hall on Tuesday afternoon for public inspection. Sixpence per head was charged for admission and the proceeds, nearly £3, were in aid of those who have suffered damage. Mr William Fayers, one of those injured by the falling house in Bentinck Street was present and two members of the National Guard, through the kindness of Major Astley, were on duty.

An interesting perspective on how the news was viewed at the time was recorded in the diary kept by a senior member of the staff, published in the *Lynn News*:

January 21: Needless to say, the London papers are crowded to-day with photographs of Lynn wreckage – many of the prints are excellent – and with full narratives, which are not so excellent. The editorial comments are scathing in the extreme, as was but to be expected [many of the reports criticised the King's Lynn engineer for opting to turn out the street lights individually rather than throwing the switches to turn off all the lights in the town be they on the streets or in people's homes].

Though the tragedy of the whole horrible occurrence weighs upon one's imagination stories of a host of humorous incidents have reached me. Many of these tales would make excellent reading if they could be published but discretion is the better part of a diarist's valour.

Talking of valour, one of two episodes brought to my knowledge confirm the theory that moral courage is a very different thing from physical pluck. As to the fact that the loudest talkers and the bullies are seldom brave when real danger of death threatens, that goes without saying, of course. On Tuesday night several individuals who in normal times are of a domineering disposition were in a pitiable state of fear. I have even heard of men who left their wives and children in order selfishly to seek a fancied security for themselves!

It is more pleasing to dwell on the stories of the quiet bravery of men who, in ordinary conditions, make no pretence of being embued with strength and, above all, the tales of the women's silent endurance thrill one. I have heard of several women who, showing hardly a sign of emotion while the bombs were raining from the skies, have suffered a physical collapse – a fact that proves the enormous mental suffering they rigidly endured on Tuesday night.

In the bombed towns, anger at their dearth of defences and the need for protection in the event of another raid, saw a Lynn editorial suggest: 'Lynn Corporation, despairing of protection by the War Office and Admiralty, should itself purchase high angle guns and an aeroplane.' On a more prosaic level, Mr R.O. Ridley, the mayor of King's Lynn, wrote to Prime Minister Herbert Asquith on 22 January 1915:

Sir,

In reference to the raid of this town by the enemy's aircraft on Tuesday evening last, the 19th inst. I am directed to inform you that considerable damage was done to property of the very poorest classes. As a result various people are homeless and others are suffering greatly in consequence of the loss and damage to their furniture.

I shall be glad of an early intimation of the Government to compensate the people injuriously affected. In the meantime, may I ask that some steps may be taken to relieve cases of immediate necessity?

I am to further call your attention to the fact that no protection of any kind is afforded to the town against raids of the above description and I beg to strongly urge that in view of the great probability of further occurrences of a like nature, some steps be devised to deal with them.

I believe I am fully alive to the difficulties of the situation and I know that great efforts are being made by His Majesty's Government in connection with the conduct of the war. I do suggest, however, that the matter referred to is of great importance and calls for some action to be taken.

The reply from 10 Downing Street was swift, dated 25 January 1915:

Dear Sir,

I am desired by the Prime Minister to acknowledge the receipt of your letter of January 22nd and to inform you in reply that it is the intention of the Government to take measures to deal with the damage suffered by reason of the recent air raid on King's Lynn similar to those adopted in the case of the recent bombardment of Hartlepool, Scarborough and other places.

The other matters referred to in your letter are receiving careful consideration.

In the immediate aftermath of the raids, the German papers were plastered with accounts of successful attacks on fortified places between the Tyne and Humber or, the 'fortified place at Yarmouth', as well as warnings to the King. Only after the German authorities gained sight of the copies of British and international newspapers did the Zeppelin crews discover the places they had actually attacked.

With their reportage of the raid, the local newspapers were quick to publish a set of 'Safety Rules' in the event of further air raids, such as this example from the *Cromer Post*, published on 22 January 1915:

It is essential in the common interest that the public should become conversant with these regulations, which are designed to prevent confusion, to assist the military and civil authorities, as well as to protect the civil population. The following summary of the official advices gives a clear idea of the best course to follow:

TO THOSE WHO HAPPEN TO BE IN THE STREET –
Take cover immediately.
There is danger from bombs from aircraft and also from fragments of shell and from bullets from the guns and against the raiders.
The assembly of large crowds might prove fatal.
The nearest basement would be the safest place.
Any fragment of shells should be handed to the police, in order that the War Office may ascertain the size and nature of the missile.

TO SCHOOL TEACHERS –
Continue the lessons as far as possible in the normal way.
Remove children from the neighbourhood of windows.
Children should not be brought from upper floors to crowd ground floor class-rooms or basement. In the event of damage to the building the children should be marched as in a fire drill.

TO THOSE IN PRIVATE HOUSES –
Stay there! Preferably in the basement.

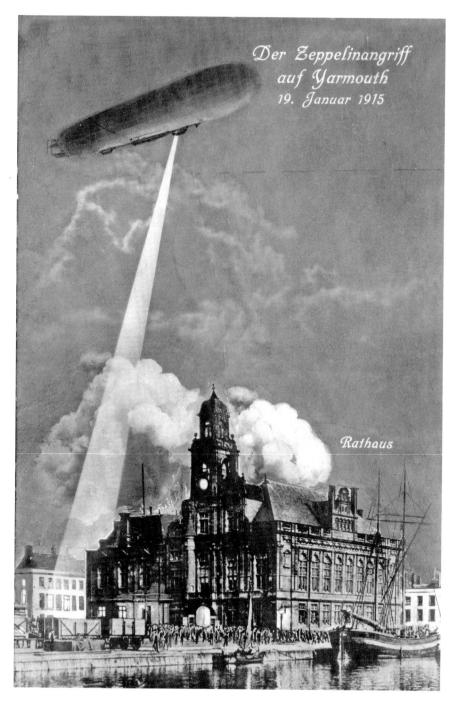

German propaganda postcard produced after the Zeppelin raid of 19 January 1915, showing Great Yarmouth Town Hall under attack and on fire; this was not exactly what happened …

DEFENCE
OF THE
REALM REGULATIONS,
1914.
NOTICE.

After this date and until further notice no Street Lights are to be lit and all House Lights as well as Lights in Business Premises are to be screened and blinds drawn down within an area of Ten Miles from the Coast, from King's Lynn to Lowestoft inclusive.

BY ORDER
ALFRED A. ELLISON,
Captain-in-Charge.
24th January, 1915.
Lowestoft and Yarmouth.
JARROLD & SONS, Ltd, Printers, King Street, Great Yarmouth.

Lighting restriction proclamation for the Norfolk and Suffolk coast issued in the wake of the first Zeppelin raid.

English coastal areas also reacted to the raid of 19 January with their own precaution notices, such as this one produced for Maidstone on 28 January 1915.

POSSIBLE AIR-CRAFT RAID OVER MAIDSTONE.

CAUTION.

IN the event of a Zeppelin or Aeroplane Raid, Maidstone must be immediately placed in total darkness.

Three Bombs will be fired from the Police Station at short intervals when the danger is imminent, and if at night the Electric Light supply will be cut off.

Inhabitants must immediately, if the raid is at night, take steps to turn out all gas and render their premises absolutely dark when viewed from outside, and should remove themselves to the basements and cellars wherever possible, and keep there till the raid is over.

Three more Bombs will be again fired when all danger from the raid is over.

Town Hall,
28th January, 1915.

W. H. MARTIN,
Mayor.

Vivish & Baker, General Printers, King Street, Maidstone.

The national press reflected the outrage of the British people with articles headlined, 'The Coming of the Aerial Baby Killers' and reports such as *The Daily Mirror*'s:

> Germany overjoyed by news of 'gallant' air huns murder raid that included 'Berlin's War Whoop: Copenhagen, Jan. 20 – I have just received a private telegram from Berlin which describes the people's joy at the success of the Zeppelin attack as being widely enthusiastic. I have an intuitive feeling that the joy could not have been greater even if Dr Barnardo's Homes had been destroyed.

The international press is typified by the *New York Tribune*, 'A Disgrace to Civilization', that spoke of a 'wanton disregard of Hague rules and humane principles. The raid belongs in the worst acts of German militarism in the present war … It is savagery which civilised opinion of the world has already condemned, which must stand condemned for all time.'

The *New York World* echoed:

> They accomplish no military purpose and the wanton slaughter of women arouses a world-wide resentment against Germany … Germany will not begin to realise what these raids cost her until she comes to make peace. Her military authorities seem to forget that the war is not going to last forever and that when it is over Germany will have to live in the world with all the other nations. How does she propose to deal with the vast body of hatred that she is building up for herself? How many years will it take for her to live down the record she is making?

The reports of the air raid in the international press stirred disquiet in the corridors of power in Germany, and Chancellor Theobald von Bethmann-Hollweg did not mince his words in a letter to Admiral von Pohl:

> According to information received, for Zeppelins to drop bombs on apparently undefended places makes a very unfavourable impression on foreign neutrals, particularly in America. Also doubt exists in reasonable circles there, as military importance and success is not readily apparent. Prompt explanation to this effect seems urgently necessary.

The Kaiser publicly praised the conduct of the raid, and the crews of L-3 and L-4 were all decorated with Iron Crosses for their part in the action, but he was quick to reiterate to his ministers that only the docks of London, military establishments of the lower Thames and the British coast could be taken as bombing targets, but royal palaces – including Sandringham – were not to be bombed.

The bombing campaign would continue, but the crews of L-3 and L-4 would have their wars abruptly ended when, less than a month later, both Zeppelins were lost during a scouting mission along the Norwegian coast.

L-3 had one engine down when she encountered a severe snowstorm and strong head winds. Fearing he would not make the return journey, Fritz put L-3 down on the beach of the North Sea island of Fanø, off the south-western coast of Denmark. Despite a hard landing the crew escaped uninjured. Fritz then burned the ship's papers and set the whole Zeppelin on fire with a signal gun. The crew were detained in Odense for the rest of the war.

L-4 was caught in that same storm of 17 February 1915, suffered the loss of electric power, the radios went down and with the engines failing Hallermund forced an emergency landing at the shore near Blaavands Huk, Denmark. The Zeppelin was wrecked beyond repair, four crew members lost their lives and the

Reportage typical of most of those found in British newspapers and magazines after the first Zeppelin raid.

remaining members of the crew were interred. Hallermund escaped at the end of 1917. He subsequently served in Finland, fighting the Bolsheviks alongside White movement forces.

In the aftermath of the air raids, it is also intriguing to read in the *Daily Telegraph* the account from the correspondent they had despatched to King's Lynn: 'That the hostile aircraft that attacked Lynn was guided by a pilot who was familiar with the countryside over which he flew there can be little doubt. The military and police authorities here are satisfied this is so.' Moreover, among all the reportage after the raid, one question was raised again and again – were spies at work? In Great Yarmouth, the *Mercury* was quick to dismiss the stories:

> It seems a great pity that the authors of such rumours as have a disturbing effect upon the community cannot be discovered and brought to book. Many of these fairy tales concern the capture of alleged spies, and the absurd stories range from the arrest of a small crowd of Germans in an empty Howard Street shop to the detention of a young girl, a German of course, in empty business premises in the Market Place in the very act of flashing signals out to sea.

It also took particular exception to a report in another local paper, which stated: 'A signal is said to have been given from Yarmouth on Tuesday evening to direct the German airship as to the best place to place bombs.' The *Mercury* concluded, with some vitriol over this matter: 'We, however, feel compelled to enter a protest when the local press give publicity to such tarradiddles.'

In the west of the county, questions over the presence of spies and stories such as there being a light shone onto the Greyfriars Tower at King's Lynn, or a similar account at Snettisham where 'the church spire was being constantly flashed upon' on the night of the raid, were taken more seriously. Above all, there were accounts of a motor car that was said to have guided the Zeppelin raiders to their targets with 'brilliant headlights' or 'that flashed upwards to the sky.'

Mr Holcombe Ingleby, MP for King's Lynn, expressed his concerns in a letter to *The Times* published on 22 January 1915:

> I have myself tested the evidence of some of the most trustworthy of the inhabitants and the evidence seems to be worth recording. The Zeppelin is said to have been accompanied by two motor cars, one on the road to the right, the other on the road to the left. These cars occasionally sent upwards doubles flashes, and on one occasion these flashes from the car on the right lit up the church, on which the Zeppelin, attempted to drop a bomb. Fortunately the missile fell on the grass meadow … After this attempt at wanton mischief the Zeppelin made for King's Lynn, and here again there is further evidence that it was accompanied by a car with powerful lights which were at one time directed

on the Grammar School. The car was stopped in the town and attention was called to the lights as a breach of the regulations. Having put them out the driver turned the car quickly round and made off at a rapid pace for the open country. Seven bombs were dropped in King's Lynn, two of them right in the heart of the crowded streets. Possibly they were intended for more important buildings, which, without the aid of the car, it was difficult to distinguish …

The official response to the concerns voiced by Ingleby was to dismiss the existence of the car, but after Ingleby's letter and the official response were also published in the local press, a number of west Norfolk residents wrote to him with their testimony about the mysterious signalling car. Having drawn together so many earnest accounts, he brought the entire matter to the notice of the Home Secretary and published the now rare and collectable booklet, *The Zeppelin Raid in West Norfolk*, in 1915.

Spies or spy scares aside, Britain had been bombed and its air defences needed to be vigorously readdressed, as Ingleby put it in the latter part of his *Times* letter: 'A couple of biplanes at King's Lynn and a couple at Hunstanton might make such a raid as to which we have been subjected impossible of success. If they could not destroy a Zeppelin they might at least drive it off.'

ONE

1915

14 April 1915

During the months of February and March, and during the first half of April, no raids actually took place. Only one unsuccessful attempt was made in March, and no extensive reconnaissances were recorded. The Zeppelins L-3 and L-4, which had carried out the raid on 19 January 1915, were lost in a snowstorm off the Jutland coast on 17 February, and no doubt imposed caution on the airship command.

Two new naval airships, L-8 and L-9, of an enlarged type, had, however, replaced these losses. On 4 March L-8, commanded by Kapitänleutnant Beelitz, attempted a raid on England from Belgium which failed. On her return the weather was cloudy and she came in too low under cloud; she was hit by gunfire off Nieuport and, losing gas, came down too low to ascertain her position. She was riddled with fire from a land battery and five of her gas bags were pierced by shell fragments. She just cleared the chimneys of the town of Tirlemont and eventually fell into some trees close by, being totally wrecked.

L-9, commanded by Kapitänleutnant Heinrich Mathy, had meanwhile made several successful reconnaissance flights from north Germany over the North Sea. Her commander was a man of great courage and resource, and soon proved himself capable of raiding England with effect. On the night of 14 April 1915, he inaugurated the series of raids on northern England which were to become a speciality of the Imperial German Naval Airship Service.

He is said, on this occasion, to have first taken his Zeppelin up the coast of Jutland to the neighbourhood of Norway, then to have crossed the North Sea to the coast of Scotland, coming southward in order to attack the industrial establishments of the Tyne, which he found with success, though the raid did little damage.

Mathy appeared off the mouth of the Tyne and coasted as far as Blyth, where he was off the harbour at 7.30 p.m. The Zeppelin appeared to take her bearings, then proceeded up the River Blyth to Cambois where she was fired at by members of C Company, 1st Northern Cyclist Battalion. She then carried on to West Sleekburn, where the first bomb, of incendiary type, was dropped in a field, doing no damage.

The Zeppelin passed on, by Bomarsund and Barrington collieries, to Choppington. Four incendiary bombs were dropped on fields on the way. A sixth incendiary bomb was dropped in front of a house at Choppington, breaking a window of the Station Hotel, just before 8 p.m. The next bomb thrown was explosive, and dropped in a field west of Glebe Farm, Choppington. Two more HE bombs were dropped in the same field. The tenth bomb, also HE, was dropped in a field west of Bedlington, followed by another HE in a second field close by. At Bedlington, L-9 turned south, passing over Crowhall Farm, where a 50kg HE bomb was dropped, again in fields doing no damage, though it fell within 30ft of two police constables, who were saved by throwing themselves flat on the ground.

The Zeppelin then proceeded towards Cramlington, dropping another HE bomb in a field, which again did no damage. At Cramlington an incendiary bomb was thrown, which fell through the roof of a warehouse. A small fire was caused but was soon extinguished by some workmen.

Proceeding south, L-9 dropped an incendiary bomb in a field near West Cramlington. The Zeppelin was then seen about 2 miles south of the latter place, apparently hovering in the air for a short time. Three more incendiary bombs were dropped in this vicinity, falling in a field west of the railway and close to the line.

L-9 then turned westward to Seaton Burn, and on the way, another incendiary bomb fell in the village but caused no damage. The Zeppelin headed further west, in the direction of Dinnington Colliery, 1 mile distant from Seaton Burn, and when about two thirds of the distance had been covered an HE bomb was dropped in a field, making a large crater. Going south-east the airship passed over Forest Hall at 8.35 p.m. and then went south over Benton, where an incendiary bomb was dropped, falling in a field a little west of the railway station.

Kapitänleutnant Heinrich
Mathy.

Incendiary bomb dropped,
by L-9 under Mathy, on
Choppington during the
Zeppelin raid of 14 April
1915 (in ring). The lady is
standing where the bomb
actually dropped.

ZEPPELIN RAID, APRIL 14th, 1915.
Incendiary Bomb dropped at Choppington.   (Z9)

L-9 dropped six incendiary bombs at Wallsend. Three did no damage, the fourth went through the roof of a cottage and slightly injured a woman and a little girl (the woman was sitting at the fireside washing the girl and they both had their hair singed, the bomb setting the floor on fire); the fifth and sixth set fire to railway sleepers.

A HE bomb then fell into the River Tyne between the electric power station and Castner Kellner's Works, the force of the explosion damaging windows at both places.

The last bomb, an incendiary, fell at Hebburn Quay, on the south side of the river, at 8.48 p.m. It struck the concrete floor of a dry dock, doing no damage. The Zeppelin then went out to sea at Marsden, between Sunderland and South Shields, persuaded by two aeroplanes, although the pilots could see nothing of her.

The HE bombs were all estimated to be around 50kg in weight. The monetary value of the damage caused by the raid was estimated at £55.

15/16 April 1915

Three Zeppelins set out to raid the Humber but ended up raiding the East Anglian coast:

L-6, under Oberleutnant zur See von Buttlar, was reported from Lydd as being out to sea, flying northward at 9.22 p.m., and reached the Essex coast at Walton-on-the-Naze at about 11.30 p.m. It passed over Frinton at 11.40 p.m., circled over Clacton at 11.45 p.m. without dropping bombs and then made across the Blackwater estuary to the Latchingdon peninsula, and passed inland over Burnham-on-Crouch at 12.10 a.m.

L-6 went north-west to Maldon, circled over the town at 12.20 a.m., and dropped two HE bombs there and two in the neighbouring village of Heybridge, accompanied by about thirty incendiary bombs in both places. One of the latter failed to ignite, while two others ignited imperfectly. A house was damaged by an HE bomb at Maldon and a fowl house destroyed. The only casualty was a girl who was slightly injured.

The airship then proceeded north-westward, passed over Smith's Green, 4 miles east of Kelvedon, and then turned eastward. It was heard and seen at Tollesbury, was over West Mersea at 12.30 a.m. and at 12.35 a.m. passed over Brightlingsea, where it was fired on by the guard at a camp of field companies, Royal Engineers.

Thence she went north towards Great Bentley and at 12.40 a.m. was over Tendring, turned south-east over Beaumont, swerved north at 12.47 a.m. at Moze Cross and at 12.53 a.m. passed over Harwich going north-east at a height estimated at 5,000ft. Three minutes later she went over Landguard Fort and

The damaged caused by an HE bomb dropped from L-6 on Maldon, 16 April 1915.

Felixstowe, attacked by a pom-pom, which fired three rounds, and by rifle fire from Landguard Fort. L-6 dropped no bombs on the fortress.

At 1 a.m. the airship passed over Shingle Street out to sea, reported to be at a height of just 500ft. At 1.10 a.m. L-6 was heard at Orford going north, then she turned south-east and, at 1.22 a.m. passed over the Shipwash, after which she disappeared out to sea.

L-5, under Kapitänleutnant Alois Böcker, was hovering off the Suffolk coast near Southwold as early as 9.40 p.m., but did not cross it until 11.50 p.m. when the Zeppelin came in at Reydon, going west. Landfall was probably given by the easily distinguishable expanse of Easton Broad. A quarter of an hour later she passed Wenhaston, going south-west, and five minutes afterwards Bramfield, where she was fired upon with rifles. She was still going south-west but soon turned north, and at 12.20 a.m. passed Halesworth.

L-5 went east to Henham Hall, where, at 12.25 a.m., she dropped one HE and twenty-three incendiary bombs near the Red Cross Hospital. No damage was done and no casualties were caused. Two minutes later, she dropped another HE and three incendiary bombs at Reydon, which also did no damage. She then circled round south-east to Southwold and, at 12.30 a.m., dropped one incendiary bomb which fell into a truck at the railway station, and another on the seashore close to the pier.

Crater left by the HE bomb dropped on Reydon, Suffolk, 16 April 1915.

Turning north-west she again passed over the railway station and, at about 12.40 a.m., dropped another incendiary bomb and another HE at Reydon, followed by two incendiary bombs at Easton Bavents, where she was attacked with rifle fire by 6th Battalion, Royal Sussex Regiment (Cyclists).

L-5 followed the coast northwards, and at 1.05 a.m. approached Lowestoft from the direction of the Herring Market. Arriving over the town, she stopped her propellers, being at a height of about 2,000ft. The siren at the electric power station was now sounded as an alarm. The airship rose to about 5,000–6,000ft, and moved off towards the Central Station and harbour at 1.15 a.m. just as the siren finished. She dropped an HE bomb and then went off along the river to Oulton Broad, dropping two HE and seven incendiary bombs on the way.

Several houses were damaged, and a fire was caused in a timber yard. Two horses were killed and four injured at the railway station; the total damage was estimated at £5,966 10s. There were no human casualties.

The Zeppelin then turned and went out to sea over Lowestoft at 1.25 a.m. A number of parachute flares were dropped by L-9 when over the land, and it would appear that some of these were regarded as incendiary bombs and are counted in the number of the latter mentioned above. How many were actually flares could not be ascertained.

L-7, under Oberleutnant zur See Werner Peterson and carrying Zeppelin chief, Peter Strasser, spent an hour skirting the Norfolk coast in high winds, coming overland at Brancaster at about 1.40 a.m. She passed south-east along the coast, going over Cromer at 2.05 a.m., Haisborough at 2.15 a.m. and passing out to sea at Gorleston at 2.35 a.m. Frustrated by the blacked-out country below, L-7 headed back to Germany without dropping a single bomb on land.

29/30 April 1915

This was the first raid on England to be carried out by a German military airship.

LZ-38 was the single army airship that carried out the raid, under the command of Hauptmann Erich Linnarz. It was first reported from the *Galloper* lightship as being 30 miles south-east of Harwich, going west, after 11 p.m. on 29 April.

At 11.55 p.m. she crossed the coast at Old Felixstowe and went straight inland, reaching Ipswich at 12.10 a.m. There, she dropped five incendiary bombs in the borough, one of which failed to ignite. One fell on a house in Brookshall Road, setting fire to it and the adjoining house; otherwise no damage was done and no casualties were caused.

Immediately afterwards, five more incendiary bombs fell at Bramford, to no effect. At 12.20 a.m. five explosive and eleven incendiary bombs fell at Nettlestead and Willisham, 7 miles north-west of Ipswich, doing no damage except for crops. The airship eventually reached Bury St Edmunds, and for ten minutes circled over the town, going round two or three times. At 1 a.m. she dropped three HE and forty incendiary bombs on the defenceless town. Luckily, most incendiary bombs simply burnt out causing no damage or were doused by buckets of water.

The most significant damage was suffered by four business premises on the Butter Market, where Day's Boot Makers and adjoining shops were gutted, and burned until morning. By some miracle, there was only one casualty – a collie dog belonging to a Mrs Wise.

The airship had now left the area of coast fog; the sky was quite clear and moonlit at Bury St Edmunds, and the LZ-38 was plainly visible at a height of about 3,000ft. She therefore hastened to return to the protection of the fog before she could be attacked, and went off eastward at high speed, dropping a single HE bomb as she went. No damage was done in either case.

At 1.15 a.m., she reached Creeting St Mary, 16 miles east-south-east of Bury St Edmunds, and dropped an incendiary bomb there, followed by another at Otley; neither causing any harm. At 1.27 a.m. another fell at Bredfield, 10 miles east-south-east of Creeting St Mary, and at 1.30 a.m. another at Melton, 2 miles from Bredfield, with the same result. The last bomb, also an incendiary, was

Some of the forty incendiary bombs droppped by LZ-38 on Bury St Edmunds, Suffolk, on 30 April 1915. Most of them simply burnt out causing no damage, or were doused by buckets of water.

Day's Bookmaker's and adjoining shops on the Butter Market, Bury St Edmunds, Suffolk, were gutted and still burning well into the morning of 30 April 1915.

harmlessly thrown at Bromeswell, and the airship proceeded out to sea near Orfordness at about 1.50 a.m.

After turning north along the coast, at 2 a.m. she passed over Aldeburgh and was last heard of at sea at 2.20 a.m., still going in the same direction. No action was taken against the Zeppelin; the mobile guns of the RNAS reached Bury St Edmunds at 1.45 a.m., three quarters of an hour after the raid, and it was by then too foggy on the coast for aircraft to go up.

10 May 1915

LZ-38, again commanded by Hauptmann Linnarz, was this time first spotted at 2.45 a.m. over the SS *Royal Edward* which was moored off Southend as a prisoner of war hulk. The raider dropped an incendiary bomb close to the port side of the ship, the flames leaping up to a height 10–12ft and lasting half a minute. The LZ-38 was travelling towards Southend and dropped two more bombs in the water between the ship and the shore.

She passed over Southend east–west at 2.50 a.m., dropping four HE and a large number of incendiary bombs on the town as she went. Two of the HE bombs failed to explode. After leaving Southend the airship went over Leigh to Canvey Island where, at 3.05 a.m., the Zeppelin came under fire of the AA (anti-aircraft) guns at Thames Haven and at Curtis & Harvey's Explosive Works, in Cliffe. There were 3in guns mounted at Cliffe and their fire, the volume of which was probably unexpected, straightaway turned the LZ-38, which appeared to be hit – although not vitally.

The Zeppelin went back over Southend, dropping more incendiary bombs there at 3.10 a.m. She headed north-east towards Burnham, and went out to sea near the mouth of the Crouch. At 4.18 a.m. she passed the *Kentish Knock* light vessel, heading east. She passed near Sunk at 4.30 a.m. and by 5.15 a.m. she had moved south of the Shipwash, after which she headed towards the Outer Gabbard and out to sea. On returning to Belgium, she was found to have been holed twice aft by AA (anti-aircraft) fire, a shell having gone through her stern.

A large number of incendiary bombs, estimated to be about ninety, were dropped on Southend but, owing to the energy of the fire brigade surprisingly little damage was done. A timber yard was burnt out and a number of small fires were started, all of which were brought under control, except for one in a dwelling house which was completely burnt out. A woman was killed and a man injured in this house. A private of the 10th Border Regiment was also injured in the town.

The very heavy load of incendiary bombs carried by the airship was remarkable, as is the time at which the raid was carried out, just before dawn. As this time was

Multi-view postcard showing the damage caused by the bombs dropped by LZ-38 on Southend, on 10 May 1915.

not again chosen for a further raid, it was evidently deemed unsuitable for some reason. The height of the airship was unusual for this period, being estimated at 9,000–10,000ft, which is much higher than previously.

During the raid a message was dropped on the town, written on a piece of cardboard in blue pencil: 'You English. We have come and will come again soon. Kill or Cure. German.'

17 May 1915

LZ-38, with Hauptmann Linnarz, was spotted again seven days later at 12.30 a.m. hovering off North Foreland for some time, and appeared to have dropped some bombs in the sea. She then approached Ramsgate. On being fired at from drifters at sea, she first went east and then northwards and was off the *Tongue* lightship shortly after 1 a.m. At 1.40 a.m. she came overland again at Margate and flew across Thanet, reaching Ramsgate at 1.50 a.m.

She dropped a number of bombs, apparently four HE and about sixteen incendiaries, on the town. One HE bomb struck the Bull and George Hotel, penetrating to the basement and blowing out the whole front of the building. A man and a woman who were on the second floor were seriously injured and died three days later; another woman was slightly injured. No other casualties were caused, though bombs fell all over the town and in the harbour. A few buildings and some fishing smacks were damaged. After throwing the bombs, the airship went out to sea under hot rifle fire and disappeared in the clouds.

She then proceeded down the coast, and came inland again about 2.10 a.m. at Deal, where she hovered for a short time with propellers stopped. At 2.25 a.m., LZ-38 approached Dover and was engaged by the AA guns of the garrison. In all, five rounds of 6-pdr and twenty-eight rounds of 1-pdr ammunition were fired. On being caught by the searchlights and fired at, the airship at once rose to a height of at least 7,000ft, dropping bombs as she did so, and emitted a dense cloud of vapour in which she disappeared (this was a discharge of water ballast).

The bombs fell at Oxney, 3½ miles from Dover. They were all incendiary, and thirty-three were found. No damage was done by any of them.

The airship carried on north, was fired on by the guard ship in the Downs at 2.50 a.m., hovered about in the neighbourhood of the North Goodwins until 3.25 a.m. and then went back to Belgium after day had dawned. She passed over the British lines at Armentières at about 4.20 a.m.

While over Ramsgate, the lights of London were discernible from the airship, but her commander's instructions expressly forbade his venturing far inland and no attempt was made to raid London. The estimated value of the damage caused by the raid was £1,600.

The aftermath of the HE bomb dropped by LZ-38 that made a direct hit upon the Bull & George Hotel, Ramsgate, 17 May 1915.

Some of the burnt-out incendiary bombs recovered from around Ramsgate after the raid of 17 May 1915.

German propaganda cartoon mocking the panic caused by Zeppelin air raids on London.

26 May 1915

LZ-38 and Hauptmann Linnarz paid their second visit to Southend on this night, several weeks later. The raid was clearly a repetition of his earlier reconnaissance of the route to London, special attention being paid to the mouth of the River Blackwater.

At 9.18 p.m. LZ-38 passed Dunkirk going west, and at 10.30 p.m. appeared off Clacton-on-Sea. She then passed south-west via Bradwell-juxta-Mare at 10.50 p.m. to Southminster at 10.53 p.m. Here, she was fired on with fifty-seven rounds from a pom-pom.

LZ-38 turned south to Burnham-on-Crouch shortly before 11 p.m., passing over Shoeburyness at 11.05 p.m., where the airship came under fire from a 3in AA gun and veered westwards to Southend. Here, at 11.13 p.m., she dropped twenty-three small HE bombs and forty-seven incendiaries. Two women were killed (one of them, unfortunately, from a fragment of AA shell), a girl was injured and several other people received minor injuries.

LZ-38 went off to the north-east, and was again engaged by Shoeburyness AA fire at 11.20 p.m. (the 3in gun at Shoeburyness fired a total of twenty-four rounds HE and thirteen rounds of shrapnel). She passed Wakering at 11.25 p.m. then left via Burnham, where she was fired on with 200 rounds of rapid fire by

A Company, 2nd/8th Battalion Essex Regiment, thence to Bradwell and out to sea at the mouth of the Blackwater at 11.45 p.m. The monetary damage caused by the raid was estimated at £947.

It was noted that the small HE bombs were more like grenades, weighing about 5lb each. The GHQ report stated:

> These clearly had no other object than the killing or maiming of as many people as possible. Owing to their small size the damage they could inflict to well-built house property was relatively slight but as the casing of the bomb was serrated in the same manner as that of a Mills grenade, the explosion of such a bomb in a crowded thoroughfare or building would cause serious casualties.

31 May 1915

The First Raid on London

LZ-38 was first reported on 31 May passing Dunkirk at 8.30 p.m. She crossed Calais at 8.55 p.m. and made for the North Foreland, passing Margate, where she was fired at with 500 rounds from the Maxim machine guns of the Southern Mobile RNAS section at 9.42 p.m. Other .45 Maxim machine guns of the Southern Mobile RNAS opened fire on her from Reculver at 9.50 p.m. and she seems to have moved over to the Essex shore. Here she was fired upon with twelve rounds of shrapnel shell by the 3in gun at Shoeburyness at 10.12 p.m.

LZ-38 passed inland between Rochford and Rayleigh at 10.25 p.m., reaching Wickford at 10.35 p.m. Later, at 10.50 p.m., she passed Brentwood and then seems to have hesitated as to her course; her commander was evidently fixing his exact position with regard to London.

LZ-38 then came straight in, passing between Woodford and Wanstead at 11.15 p.m. The airship was seen over London for the first time, about 400 yards away from Stoke Newington Station and, at this point, commenced dropping bombs at 11.20 p.m.

The first bomb to drop in the Metropolitan Police area was an incendiary. It fell on 16 Alkham Road, Stoke Newington, penetrating into two bedrooms and destroying their contents by fire. The spot where this bomb fell is about 300 yards south-east of Stoke Newington Station and it may possibly have been aimed at the station. The next bomb was also an incendiary and it fell on 8 Chesholm Road, falling through the roof of the back bedroom but without doing any further damage. This was followed by three HE grenades that fell on 41, 43 and 45 Dynevor Road, Stoke Newington. At the first house the windows and doors were blown out, but nos 43 and 45 also had the back extensions of each house practically demolished and the rest of the doors and windows blown out.

The airship then steered a course due south about 500–600 yards west of the main Kingsland–Stoke Newington road, which was doubtless visible. Bombs were then thrown in rapid succession, the next being an incendiary which fell at 27 Neville Road, Stoke Newington, completely gutting the premises. This was followed by an HE grenade, which landed in the roadway of Neville Road and failed to explode.

An incendiary bomb fell on a shed at the rear of 21 Neville Road, but caused no fire, and another incendiary followed this at 47 Neville Road, falling through the roof to the floor below without causing any fire. Another incendiary bomb was dropped at 6 Allen Road, and this went through the roof of the house to the ground floor, gutting two rooms and injuring four children slightly. At 69 Cowper Road, an incendiary bomb fell into a small water tank without causing any serious damage, while at 71 Cowper Road another incendiary caused a small fire.

The next bombs to fall were two grenades at 102 Shakespeare Road. One struck the coping of the house and another fell on the front steps. Considerable damage was done to no. 102 and the adjoining houses. Three more incendiary bombs were thrown into Barrett's Grove, Arundel Grove and St Matthias' Road, Stoke Newington, but no damage was caused.

Two HE grenades were dropped on Woodville Grove. These fell into gardens and did not explode. They were followed by three incendiaries and an HE grenade dropped in Mildmay Road, which caused very slight damage. From this point onwards to the Shoreditch Empire Music Hall no grenades were thrown. More incendiaries fell in Queen Margaret's Grove and King Henry's Walk without doing any damage; two, however, caused a fire in Ball's Pond Road in which two people were burnt to death, and a man and four women injured.

Incendiary bombs were dropped all the way down Southgate Road at close intervals, but fortunately they all fell into gardens or onto roadways and caused no damage. After crossing at Regent's Canal, an incendiary bomb was dropped at 6 Witham Street but only caused a slight fire, which was extinguished by the occupier.

The airship now veered more to the south-east, and dropped several incendiary bombs, causing only slight damage, until at 28 Hemsworth Street, Hoxton, where the premises were gutted, as were those at 31 Ivy Lance, Hoxton, where an incendiary bomb caused severe damage by fire and slightly injured a child. The next bomb, dropped at Bacchus Walk, Hoxton, destroyed the premises, and hit and seriously injured a soldier. Between this point and the Shoreditch Empire, three more incendiary bombs were dropped without causing any serious damage, two of them falling onto stone pavements.

Subsequently, four incendiary bombs were dropped together, three falling on the Shoreditch Empire Music Hall and the other on the house next door; the damage in both cases was slight. A grenade was also thrown at this point, and this fell onto the pavement in front of the music hall without causing any casualties. The audience was in the building at the time, and any tendency to panic was averted by the promptitude of the manager in addressing the audience from the stage.

The next bomb was an incendiary, and fell on the premises of Hopkins & Figg's, drapers of Shoreditch, without causing any serious damage. There were about thirty female assistants sleeping on the premises and the consequences might have been very serious had the bomb set the building on fire.

Three incendiary bombs fell on Bishopsgate Street Goods Station, but the fires were promptly extinguished by the men on duty. Two incendiaries and one HE grenade fell into Pearl Street, Shoreditch, but did no great damage. These were followed by three incendiary bombs in Princelet Street, and an HE grenade in Fashion Street, none of which caused any serious damage.

Altogether, four men (including two soldiers), two women and two children were injured in Hoxton and Shoreditch. Fortunately there were no fatalities.

The airship then passed over Whitechapel. Incendiary bombs which fell on Osborn Street, Whitechapel and near Whitechapel Church did no damage. A HE grenade fell into a large tank of water at the whisky distillery of Johnnie Walker & Sons, Whitechapel, followed by three incendiaries in Commercial Road East, none doing any harm.

LZ-38 began to turn due north-east and, at the same time, dropped seven HE grenades in close succession. The first of these fell in a yard at 13a Berners Street, injuring a horse, followed by two on the same spot in Christian Street, Whitechapel. The casualties here were severe, as two children were killed, with five people seriously hurt and five slightly injured.

Another HE grenade fell in Burslem Street, St George's, but failed to explode; followed by another in Jamaica Street and another in East Arbour Street, with similar results. The next bomb was also a grenade that fell on Charles Street, Stepney, but only broke some glass. The next two bombs were incendiaries at 130 Duckett Street and 16 Ben Jonson Road, Stepney, and these caused slight fires in both instances. They were closely followed by a grenade, which also fell on Ben Jonson Road, but caused no damage.

It is worthy of note that the commander of LZ-38 made no attempt to attack the docks which, at this point in the raid, lay only about 1 mile away from him to starboard.

A relatively large distance of 3 miles now elapsed before the next bombs were thrown. The fact that the airship was passing over the relatively thinly inhabited areas on each side of the River Lea could apparently be seen from the airship and, for this reason perhaps, no bombs were thrown hereabouts. The next bombs released were two incendiaries, which fell at Wingfield Road and Colgrave

Road, West Ham, but did no damage. A grenade in Florence Street, Leytonstone, caused little harm, as did others which fell at Park Grove Road, Cranleigh Road, Dyer's Hall Road and Fillebrook Road. The last bomb in Fillebrook Road, fell at about 11.35 p.m. The casualties at Leytonstone amounted to three people being slightly injured.

The total number of bombs dropped in the Metropolitan Police area were:

30 HE grenades at 5lb each = 150lb
89 Incendiary bombs at 25lb each = 2,225lb
This weight represents a total of 1 ton 1cwt and 23lb of bombs dropped on London.

LZ-38 now went off east, passing Brentwood at 11.55 p.m. and was spotted between Burnham and Southminster at 12.30 a.m. Her commander clearly had some difficulty in fixing the locality of his point of departure – no doubt the mouth of the Crouch – and hesitated for a few moments just as he had outside London, before deciding his position with regard to the coast. The airship was fired upon by an AA gun at Southminster and the mobile guns of the RNAS at Burnham. LZ-38 went out to sea at the mouth of the Crouch about 12.40 a.m. The estimated monetary value of the damage caused by the raid was estimated at £18,596.

The question of the height at which this airship was travelling was of some importance. At Shoeburyness her height was estimated at 7,500ft by the military authorities. The reports of the RNAS, however, speak of her as having passed near that place (probably on her return) at 10,000ft and, 'at no part of its journey does it appear to have descended much below this elevation'.

No action against the airship was taken by the AA guns in London, then controlled by the RNAS. The reason given for this inaction was the airship was so high that it was neither seen nor heard: 'There is no authentic case of anyone having been able to see it during its passage over London … it was faintly heard by the gun-station at Clapton.' This statement appears to be substantiated, but at the same time the accurate manner in which the airship followed the straight line of the Kingsland Road from Stoke Newington to Shoreditch, at a height of 10,000–11,000ft, even by moonlight, is remarkable.

So great a height was not attained by any of the other airships, whether army or naval, which raided London later in the year. Their height was between 7,000 and 10,000ft during the raids of 7 and 8 September, until they had got rid of their bombs and were going off. On 13 October, however, the height of 12,000ft was attained by a naval airship, but it was not until autumn 1916 that this became the normal raiding height.

One enterprising postcard manufacturer devised a card to help the population gauge the Zeppelins' range:

The 'Coin of the Realm' Zeppelin Range Finder –
When the diameter of the following coins held at arm's length subtends the Zeppelin broadside on:

The range of the Zeppelin is:

Threepence	4 miles
Sixpence or Half Sovereign	3½ miles
Sovereign	2¾ miles
Halfpenny	2½ miles
Half Crown	2 miles
Crown	1/3-5 miles
Penny and Farthing alongside	1/1-5 miles

In the aftermath of the first raids on London, the Metropolitan Police Commissioner issued the following advice published in leaflet form:

POLICE WARNING

WHAT TO DO WHEN THE ZEPPELINS COME

Sir Edward Henry, Commissioner of the Metropolitan Police, has issued a series of valuable instructions and suggestions as to the action that should be taken by the ordinary householder or resident in the event of an air raid over London.

New Scotland Yard, S.W.

In all probability if an air raid is made it will take place at a time when most people are in bed. The only intimation the public are likely to get will be the reports of the anti-aircraft guns or the noise of falling bombs.

The public are advised not to go into the street, where they might be struck by falling missiles; moreover, the streets, being required for the passage of fire engines etc should not be obstructed by pedestrians. In many houses there are no facilities for procuring water on the upper floors. It is suggested, therefore, that a supply of water and sand might be kept there, so that any fire breaking out on a small scale can at once be dealt with. Everyone should know the position of the fire alarm post nearest to his house.

All windows and doors on the lower floor should be closed to prevent the admission of noxious gases. An indication that poison gas is being used will be that a peculiar and irritating smell may be noticed following on the dropping of the bomb.

Many inquiries have been made as to the best respirator. To this question there really is no satisfactory answer, as until the specific poison used is known an antidote cannot be indicated. There are many forms of respirator on the market, for which special advantages are claimed, but the Commissioner is advised by competent experts that in all probability a pad of cotton waste contained in a gauze to tie round the head and saturated with a strong solution of washing soda would be effective as as filtering medium for noxious gases, and could be improvised at home at trifling cost. It should be damped when required for use and must be large enough to protect the nose as well as the mouth, the gauze being so adjusted as to protect the eyes.

Gas should not be turned off at the meter at night, as this practice involves a risk of subsequent fire and of explosion from burners left on when the meter was shut off. This risk outweighs any advantage that might accrue from the gas being shut off at the time of a night raid by aircraft. People purchasing portable chemical fire extinguishers should require a written guarantee that they comply with the specifications of the Board of Trade, Officer of works, Metropolitan police or some Fire Prevention Committee.

No bomb of any description should be handled unless it has shown itself to be of incendiary type. In this case it may be possible to remove it without undue risk. In all other cases a bomb should be left alone and the police informed.

E.R. HENRY

4/5 June 1915

SL-3, commanded by Kapitänleutnant Fritz Boemack, was first reported 85 miles east of the Humber at 7.30 p.m. She came in very slowly, lying off the Yorkshire coast until dark, and it was not until 10.55 p.m. that she was off the coast at Ulrome. She then moved northwards to Fraisthorpe, after which she returned southwards to Ulrome at 11.20 p.m. where she made landfall at 11.45 p.m. She proceeded out to sea again, her commander being apparently uncertain of his position, and slowly followed the coast northward until she made Flamborough Head.

This landmark appears to have been recognised, since at 12.30 a.m. when 1 mile north of the head, the Zeppelin came overland again and went inland south-west, dropping an incendiary bomb between the villages of Kilham and Langtoft, which did no damage.

SL-3 carried on south-west to Driffield where, after hovering for a time, at 1.05 a.m. she dropped two HE bombs, one in a field and the other in a garden, both only causing minor damage.

She attempted to turn southward to Hull but, probably owing to a ground mist rendering visibility difficult, her commander gave up and turned back, passing Fraisthorpe again at 1.10 a.m. and Bridlington at 1.15 a.m., going north.

Ten minutes later, the Zeppelin fled seaward at Flamborough Head under rifle fire from the coastguard.

The raider flew very low. She was clearly seen along the coast between Ulrome and Bridlington. At Flamborough, when going out to sea, she was claimed to have been only about 1,000ft up. This low altitude clearly marks the raid as, in fact, a reconnaissance – the first visit of a Zeppelin to that part of the coast, which would afterwards become so well known to the German airship commanders.

L-10, under Kapitänleutnant Klaus Hirsch, was sighted between 9.35 and 9.55 p.m., south of the *Sunk* lightship, by four armed trawlers *Resono*, *Lord Roberts*, *Cygne* and *Zephyr* – the latter opened fire on the Zeppelin. At 10.45 p.m., she was between Gunfleet and Foulness, going south-west towards Shoeburyness. At 11 p.m. she passed over Sheerness and turned south-east over Sheppey to Whitstable where, at 11.10 p.m., she turned again north-west, circling in the direction of Faversham, over which she passed five minutes later, going north-east.

Between 11.20 p.m. and 11.30 p.m. she was over Sittingbourne. L-10 circled twice over the town and camp, dropping three HE bombs and eight incendiaries. One house was burnt out, two others damaged, and a man and a woman injured.

The raider then went off south-west towards Maidstone, and eventually turned towards Chatham where she was reported overhead and fire was opened on her at 11.45 p.m. One HE bomb was dropped in reply at Rainham, but there were no casualties.

She went on westwards to Tilbury, where she was again fired upon at 11.57 p.m., here she hovered for half an hour, which was no doubt utilised in careful observation, the Zeppelin making two circuits of figures of eight.

Meanwhile, at 12.25 a.m. she dropped five HE and three incendiary bombs on Gravesend. The yacht club, which was being used as a military hospital for wounded was burnt, but all patients were safely removed to Chatham. Other considerable damage was done to property; two houses were demolished and a third badly damaged; a stable with four horses in it was also destroyed. Six people were injured (two men, three women and one child), none of them seriously.

The Zeppelin then went off towards the north-east before the wind, passing Laindon at 12.45 a.m., Manningtree at 1.05 a.m., Holbrook at 1.10 a.m., Wherstead at 1.12 a.m., Ipswich at 1.15 a.m., Woodbridge at 1.20 a.m., north of Orford at 1.25 a.m., Aldeburgh about 1.28 a.m. and Leiston at 1.30 a.m., before going to sea near Dunwich at 1.35 a.m. She was last heard of leaving Southwold at 1.40 a.m. heading north-east.

She had been lying low over Suffolk as if trying to get bearings on the coast, and was either seen or distinctly heard at all the places mentioned above.

The total monetary value of the damage caused by the raid was estimated at £8,740.

6/7 June 1915

On the afternoon of 6 June 1915 two naval Zeppelins, L-9, commanded by Kapitänleutnant Mathy, and probably L-10 under Kapitänleutnant Hirsch, left their sheds in northern Germany to bomb England. The army airships LZ-37, LZ-38 and LZ-39 also rose from their sheds in Belgium, but only LZ-37, commanded by Oberleutnant van der Haegen, and LZ-39, commanded by Hauptmann Masius, came over the sea.

LZ-38, under Hauptmann Linnarz, descended almost immediately. LZ-37 and LZ-39 were unable to find the English coast, probably owing to fog.

On her return journey, LZ-37 was destroyed at 3 a.m. by Flight Sub Lieutenant Reginald Warneford, RNAS, over Mont St Amand, near Ghent; the crew were killed with the exception of one man who had a miraculous escape. The burning airship tragically fell onto a nunnery and two of the nuns were killed.

LZ-39 returned safely to her shed, but LZ-38 was also destroyed, in an aeroplane attack on her shed at Evere, near Brussels at 2.30 a.m. the same morning.

After Flight Sub Lieutenant Reginald Warneford successfully brought down LZ-37 on 7 June 1915, he was awarded the Victoria Cross and enjoyed great celebrity, even after his death in a tragic flying accident ten days later on 17 June 1915.

L–9 and L–10 both reached the Norfolk coast at about 7.30 p.m., when L–10 had to abandon the raid and turn back after suffering engine trouble. Mathy, in L–9, which was clearly identified 12 miles north-east of Mundesley at 8.15 p.m., went on.

At 8.40 p.m. the Zeppelin was reported from Cromer as going north out at sea. Five minutes later, she was steering south-west, 2 miles out. At 9.40 p.m. L–9 was 2 miles from the Lincolnshire coast, off Sutton-on-Sea, still steering north-west. Apparently, after reconnoitring the Norfolk coast, Mathy had given up the idea of raiding in that direction and determined on visiting the Humber instead.

Arriving off the Lincolnshire coast, he employed the same reconnoitring tactics before deciding his course. At 9.50 p.m. he moved over the coast at Theddlethorpe, flying west, and at 10.10 p.m. was reported by HMS *Thrasher* patrolling off the Spurn. At 10.40 p.m. he was heard off the Yorkshire coast at Withernsea, where he dropped a flare, and five minutes later at Hornsea.

There was a good deal of fog off the mouth of the Humber, and Mathy was evidently unwilling to try the approach to Hull by way of the river. He therefore went north along the coast instead, to Flamborough Head, which he found and identified at about 11.10 p.m. A flare was dropped at Bridlington.

L–9 then turned and proceeded straight down towards Hull, passing south-west of Hornsea, and dropping two incendiary bombs which did no damage at Wyton Bar. She reached the neighbourhood of Hull and Sutton at about 11.35 p.m.

Mathy reconnoitred in a south-easterly direction for ten minutes, evidently trying to fix his position. The river was enshrouded in a thick mist, which explains the time taken up in finding the exact position of the city. At this moment, however, the mist seems to have cleared from over Hull itself and, taking immediate advantage of the opportunity, L–9 bore directly down upon the city from the direction of Hedon, over which she passed. At 11.48 p.m., the first bombs were dropped.

One HE and one incendiary bomb fell in the Alexandra Dock, a second incendiary fell on the stern of the Swedish SS *Igor* lying in the Victoria Dock basin, doing slight damage. A third fell through the roof of a timber shed on the quay at the basin, damaging it and causing a small fire, which was extinguished by NER (North Eastern Railway) firemen. This was followed by an HE bomb on the railway lines close by, making a crater which destroyed about 12ft of track.

Another incendiary bomb was promptly extinguished in Alexandra Terrace, Delapole Street, while an HE bomb exploded in St Paul's Avenue, Church Street, partially destroying three houses. Several people were injured.

An incendiary in Feather Lane dropped on the footpath, doing no damage, but an HE in East Street then demolished two houses, killing a man, a woman and a child. Two incendiaries in Church Street and Clarence Street had no effect, and a third in a grain warehouse in High Street was smothered by the loose grain and extinguished. Two more incendiaries and an HE bomb in the same street injured five people and damaged property within a radius of 500 yards.

Three incendiary bombs then set fire to a large drapery establishment in the Market Place, which was completely burnt out, and to a public house at the Old Corn Exchange, causing a small fire.

Hull Market Place after the Zeppelin air raid of 6 June 1915. Holy Trinity Church was saved by the excellent work of locally billeted troops and the fire brigade.

Holy Trinity Church, which stood a short distance from the drapery shop, was saved by the excellent work of the troops and fire brigade, only some of the outer protective glass of the windows was destroyed by the heat. An incendiary bomb in Mytongate did no damage, but another at the Golden Gallon Inn dropped through the roof and did considerable harm before it was extinguished, and two incendiaries landing in Blanket Row killed a man and two boys, and injured a third boy.

Three HE bombs fell together into the Humber Dock; the concussion of the explosion broke windows in the surrounding buildings. Two incendiaries then caused a small fire in Myton Place, north-west of the dock.

The Zeppelin, which had hitherto pursued a sinuous course over the docks and the old town of Hull, following to a great extent the line of the quays in

The damage inflicted by the HE bomb dropped on Edwin Place, Porter Street, Hull, where four houses were demolished and several damaged. Four people were killed in this area and several injured on 6 June 1915.

order to bomb warehouses, now went off over the western portion of the city as far as the line of the North Eastern Railway, dropping bombs as she went.

One HE bomb fell in Edwin Place, Porter Street, demolishing several small houses and killing three people. In nearby Sarah Ann Place, an incendiary bomb killed another person. In Albert Terrace, Pease Street, an HE bomb smashed several windows and did other damage. In Great Thornton Street, Walker Street, Goodwin Street and Campbell Street five incendiary bombs did slight damage to houses, set a hayloft on fire and injured two people.

In St Thomas's Terrace, an HE bomb totally demolished two houses and shattered the roof of St Thomas's Church, killing two women and a man and injuring two further women. In South Parade, Regent Street, Coltman Street, Gee Street, Cholmley Street and Constable Street, seven incendiary bombs killed two girls and burnt two motor cars, but only did very slight damage to houses.

The last bomb, an incendiary, dropped on Selby Street at the west end of the city, close to the railway. It fell through a roof, but caused no other harm. The Zeppelin then turned east and flew back the way it had come, but this time passing north of the Paragon Station, and dropped no more bombs until it was north of the Victoria Dock.

Two incendiary bombs were then dropped in Danson Lane; one fell in the playground of St Mary's School, wrecking all the windows of the school, and the other set fire to a furniture warehouse. A third incendiary bomb was dropped, along with an HE bomb, on a large sawmill in the same street, destroying it. The fire was extinguished by NER firemen.

An incendiary bomb set fire to a house in Bright Street, which was burnt down. Another fell on the railway lines near Waller Street, slightly damaging the track and breaking the windows at Southcoates Station, followed by an HE bomb in the same street, which entirely destroyed three houses and damaged others, killing a man, three women and one child. A further incendiary bomb was dropped on the Rovers' football ground, followed by another in Woodhouse Street, doing no damage. These were the last bombs dropped. Three or four of the incendiary bombs had not ignited.

L-9 had approached Hull at a great height, probably 8,000–10,000ft, but her commander brought her down much lower as soon as he realised that no anti-aircraft gun defence was to be feared, and bombed the city from a height of about 5,000ft. No anti-aircraft guns were mounted at Hull and the only defence that could be made was by the guns of HMS *Adventure*, which was under repairs at Earle's Yard.

In all, five men, thirteen women and six children had been killed, and twenty men, thirteen women and seven children injured during the raid – a total of sixty-four casualties.

Those who lost their lives as a result of this raid were:

- Georgina Cunningham (27), 22 Edwin's Place, Porter Street
- Elizabeth Picard Foreman (39), 37 Walker Street (Died from shock 7 June 1915)
- Joanna Harman (67), 93 Arundel Street (Died from shock 7 June 1915)
- George Hill (48), 12 East Street
- Jane Hill, (45), 12 East Street
- A. Johnson, Campbell Street
- Edward Jordan (10), 11 East Street
- Alfred Matthews (50), 11 Waller Street
- Hannah Mitchell (42), 5 Alexandra Terrace, Woodhouse Street (Died from shock 7 June 1915)
- George Mullins (15), 39 Blanket Row
- Norman Mullins (10), 39 Blanket Row
- Emma Pickering (68), Sarah Ann's Terrace, Porter Street
- Maurice Richardson (11) and Violet (8), brother and sister, 50 South Parade
- Sarah Ann Scott (36), The Poplars, Durham Street (Died from shock 8 June 1915)
- Eliza Slade (54), 4 Walter's Terrace, Waller Street
- Tom Stamford (46), killed at 5 Blanket Row
- Ellen Temple (50), 20 St James' Square, St James' Street (Died from shock 8 June 1915)
- Alice Priscilla Walker (30), 2 St Thomas's Terrace, Campbell Street
- Millicent Walker (17), 2 St Thomas's Terrace, Campbell Street
- William Walker (62), 2 St Thomas's Terrace, Campbell Street
- Annie Watson (54), 21 Edwin's Place, Porter Street
- William Watson (67), 21 Edwin's Place, Porter Street
- Florence White (30), 3 Waller Street
- George White (3), 3 Waller Street

In going off, the Zeppelin followed the line of the Humber to the sea, passing over Paull at about 12.10 a.m. and over Sunk Island ten minutes later. At 12.20 a.m. the Maxims at Sunk and Stallingborough opened fire, supplemented by rifle fire at the latter place. A minute later, the pom-pom at Immingham and a Maxim on the SS *Kale*, lying off the coast, joined in the fire. L-9 veered off eastwards and then went on, flying high, towards Grimsby, which she reached at 12.25 a.m. She stopped her propellers and hovered for a short time, after which she circled over the docks and part of the town. Seven incendiary bombs, one of which did not ignite, were dropped on the docks, damaging three railway trucks. No casualties were caused.

Rifle fire was aimed at her from New Clee. She then left in a south-easterly direction, but was met with pom-pom fire at Waltham at 12.30 a.m. and was last seen at 12.40 a.m. from Donna Nook, flying south-east.

The total monetary value of the damage caused by the raid was estimated at £44,795.

PRECAUTIONS IN THE EVENT OF ZEPPELIN AIR RAIDS – 1915

Further air raids during the war saw new and renewed precautions announced to the public in the press, on posters and on hand bills. Lighting restrictions had been employed to reduce the danger of bombardment from the sea, and had been in place at the request of the Admiralty since September 1914. These had been tightened up and publicity renewed along the coast in the immediate aftermath of the Zeppelin raid of 19 January 1915.

A month after the first raid, the dangers of moving around streets in darkness and reduced lighting led to the edges of street kerbs at street corners being whitened with lime wash, along with the carriageway crossings of footpaths.

In many areas, emergency plans for the care of air raid casualties were made and published. A typical example comes from Norwich, Norfolk, in May 1915, when Dr H.J. Starling, chairman of the Norwich division of the RMA, circulated printed notices of the medical arrangements for casualties occurring in the city of Norwich caused by enemy aircraft. The climate of fear of air attacks during 1915 was such that, to avoid any panic, the doctor was quick to point out first in his announcement: 'It cannot be strongly emphasised that the existence of such a scheme in no way adds to the probability of such air raids taking place and is drawn up for the sole purpose of reassuring the public and to avoid congestion at the Norfolk & Norwich Hospital'.

The public were thus requested to direct all casualties in the first instance to the nearest dressing station. Seven schools were set up as dressing stations for the city of Norwich, namely Bull Close, Avenue Road, Crook's Place, Wensum View, Heigham Street, Surrey Road and Thorpe Road. Each school was provided with 'all appliances for first aid, dressings, splints, stimulants etc.' At least three doctors were appointed to each school, and would give first aid to all casualties brought there.

The Red Cross would also provide two sections, each with its own doctor and stretcher bearers to attend each school. They would work in the streets of the neighbourhood to search for and recover the injured, then direct, aid or remove them by stretcher to the nearest dressing station. The chief constable would also direct appointed members of the Norwich War Emergency Corps to assist those whose houses had been damaged by incendiary or explosive bombs.

Lighting restriction order notice for Norwich, 1915.

CITY OF NORWICH.

IMPORTANT MILITARY ORDER.

The Military Authorities Order that all Lights other than those not visible from outside of any House, Premises or Buildings, whatsoever, must be Extinguished in Norwich, between the hours of 5 p.m. and 7.30 a.m. to-day, and until further notice.

Signed **E. F. WINCH,**

Norwich, January 26th.   Chief Constable.

15/16 June 1915

Zeppelins L-10 and L-11 set out from their sheds at Nordholz, but L-11, under Oberleutnant zur See von Buttlar, had its crankshaft break 90 miles north-west of Terschelling and had to turn back.

L-10, commanded by Kapitänleutnant Hirsch, crossed the coast of Northumberland between Newbiggin and Blyth at 11.25 p.m., proceeding south-west. She came in at a height of about 6,000ft, flying very fast.

The objectives of this raid were the shipbuilding and engineering works on the Tyne. The airship made the same landfall as on 14 April, and pursued the same course to strike the Tyne – only this time far more certainly and directly. In both cases, the direct approach up the Tyne from its mouth was purposely avoided, mainly in order to elude attack by the batteries at the mouth of the river. The course taken passed over Newsham, Gosforth and Benton.

No bombs were wasted, as they had been on the former occasion, in the open country north of the river. The first were thrown at Wallsend. The course steered would have brought the airship direct to Newcastle but, when north-west of Wallsend, it was altered several points to the south-east and the airship bore directly down upon Palmer's Works at Jarrow.

Wallsend was reached at 11.40 p.m., the Zeppelin travelling at a speed of 60mph. Owing to the sudden appearance of the airship, and the fact that the telephone arrangements were not yet fully developed to deal with such emergencies, many of the Tyneside industrial establishments had not received the order to extinguish lights until the Zeppelin was upon them, and consequently showed a full blaze of light. Siren blasts sounded by HMS *Patrol* as a warning were also not understood, and Palmer's Works presented a perfect target to the enemy.

At Wallsend, one incendiary and two HE bombs were dropped south of the Infectious Diseases Hospital, one HE near Holy Cross Church and another at Burn Closes, none of these doing any damage. Two incendiary bombs then fell near the secondary schools at Church Bank, followed by six HE and three incendiary bombs on the North-Eastern Marine Engineering Works. Houses in Wallsend were affected and, though there was no fire, the engineering works suffered heavily, machinery being displaced and damaged to the value of £30,000. No casualties occurred. Slight damage was also done to Wallsend Colliery.

The Zeppelin passed over the Tyne to Hebburn, where one HE and one incendiary bomb fell on Hebburn Colliery, having only slight effect. These were followed by one HE and three incendiary bombs on the Ordnance Works football field in Jarrow, which broke windows in a large number of houses in the neighbourhood. One HE and two incendiary bombs then fell in Blackett Street, breaking more glass.

The airship turned north-east and went over Palmer's Works, dropping seven HE and five incendiary bombs which did considerable damage and injured seventy-two, besides breaking more glass in the houses of the town. One incendiary bomb then fell at the back of Berkley Street and another in Curlew Road to minor effect.

The Zeppelin recrossed the Tyne to the north bank and dropped bombs on Willington Quay and East Howdon. Two incendiaries and one HE bomb fell inside Cookson's Antimony Works, destroying a shed crane and weighing machines etc., followed by an incendiary bomb in Pochin's Chemical Works, two incendiaries in Stephenson Street and one HE bomb in Coach Open. The latter killed a police constable.

An incendiary bomb was dropped in the Tyne Commissioners' yard, another in Dock Street and two in Tyne View Terrace. A considerable amount of damage was done to small houses by these bombs.

The Zeppelin then pursued a straight east-north-east course to the sea, crossing the river again between Albert Edward Dock and Haxton Colliery staithes, where two HE bombs were dropped, with one falling into the river and the other on a fairground adjoining the staithes. The explosion set fire to a scenic railway and did considerable damage to the windows of surrounding houses.

L-10 went right over South Shields, and dropped another HE bomb between the Tynemouth Battery and Frenchman's Point, on the Bents Ground, just outside the area of the South Marine Park. Slight damage was done to the park railings and to windows in the neighbourhood at South Shields. No casualties were caused. Finally, an incendiary bomb fell on the sands. The Zeppelin went out to sea at 11.52 p.m. on an east-north-east course, flying at an estimated height of 6,000ft.

The response from the ground had seen the AA gun at Low Walker open fire when L-10 was at 3,000ft, and increased the range to 3,500ft for the second shot which, it was claimed, was very near. The guns at Carville and Pelaw opened fire at 4,000ft range and the pom-pom at South Shields at 5,000ft. In addition to these guns, a 4.7 gun of the Heavy Battery Durham Royal Garrison Artillery, stationed at Cleadon, fired five rounds of shrapnel at L-10. Rifle fire was lso opened on her at East Boldon and by men of No 3 Supernumerary Comp v, Durham RGA, at Jarrow. One house was damaged by an AA shell at Wallsend.

From a military point of view, this raid was one of the most effective that had been carried out by Imperial German airships or Zeppelins to date. Although the total monetary value of the damage, some £41,760, was somewhat less than the raid of 6/7 June on Hull, considerable military damage was inflicted, which was not the case in Hull. Owing to the low height at which she flew, the Zeppelin was able to choose her course carefully and bomb her objectives with accuracy.

9/10 August 1915

Five of the newest German naval airships (L-9, L-10, L-11, L-12, and L-13) set out from their north German sheds to raid England on 9 August.

L-9, under Kapitänleutnant Odo Loewe, was first sighted off Ulrome on the Yorkshire coast at 8.15 p.m. Five minutes later she passed over the Naval Air Station at Atwick, going south at a height estimated at about 3,000ft. The Zeppelin was fired upon by rifles from Skipsea. Two aircraft took off at 8.25 p.m. in pursuit, and the airship, which was then off Hornsea, immediately rose to 10,000ft, stopped her engines and drifted, evidently listening for the aeroplanes. On their approach she steered eastward out to sea and was lost in the fog. Ten minutes later, she reappeared to the northward at Fraisthorpe, 6 miles south of Bridlington, but again went out to sea and was next seen off Hornsea at 9.10 p.m., going north.

A naval aeroplane again rose from Atwick, and the airship promptly went out to sea again, followed by the aeroplane for some thirty-five minutes. The Zeppelin kept out to sea for an hour, finally coming in at Aldborough at 10.05 p.m., from where she passed over Atwick at 10.15 p.m. going west-north-west. Ten minutes later she struck the railway at Hulton-Cranswick and turned south along it, evidently making for Hull.

Before he had got far, the Zeppelin commander became uncertain of his position and, between 10.30 and 10.50 p.m., circled twice in the neighbourhood of Leconfield. Finally, he went off north-west of Beverley in a south-westerly direction and at 11.03 p.m. passed South Cave. At 11.10 p.m. he reached the Humber and followed it westward to Goole, which was showing lights and was severely bombed at 11.15 p.m. An empty goods train was travelling along the line from Hull to Goole and reached the Hook Bridge over the Ouse at Goole at 11.15 p.m. just at the same time as the Zeppelin, which apparently saw it and dropped three incendiary bombs just east of the bridge. Two more fell in the river.

L-9 then flew over the town from east to west, dropping eight HE and thirteen incendiary bombs on the town and docks. One of the HE bombs appears to have been of the large type, weighing 200kg, and two of the others were estimated to have been 100kg in weight. Ten houses were demolished, and several others badly damaged by explosion and fire, a dozen people were buried in the debris. A large shed on the dockside was badly damaged, and in Aire Street, which was exclusively shop property, although no building was actually struck, a great amount of damage was done by the concussion of a large HE bomb which fell on the dock wharf opposite. Almost every window in the street was destroyed and doors were torn from their hinges. An incendiary bomb went through the roof of the Exchange but to no effect. The total damage to the town was estimated at £7,000.

The fires were quickly brought under control, but sadly one man, nine women and six children were killed. Two women were seriously injured, and two men, two women and five children received minor injuries. All of them were civilians living in poor circumstances.

L-9 headed off westward, throwing sixteen incendiary bombs as she went, which fell between the Lancashire & Yorkshire Railway goods sidings and the north-eastern loop line between the docks and the Dutch River Bridge. No damage was done. The Zeppelin passed Snaith at 11.25 p.m. and then went north to Selby, where she passed over the Olympia Company's oil tanks and turned eastward at 11.30 p.m.

At 11.45 p.m. she dropped three incendiary bombs, one of which failed to ignite, in an open field at Hotham, 7 miles south of Market Weighton. No damage was done. Passing north of Beverley again, about midnight, she went south and then east, was seen at Sigglesthorne just before 12.10 p.m. and finally proceeded out to sea at Hornsea at 12.12 a.m.

There was a heavy ground mist the whole time the Zeppelin was over the East Riding, which was probably the chief cause of her inability to find Hull, after she had been confused as to her exact landfall by the action of the aeroplanes at Atwick.

L-11, commanded by Oberleutnant zur See von Buttlar, was first reported by HMS *Cynthia* 6 miles east of the *Shipwash* lightship at 9.50 p.m. She appears to have been in company with L-10. She made landfall at Dunwich at 10 p.m., swung north, passed over Wrentham at 10.10 p.m. then over Pakefield, where an HE bomb was dropped on a farm, doing slight damage to fences and windows. She then made for Lowestoft from the south-west.

At 10.18 p.m. L-11 began bombing the town. A HE bomb first demolished a wall and two greenhouses between the Avenue and the London Road. The airship then went out to sea near the Claremont Pier, and circled to the south, dropping four water flares, which fell in the sea and ignited. Having thus obtained the exact position of the town she came in again near the Grand Hotel and, circling northward, dropped six more HE bombs. The first wrecked two small houses and a shop in Lovewell Road, killing a woman and burying another woman and two children, who were rescued from the debris unhurt. The second landed in a back garden in Lorne Park Road, damaging walls and outbuildings. The third also fell in a back garden in Wellington Esplanade, doing similar damage, as did the fourth which wrecked another house. The fifth destroyed the roof and two of three rooms of a house, injuring a woman and three men of 2/4th Battalion, The Norfolk Regiment TF, who had been billeted in the house. The sixth fell on a pavement in Wellington Road, causing only slight harm. The total damage was estimated at £4,971.

Naval 6-pdr AA guns came into action against the Zeppelin, firing twelve rounds without visible effect. L-11 went out to sea in a north-east direction about 10.25 p.m. and an aeroplane of the RNAS rose in pursuit from Yarmouth at 10.31 p.m., but the pilot, who cruised at 5,000ft for about an hour, could see absolutely nothing. There were patches of dense fog and his aeroplane, which had to rise and land in fog was damaged on landing.

L-10, under the command of Oberleutnant zur See Friedrich Wenke, approached the coast in company with L-14 and was off Aldeburgh at 9.40 p.m., making landfall about ten minutes later. She passed over Leiston at 10 p.m. and reached Dennington at 10.10 p.m. Turning south-west, she passed Framlingham at 10.15 p.m., and Otley at 10.20 p.m. before turning westward at Tuddenham, just missing Ipswich, where the Zeppelin was heard on the north side of the town at 10.30 p.m. At 10.35 p.m. she turned south-west at Bramford and was next heard of at Coggeshall, where she suddenly appeared at 11.15 p.m., heading south.

She passed near Kelvedon, and over Witham and Wickham Bishops to Maldon, at about 11.30 p.m. Around 11.45 p.m. she dropped a flare at Little Stambridge, north of Rochford, and five minutes later was heard between the latter place and Leigh. At 11.57 p.m. L-10 passed over Shoeburyness, out into the Thames estuary and dropped her first bombs, evidently aimed at vessels in the river about the Nore. Five or six were heard to explode.

L-10 next appeared off Sheppey and passed over Eastchurch village at 12.09 a.m. Three minutes later she bombed the aviation ground, dropping six HE and six incendiary bombs but causing only slight damage to glass and no casualties. One of the HE bombs was of large size, apparently 100kg, and this appears to be the first instance of the use of such a large bomb in the aerial bombing campaign. The HE bombs fell close to the sheds and were extinguished by an infantry detachment.

At 12.08 a.m. two aeroplanes had risen from the aerodrome to engage the Zeppelin, but although they were in the air whilst she was dropping bombs, they could not see her on account of the darkness and fog, so they landed just an hour later. The raider passed on southwards, dropping two more HE bombs at Pumphill, ½ mile south of Eastchurch at 12.15 a.m. These did no damage. At 12.25 a.m. she passed Teynham and turned eastward along the Swale, passing between Leysdown and Faversham. She went out to sea near Whitstable about 12.30 a.m., passed near the *Edinburgh* lightship at 12.50 a.m. and the *Kentish Knock* at 1.05 a.m., being last reported passing the *Galloper* at 1.25 a.m.

L-12, under Oberleutnant zur See Werner Peterson, was first reported, in company with L-13, as approaching the *Edinburgh* lightship from the east at 10.15 p.m. Both Zeppelins, which were flying on parallel courses, were off the coast between Reculver and Birchington between 10.30 and 10.40 p.m. L-13 then apparently parted company, going on up the estuary while L-12 turned eastward, appearing over Westgate at 10.48 p.m. A naval aeroplane went up from Westgate on the approach of L-12, but failed to find her in the fog, and on returning to the aerodrome it crashed, killing the pilot, Flight Sub Lieutenant Reginald Lord (23). He was buried with full military honours in Margate cemetery.

The Zeppelin went on along the coast, being off North Foreland at 10.53 p.m., when she turned west and apparently went over the land to Ramsgate. She now hovered over the downs for over an hour and, at 12.05 a.m. was spotted over Kingsdown, drifting seawards. At 12.28 a.m. she approached Dover from the south and passed over the harbour at a height of about 5,000ft, dropping bombs while under fire from the anti-aircraft guns of the defences. The accuracy of this fire was evidently unexpected; as soon as it opened the Zeppelin endeavoured to get out of it and swung round to the eastward. The second or third round from the 3in gun (the only one at that time mounted) hit her fairly, and she was seen to reel on being struck and her bows lifted slightly. She then dropped all her water ballast, rose swiftly and, at 12.35 a.m., disappeared in a south-easterly direction. She was pursued by both a RNAS and a RFC aeroplane but, owing to her banking to a great height, they were unable to engage her and soon lost sight of the Zeppelin in fog.

L-12 had dropped about a dozen bombs, all falling in the harbour, apart from three incendiaries which fell on Admiralty Pier. Two of these fell on the parapet and burnt themselves out, one fell through a corrugated iron roof near the transport office, setting fire to the platform, but was soon put out. A HE bomb burst under the bows of the trawler *Equinox*, wounding one man severely and two slightly. No other casualties or damage were caused.

The injury to the Zeppelin had proved serious. Making for Belgium, L-12 was losing gas and was compelled to descend gradually to the surface of the water when only a few miles out of Ostend. She was taken in tow by German torpedo boats which she had summoned to her assistance. She was located at about 9 a.m. by a British naval aircraft from Dunkirk, 3 miles west of the port, who bombed L-12 from as low as 500ft and then returned. Two other aeroplanes arrived and continued to bomb the airship, which was by then low in the water, still being towed. By midday she made it into Ostend harbour alongside the quay, where she was bombed again. At about 2 p.m., while workmen were attempting to haul her up onto the quay, she was found to have been so severely damaged by bombs that her back broke and her rear half slipped back into the water. Her hull was salvaged but the rest of the airship was scrapped on the spot.

L-13, under the command of Kapitänleutnant Heinrich Mathy who had taken the Zeppelin over with his old crew from L-9, was seen first in company with L-12 off the *Edinburgh* lightship at 10.15 p.m., and appeared off Reculver with her half an hour later. There they separated and she went up the river, being reported from Wakering Haven and off Shoeburyness at 11.10 p.m. She returned to the mouth of the Thames and, at 11.55 p.m., was off Foreness. There, her forward engine broke down and she drifted for an hour.

By 12.48 a.m. she was off the North Foreland, going south and, at 12.59 a.m. began dropping bombs in Sandwich Bay. She then went off eastward, going home over Belgium, where it is known she had a further mishap to her engines and her radiator burst. The official report issued at Berlin with regard to the raid stated:

> On the night of 9–10 August, our naval airships raided several fortified places and ports on England's east coast. In spite of violent efforts to offer resistance, bombs were dropped on British warships in the Thames, on the London Docks, the torpedo-boat base at Harwich and important establishments on the Humber. Good results could be observed. The airships returned from their successful enterprise.

From this statement it may be deduced that the commander of L-9 eventually realised that he had not found Hull; that the commander of L-11 mistook Lowestoft for Harwich; and that the commander of L-10 either mistook Sheppey for London or, as is more probable, was deliberately lying in his claim to have bombed the London Docks. Neither the attack on Dover nor the destruction of L-12 was mentioned.

12/13 August 1915

On the night of 12 August, three Imperial German Navy Zeppelins, L-9, L-10 and L-11, were reported approaching the British coast. Only L-10 made landfall and attacked Woodbridge and Harwich.

L-10, commanded by Oberleutnant zur See Wenke, came over the coast at Pakefield at 9.25 p.m. and flew south-west, passing between Kessingland and Carlton Colville at 9.30 p.m. and was fired on at Benacre by Maxims of the Shropshire Yeomanry, without result. She passed near Covehithe at 9.35 p.m., over Brampton at 9.45 p.m., Halesworth at 9.50 p.m., Walpole at 9.55 p.m., Peasenhall at 10 p.m., and on to Bassingham where, about 10.05 p.m., she dropped a couple of petrol tanks. Turning south she passed between Great Glemham and Framlingham, then hovered over Wickham Market at 10.15 p.m., where she dropped two flares.

Going southward to the railway, at Pettistree she dropped an incendiary bomb, which did no damage, followed by another, which fell on the main road at Melton, also to no effect. The airship now turned along the line of the railway to Woodbridge and, just outside the town, she was attacked by machine guns and rifle fire by troops of 2/3rd Battalion, London Infantry Brigade. This drew the return fire of the airship which, at about 10.25 p.m., dropped four HE and twenty incendiary bombs on the town. Five cottages were destroyed and one partly wrecked; sixty-four other buildings and a church were damaged.

The wreckage caused on St John's Hill, where the first bomb fell.

The house on St John's Hill where Mrs Bunn was killed.

The 'x' shows the house where Mr and Mrs Tyler were killed.

The funeral of the air raid victims. A total of six people were killed and seven injured during the raid.

The total amount of damage was estimated at about £2,250. Six people were killed and seven injured. In the German communiqué it was stated that 'blast furnaces' at Woodbridge were bombed. Woodbridge was at the time a purely agricultural market town, without factories of any kind.

Leaving Woodbridge at about 10.30 p.m. the airship went on towards Ipswich, dropping three HE and three incendiary bombs at Kesgrave, which did no damage, until at 10.40 p.m. she was attacked over Rushmere Heath by AA motor machine guns of the RNAS. She dropped two HE bombs in reply, which exploded near the guns but did them no injury, and then turned away from Ipswich towards the River Orwell, which she crossed at 10.55 p.m. near Pin Mill. The Zeppelin passed westward for a short space and was over Woolverstone at 11 p.m.

L-10 then turned south, passing Holbrook at 11.05 p.m. and slowly veering round towards Harwich. Around 11.10 p.m. she dropped one HE bomb on Shotley Marshes, which naturally did no damage, and then crossed the River Stour. Shortly before 11.15 p.m. eight HE and four incendiary bombs fell on Parkstone. Two of the HE and two of the incendiary bombs fell in the river and one HE in the mud. Two houses were wrecked and a number of windows broken by the others. Seventeen people were injured, and damage was done to the telephone wires and communication with the Ray Hill gun cut off. This gun, and that at Beacon Hill, fired without result. An aeroplane was ready to go up, but did not do so as it was considered that the searchlights would have dazzled the pilot.

L-10 was claimed to have been attracted to Harwich by lights shown in the harbour by the naval authorities in order to allow a squadron of destroyers to enter and take up their moorings. After bombing Parkstone, L-10 went over Fagbury Point and Trimley, behind Felixstowe and out to sea near Bawdsey at about 11.25 p.m. after being fired at by Maxim gun cars of the RNAS at Shingle Street. She went along the coast, came in over Orford at 11.35 p.m. and finally went out to sea north of Orfordness shortly afterwards.

L-11, commanded by Oberleutnant zur See von Buttlar, was first heard off Harwich at 9 p.m., then over the Thames estuary and Thanet at about 10.10 p.m. She appears to have gone east into the Straits of Dover and was seen off Deal heading towards land at 10.45 p.m. She was off Westgate at 11.22 p.m., then seems to have wandered around over the estuary for a time, before being heard from Wakering Haven in Essex at 11.45 p.m., from Warden Point in Sheppey at 11.50 p.m., and from the Nore at 11.55 p.m. She finally turned eastward, was heard between Broadstairs and Ramsgate at 12.30 a.m. and seems to have gone out to sea south-east of Deal at 12.48 a.m.

L-9, commanded by Hauptmann Stelling, seems to have followed L-10 to the neighbourhood of Lowestoft and was vaguely reported in the neighbourhood of Corton, but a propeller shaft became loose and she was forced to return without coming overland at all.

The Intelligence report remarked of Stelling that he was:

An officer who had been a well-known pre-war pilot of the Parseval Airship Company. Shortly before the war he had piloted a Parseval airship to England for the British Government, which had acquired it, and flew it over London.

During the war, however, Stelling would never reach the capital. The report also commented:

The inactivity of L-11 was very characteristic of her commander. All three airships were probably intended to come to London. L-10 certainly gave up the attempt owing to the mist. Her commander found a very good substitute in Woodbridge.

17/18 August 1915

Zeppelins L-10, L-11 and L-14 crossed the North Sea on this evening in August. Only two actually raided England: L-10 came over the coast of Suffolk, attacked London and dropped eleven bombs in Kent; while L-11 headed straight for Kent. L-14, on her maiden voyage, merely approached the Norfolk coast and then went away after dropping her bombs in the sea.

L-14, commanded by Kapitänleutnant der Reserve Alois Böcker, was first reported from the *Would* lightship at 8.20 p.m. At 9.45 p.m. she passed the *Haisborough* lightship, going north and then seems to have turned, passing the lightship again at 10 p.m. and going down the coast. At 10.40 p.m. she passed the *Cross Sand* lightship heading north-north-west, and at 10.50 p.m. returned at a height of only 200ft and dropped an incendiary bomb over the lightship, and missed. She was near Yarmouth at 11.10 p.m., and at 11.25 p.m. was off Caister, going seaward. She was next reported from the *Cross Sand* lightship again, south of which she dropped a number of bombs, estimated at fifteen or twenty, into the sea and fired machine guns at a steamer. At 11.40 p.m. she was reported from the *Corton* lightship, and thereafter no further reports of her were received; she no doubt made off directly homewards.

L-10, under the command of Kapitänleutnant Klaus Hirsch, made the English coast at 8.25 p.m. at Sizewell Gap, and then rounded Orfordness into Hollesley Bay, by 8.50 p.m. At 8.56 p.m. she came over land at Shingle Street, where she was fired at by two Maxim gun cars of the RNAS. She was feeling her way cautiously, no doubt in order to be certain of her landfall and to avoid the Harwich defences, which her commander did not intend to attack. At 9.05 p.m. she passed Shottisham, crossed the River Deben and the Orwell, behind Harwich, and at 9.15 p.m. was at the back of Shotley. She passed Holbrook and, at 9.35 p.m. reached Manningtree, there striking the railway, which she followed for the greater part of the journey to London.

At 9.32 p.m. she turned south of Stratford St Mary and at 9.40 p.m. passed Colchester. Ten minutes later she was at Kelvedon and, after that at Witham. At 10.08 p.m. she went over Writtle and at 10.17 p.m. was at Ongar, where she dropped a petrol tank. She then sighted and followed the railway to Epping and went on towards Waltham Abbey, where she was fired at by a 1-pdr AA gun.

L-10 turned south-west to Chingford, where she arrived at 10.30 p.m., and at about 10.34 p.m. dropped her first bombs at Walthamstow. The first, an incendiary, fell in Lloyd's Park, doing no damage. A third, that fell close by, on the Stratford Co-operative Store buildings in Hoe Street, penetrated the roof and floor of a shop but the fire was quickly extinguished by a watchman and did not set fire to the premises.

The Zeppelin had now picked up the Midland Railway line from Tottenham to Forest Gate, and followed this more or less throughout her course over London. The first HE bomb now fell on Baker's Avenue, Leyton, demolishing four flats and shattering all the windows in the street. Four men, two women and two children were injured. Three more HE bombs then landed close by in the grounds of the Baker's Almshouses, Lea Bridge Road, shaking the buildings, breaking the windows and seriously injuring one woman. Another HE bomb landed in the roadway of the Lea Bridge Road, destroying the tramlines and pavements. One HE and three incendiary bombs fell in Lea Bridge Road, badly damaging the buildings of the tramway depot, causing a slight fire and breaking the windows of seven shops. One man was slightly injured.

The Zeppelin then dropped an HE bomb in Dunton Road, badly damaging a house and breaking the windows of many others. This was followed by an HE bomb in Parmer Road, south of the railway, which also severely affected a house. L-10 dropped four HE bombs on and around the railway station at Leyton, wrecking the station booking office and a billiard saloon, as well as seriously damaging one house and breaking windows in a large number of others, and a Wesleyan Chapel. It also badly damaged a steam roller. Three men and one woman were killed; seven men, four women and three children were injured.

Two incendiary bombs were dropped in Grosvenor Road, to very little effect, while a single HE bomb fell in Claude Road, wrecking one house and breaking glass in 175 others in the neighbourhood. A man, a woman and a child were killed.

An incendiary in Murchison Road did scarcely any damage, but the HE bomb and two incendiaries which fell in Albert Road badly damaged one house and caused small fires. A woman and a child were slightly hurt. Another incendiary in Twickenham Road also caused a fire but no casualties.

There were two HE bombs in Oakdale and Ashville Roads respectively, which did a great deal of damage, badly wrecking thirty houses and breaking the windows of 123 others, besides killing two men and injuring four men, eleven women and six children.

The Zeppelin was now ¼ mile south of the railway line which she had been following, and bore eastward in order to regain it. Crossing the Great Eastern Railway line to Chingford, between Grove Green and Norman Roads, she dropped an incendiary bomb on the permanent way with no destruction. Serious damage was now done by a couple of incendiary bombs which fell on St Augustine's Church, Lincoln Street, in Leytonstone, gutting the building.

Another incendiary bomb in Mayville Road caused a fire in a kitchen, and an HE bomb in Southwell Grove Road demolished the backs of two houses and broke windows in 132 others, and a man was killed. Another HE and two incendiary bombs, dropped on Wanstead Flats just north of the Midland Railway and east of Montague Road, broke the windows of seventy-three houses.

The last HE bomb dropped on London by L-10 fell near the model yachting lake in Wanstead Flats. The concussion broke windows in seventy-five private houses in the neighbourhood, but caused no other damage. Four incendiaries followed on the Flats, one near the Woodford Road, two close to the bandstand and one at the head of Tylney Road, all having no effect. Two more incendiary bombs then fell, one in Aldersbrook Road and one in a pond near the terminus at the junction of that road with Wanstead Park Avenue. The former broke a fence, the latter blew all the water out of the pond but did no further damage. These bombs, which fell about 10.45 p.m., were the last of those dropped by the Zeppelin in the London district, but it seems that the airship was not seen when she was over Leyton by any gun crew, except one near Waltham.

She now went off parallel to, and on the north side of, the Great Eastern main line, passing between Ongar and Brentwood at 11 p.m. and reaching Chelmsford at 11.15 p.m. Here, two HE bombs were dropped, one exploding in a field west of the town without doing any damage, the other passing through a house in Glebe Road near the Marconi Works and embedding itself in the earth beneath the basement without exploding.

The airship, having passed north of Chelmsford, now crossed to the south of the railway and was south of Tiptree about 11.30 p.m., and ten minutes later, after passing north of Mersea Island, reached Wivenhoe, where she turned northward, passed to the east of Colchester and dropped a flare, which lit up the whole country with an intense glare, at Ardleigh about 11.43 p.m. At 11.50 p.m. she turned north-east near Stratford St Mary.

At midnight she was near Holbrook, and reached the Orwell at Wherstead five minutes later. She turned down the river towards Harwich, but at Trimley bend went off directly south-east to Felixstowe, thus again avoiding the guns of the fortress. She passed Felixstowe at 12.12 a.m., crossed the mouth of the Deben and went out to sea at 12.20 a.m. at Shingle Street, at the same spot where she had crossed the coast on her inward journey and under the fire of the same Maxim gun cars of the RNAS, again without detriment to herself.

On her return to Germany she appeared over the Dutch island of Vlieland early in the morning and was received by the Dutch with both gun and rifle fire. L-10 returned to base after this raid, but was destroyed in a storm three weeks later on 7 September 1915, when she was struck by lightning and burnt with all hands on board.

L-11, under the command of Oberleutnant zur See von Buttlar, passed the *Gunfleet* light vessel shortly before 9 p.m. and crossed the coast at Herne Bay about 9.30 p.m. She came in low over the pier from the sea, but rose to 7,000ft immediately on being fired at with rifles by men of the 42nd Provisional Battalion. She went directly over Herne to Canterbury, and at 9.35 p.m. crossed the city from east to west towards Chartham, following the railway to Ashford. At Chartham, she diverged from the line southward, and proceeded parallel with the straight line of Stone Street, which goes directly south from Canterbury with woods on either side and, from the air, would probably look like the straight line of a railway.

She passed over Hastingleigh about 9.45 p.m. and at Smeeth turned west along the railway. On approaching Ashford at 9.55 p.m. she first made a half circle round the town to the north, and went off directly overhead until she reached the Canterbury road where she turned north-west, dropping bombs. Eight incendiary bombs fell in gardens on the east side of the road and in a cemetery close by. No damage resulted, beyond killing a couple of fowls. These missiles were followed by ten incendiary and two HE Bombs which fell on fields near the sanatorium in the Maidstone road. A sheep was killed, but no other damage done.

L-11 then went away at about 10.15 p.m. in the direction of Faversham, turned north and when 2–3 miles east of Faversham, turned south again and retraced her course as far as Molash. Here, she circled in various directions, evidently seeking an objective and finally at about 11 p.m. dropped sixteen HE and twenty-five incendiary bombs in open country in the parishes of Badlesmere and Sheldwick near the Leeds Press VAD Hospital. Some injury was done to windows in Badlesmere Church and a few fruit trees, but otherwise no damage was reported and there were no casualties.

The AA gun at Faversham was unable to come into action owing to the electricity for the searchlight having been cut off from the Powder Works, since the managing director considered that the light was too near the works and betrayed their position.

At about 11.15 p.m. the Zeppelin went away in a north-east direction, passed over Hernhill and out over the River Swale at Graveney. She was heard over the Swale from the Isle of Sheppey at 11.30 p.m., then passed over the coast again and south of Whitstable, over the Blean Woods and north-west of Herne, where she was again subjected to rifle fire at about 11.40 p.m. by the 57th (West Lancashire) Divisional Cyclist Company. On being fired upon she veered off to the northward and finally passed out to sea about 11.45 p.m. at Herne Bay, being last heard from Shellness, going away into the Thames estuary.

The speed of L-11 varied greatly. From the Gunfleet to Ashford she flew at 60mph with the favourable wind, but when working northward to Faversham she went very slowly with many tacks and much circling.

The commander's orders were no doubt to attack London. The load of incendiary bombs carried can leave no doubt that the raider's intention had been to attack London in concert with L-10. On making land he abandoned the attempt, and it is not improbable that he next made up his mind to attack Dover from the land side. His courage failed him; doubtless the memory of the destruction of L-12 on 10 August, when attacking the same town, contributed to his pusillanimous behaviour. After circling over Ashford it would seem he made an attempt to find Faversham munition factories or Chatham, and failed to locate his objectives. On his return to Germany, von Buttlar claimed to have bombed London as he had been expected to do.

The total estimated value of the damage caused by this raid was £35,750.

7/8 September 1915

Three Imperial German Army airships set out from their sheds in Belgium to bomb London on 7 September.

LZ-77, under the command of Hauptmann Horn, was first seen from the *Kentish Knock* lightship going west at 10.05 p.m., and was next sighted over Clacton-on-Sea at 10.40 p.m. going due west. The coast-watching troops at Clacton fired a few rounds of small arms ammunition as she passed. She was over Brightlingsea about 10.50 p.m. and, changing course slightly to the north-west, passed south of Wivenhoe at 10.55 p.m. to Coggeshall which was reached at 11.10 p.m. She then turned south, was west of Kelvedon about 11.15 p.m. and over Witham at 11.20 p.m. Thence she went west to Great Walsham at 11.30 p.m. and, following the River Chelmer, north-west to Felsted at 11.40 p.m., then south-west again to Leaden Roding, where she was at 11.57 p.m. At about 12.15 a.m. she was over Hatfield Broad Oak, the most westerly point to which she penetrated.

The airship seemed to be wandering slowly, being perhaps in search of a suitable target or in a quest for some navigational marks. Her commander may either have realised that he had lost his way, or he may have thought that he had ventured far enough inland. The former supposition is rendered improbable by the fact that the glare of London was assuredly visible to him at this moment. It is not unlikely that he actually saw the searchlights sweeping the sky for LZ-74 and SL-2; this spectacle may have caused him to hang back.

He sharply turned his ship east-north-east and passed at a moderate speed over Dunmow at 12.25 a.m. and on to Bocking where, at around 12.35 a.m., he turned north-east passing between Cavendish and Sudbury at 12.50 a.m. and between Bury St Edmunds and Stowmarket about 1.05 a.m., arriving in the neighbourhood of Walsham-le-Willows at 1.15 a.m. Here, he altered his course again east and slightly south and dropped his first bomb, an incendiary, at Monk Soham at 1.30 a.m. The bomb failed to ignite and no damage was done.

Persisting on the same course, LZ-77 deposited its first HE bomb on Framlingham at 1.35 a.m. The bomb fell on farmland covered in wheat stubble and failed to act. The next bomb fell on Great Glemham at 1.40 a.m., where two incendiary bombs dropped among some farm buildings. Here a reaper and a self-binder were destroyed, the total loss being estimated at £7.

Passing over Benhall and circling over Saxmundham between 1.45 a.m. and 1.50 a.m., the airship followed the railway for about 2 miles in the direction of Leiston, turned north-north-east and passed over Theberton at 1.53 a.m. It was south of Southwold at 2.05 a.m.

LZ-77 followed the coast to Lowestoft and went out to sea there, in a north-east direction at 2.20 a.m. At 4.05 a.m., explosions were heard out to sea from the direction the airship had been taking. These were believed to be LZ-77 dropping its bombs in the sea, possibly directed at shipping.

SL-2, commanded by Hauptmann von Wobeser, was first heard at 10.15 p.m. north of the *Tongue* light vessel, going due west. At 10.50 p.m. she struck the coast at the mouth of the Crouch. She was flying very low, and was fired at with rifles from the lookout post at Holliwell Point. She immediately rose and went on westward, following the river to Burnham.

Going thence south-west, she reached Rochford soon after 11 p.m. and Wickford at 11.10 p.m., Billericay at 11.24 p.m., Brentwood at 11.27 p.m., Harold Wood three minutes later, Romford at 11.36 p.m. and finally Leytonstone at 11.40 p.m. Here, she turned south following the left bank of the River Lea and making straight for the Isle of Dogs.

Now she began to release her bombs, seven incendiary and one HE bomb falling at 11.45 p.m. in a line between Havannah Street and Gaverick Street, Millwall. The first failed to ignite. Eleven people were injured by the HE bomb which fell in Gaverick Street, demolishing three houses and severely damaging two others. She then dropped an incendiary bomb which hit a barge in the river off Snowden's Wharf, damaging it and severely burning two men. This was followed by an HE bomb which fell in the Thames mud above low-water mark on the Millwall side, and three incendiary bombs at Deptford at 11.50 p.m. The first fell in the Royal Victualing Yard, destroying a number of barrels of wine; the second fell on the Supply Reserve Depot, Army Service Corps, breaking the glass roof of a store but otherwise doing no damage; while the third fell in Hughes Fields, setting fire to a small dwelling house and killing the occupants, two adults and three children.

The airship then followed the Thames in the direction of Woolwich. She crossed Deptford Creek and dropped two incendiary bombs in Norman Road, one doing slight damage, and one in Greenwich Road, which caused none. Another bomb was dropped on South Street and one in Brand Street, neither of which ignited, and one on Royal Hill, which did no harm. A further bomb followed in Greenwich Park close to the Observatory, doing only slight damage to a fence, and three more in the park, one of which did not ignite. A bomb at Woodlands, Beaconsfield Road also did no damage.

A HE bomb that fell in the garden of Mr Justice Scrutton's residence at Mycenae Road did not explode, and was followed by an incendiary in Glenluce Road, and one incendiary and one HE bomb in fields near Victoria Road; the explosion of the latter broke a few windows but did no other damage. An incendiary bomb then fell in Charlton Lane, near the Gorse & Groom pub, slightly damaging a fence, followed by another in Pett Street, Woolwich. Finally, an HE bomb fell in Kingsman Street, Woolwich, which broke a number of windows, three people being slightly injured by fragments of glass.

Then the airship crossed the river. Observed in Woolwich, a 6-pdr gun in the arsenal opened fire at 11.52 p.m. with three rounds of common shell without effect. The airship was reported as flying very high (over 8,000ft) and fast. When across the river, she turned west again, appearing at 12.05 a.m. over Southwark, where she dropped two HE bombs in Keetons Road, damaging two houses and wrecking a greenhouse, followed by one HE in Ilderton Road, where a house was wrecked and six people killed.

Crossing the Surrey Canal, the airship dropped her next bomb, an HE, in Sharratt Street, where a house was badly damaged and five people seriously injured. Another HE bomb fell in Hunsdon Road, New Cross, injuring a policeman, followed by another in Monson Road, which damaged a school but caused no casualties. Two HE bombs then landed on sidings of the London, Brighton & South Coast Railway at New Cross, causing slight damage and injuring a policeman, followed immediately by another HE in Monson Road, which damaged a school but caused no casualties. Two more HE bombs again fell on sidings of the London, Brighton & South Coast Railway, this time damaging the permanent way and some rolling stock.

Two HE bombs destroyed a house and severely damaged two others in Childeric Road, killing a man, a woman and one child, and injuring two women and four children. These were followed by two more HE bombs, which fell on Cliften Hill, demolishing one house and damaging two others and a school in Angus Street. Two adults were killed, three adults and three children injured. The last bomb, also HE, fell in Edward Street, damaging 100 houses but causing no casualties.

The SL-2 then passed on between Brockley and Lewisham at 12.27 a.m. and apparently followed the railway to Chislehurst, where two airships were seen at 12.33 a.m. The other was LZ-74, which was now following SL-2. The latter ship passed over Bexley Heath and reached Erith at 12.40 a.m. Having found the Thames again, she followed it to Purfleet, where she crossed to Essex at 12.45 a.m. and steered a steady north-east course for the coast via Pitsea, Rayleigh at 1.05 a.m. and Mundon about 1.20 a.m. to Maldon at 1.25 a.m., past Thorrington at 1.50 a.m., Walton-on-the-Naze about 2 a.m. and Harwich at 2.10 a.m. Here she was fired at by pom-poms on Beacon Hill and Landguard, without result. At 2.15 a.m. she passed Felixstowe, and at 2.20 a.m. was out at sea, flying at an estimated height of 5,000–6,000ft and at a speed of 40–50mph.

Unfortunately the searchlights of Harwich could not be used as a naval aeroplane went up in pursuit from Felixstowe and it was considered that the light would have dazzled pilot Lieutenant Commander Hope Vere. The aircraft subsequently landed at Trimley and was wrecked, Hope Vere escaping unhurt.

SL-2 came to grief on her return to Belgium. After suffering a failure to one of her engines and being low on fuel she attempted to land and, whilst descending, went out of control and fell, striking a house and severely damaging herself. It was supposed, at the time, that she had been damaged by gunfire as a large hole was observed in one of the gondolas, but it is probable that this was a result of her accident since nothing seemed to be wrong with her until she reached Belgium.

LZ-74, commanded by Hauptmann George, came in over Bradwell-juxta-Mare at 10.55 p.m. and followed the course of the River Blackwater to Maldon, where she was heard at about 11.10 p.m. She was at Danbury at 11.15 p.m., was seen at Great Baddow at 11.20 p.m. and was near Chelmsford five minutes later. She then set her course for the valley of the Lea, via Ongar at 11.35 p.m., North Weald Bassett at 11.40 p.m. and Broxbourse at 11.55 p.m. There she turned south along the river and railway, and reached Cheshunt at about midnight, flying at an estimated height of 9,000ft.

Here, she threw her first bombs. Eighteen HE Bombs (one of which failed to explode) and twenty-seven incendiaries fell between Windmill Lane and Turkey Street. Astonishingly, no casualties were caused. Minor damage to property resulted in Windmill Lane, where the first HE bomb fell, wrecking six houses and slightly damaging five; while at the rear of Turner's Hill, nine HE and five incendiary bombs fell, seriously damaging a coach house and garage and breaking the windows of a large number of cottages.

On the Great Eastern Railway, Cheshunt and Edmonton branch, two HE bombs damaged the permanent way; and glass greenhouses were wrecked at nurseries in Park Lane and Bullsmoor Lane by six HE bombs, with three small fires caused by a further twenty-one incendiaries. No damage was caused by the last bomb, an incendiary, which fell in a field near Turkey Street, on the north side of the New River.

The airship was now engaged by AA guns at Waltham Abbey, which fired one round from a 3in gun and two from a 6-pdr without result. She followed the right bank of the River Lea towards London. Edmonton was passed at about 12.10 a.m. and Finsbury Park at around 12.15 a.m.

LZ-74 then crossed the Thames, travelled slowly south-east and was next reported over Peckham Rye, then at Honor Oak Park and Bromley. Turning north-east LZ-74 was again in company with the airship SL–2, at about 12.35 a.m. near Chislehurst, but at a considerably greater height. Here, following her sister ship, LZ-74 set her course for the north-east, passed Bexley Heath and crossed the river to Purfleet at 12.55 a.m.

She was then engaged by a 3in gun, which fired two rounds, and a pom-pom which fired nine. The raider was flying at an estimated height of 10,000ft and a speed of 40mph. No hits were observed, and the ship went on via Laindon at 1.07 a.m. to Chelmsford at 1.12 a.m. She was seen at Broomfield at the same time turning from north to north-east to follow the railway. She went on parallel with the line at a distance of a mile or two south of it, past Witham at about 1.40 a.m., Tiptree at 1.45 a.m., Colchester at 1.55 a.m. and Mistley at 2.05 a.m. Here she followed the railway and the River Colne to Harwich, where she passed at 2.12 a.m., two minutes after SL-2. She was only heard but not seen, and was not fired at.

LZ-74 then went out to sea, over the *Cork* lightship at 2.18 a.m. and returned to Belgium, passing over part of Holland.

8/9 September 1915

Naval Zeppelins L-9, L-13, L-14 and army airship LZ-77 were all involved in sorties on 8 September.

At approximately 1 p.m. messages were intercepted from four German airships that they only had 'HVB' on board. Intelligence knew what that meant – a copy of a German code book known as HVB (*Handelsschiffsverkehrsbuch*) that served as the official instrument of correspondence in the German Mercantile Marine had been captured in Austria early in the war, and another copy had been fished up later from the sea by a Lowestoft fishing trawler. The Germans knew that the code book had been compromised but continued to use it, with frequent changes of ciphering key, in official correspondence. When Zeppelin commanders set off on a raid, however, they always left behind the more confidential naval signal book, a fact they reported by wireless message, using the phrase 'Only HVB on board.' These signals were frequently intercepted and were seen as a good indicator that Zeppelins were then setting out from Germany to raid Britain.

L-9, commanded by Kapitänleutnant Odo Loewe, came in over the Yorkshire coast about 9.15 p.m. at Port Mulgrave, between Whitby and Kettleness. She went westward to near Mickleby and then north over Hinderwell to Staithes, where she circled and carried on westward to Ings House. Arriving there at about 9.30 p.m., she dropped her first bomb, an incendiary, which caused no damage. A petrol tank was dropped at the same time.

A HE bomb was dropped at West Loftus with the same negative result, and the Zeppelin turned northwards, dropping four incendiary bombs near Carlin How, two in Scarfe's Field and two in Watson's Garden. None caused any harm. At 9.35 p.m. L-9 was over Skinningrove, where she dropped her main load of bombs, nine HE and twelve incendiaries, aiming them at the iron works. The water main and electric light cables were broken, two railway trucks partly burnt, a concrete step way at the jetty broken and other minor damage done. An incendiary bomb hit the top of the benzol house, which fortunately was composed of concrete, and so it did not penetrate, while an HE bomb dropped within 10ft of it, doing much of the damage mentioned above, but not injuring the house. Had this bomb hit the benzol house or tanks, which contained 45,000 gallons of benzol the greater part of the works would, in all probability, have been destroyed. Another HE bomb narrowly missed the TNT stores. No casualties were caused, the workmen having taken refuge in the neighbouring mines and elsewhere on the first alarm.

After circling above the works while dropping the bombs, L-9 then made off in the direction of Hinderwell, which she passed at 9.40 p.m., going towards Kettleness. There, she turned south and finally went out to sea at 9.45 p.m. near Sandsend. Three RNAS aeroplanes went up from Redcar Aerodrome in pursuit. Unfortunately, the atmosphere was hazy with Middlesbrough smoke, and the Zeppelin could not be seen, though one pilot searched for L-9 for 1½ hours.

The Intelligence report noted, 'that the benzol and other plant at the Skinningrove Iron Works was largely erected by German contractors and German workmen; there can be little doubt, therefore, that the enemy possessed full information as to the nature and topography of the works and that the raid was aimed at Skinningrove only.'

L-13, under Kapitänleutnant Heinrich Mathy, was first heard at sea off the Norfolk coast, north of Holkham at 7.35 p.m. where some armed trawlers opened fire upon her. She went out to sea again, and was seen coming back overland at Brancaster at 8.05 p.m. Mathy was feeling his way to the mouth of the Wash, which he did not reach until 8.25 p.m. He then turned, followed the coast and passed over Hunstanton at 8.30 p.m., near Sedgeford at 8.35 p.m., then went out to sea again and was seen by a fishing boat.

At 8.45 p.m., L-13 was over the mouth of the Lynn Cut, where the Zeppelin was observed by the SS *Annandale* going slowly up the river. For a full four minutes she stopped at an elevation of 1,500–2,000ft to observe her position. At 8.49 p.m. she started again, passing between King's Lynn and Terrington St Clement where she picked up the course of the Ouse near Wiggenhall St Germans. Mathy then set his course following the river and railway to Downham Market at 9 p.m., he followed the Bedford Level and headed for London where he dropped fifteen high explosive bombs across Golder's Green, in Middlesex. These bombs were probably intended for the aviation ground at Hendon. Five incendiary bombs were first thrown all together in a field adjoining Decoy Farm, followed by an HE bomb in another field close by, owned by the Express Dairy Company. No real damage was done.

An incendiary bomb fell on waste ground at the corner of Leeside Crescent and Prince's Park Avenue, and two in a wood adjoining Leeside Crescent, about 70 yards from Alba and Russell Gardens. These were followed by an incendiary bomb through the roof of an empty house in Highfield Road and one in the garden at the rear of 19 Alba Gardens. The last bomb thrown here was an HE that fell into a garden between Alba Gardens and Russell Gardens. The incendiary did very little damage but the second HE blew out the windows of a large number of houses. There were no casualties.

L-13 pursued her course over north-west London at a slow speed, passing presumably over Child's Hill, Frognal, Belsize Park, Primrose Hill and the north-east corner of Regent's Park to the neighbourhood of Euston Station, where she resumed her bombing with serious results. At 10.45 p.m. she dropped three incendiary bombs, two of which fell between Woburn Place and Upper Bedford Place and one in Russell Square, at the junction of Woburn Place and Southampton Row. A hotel in Bedford Place was slightly damaged by fire.

A HE bomb was dropped in the centre of Queen's Square breaking all the windows in the square, but luckily causing no serious harm to the buildings which included several hospitals. Five incendiary bombs fell in Osmond Yard, followed by others in East Street, Emerald Street, Theobald's Yard and Lamb's Conduit Passage where, at 10.49 p.m., an HE bomb was also dropped. A good deal of damage was done until the fires caused by them could be extinguished. One man was killed and sixteen other people injured in Lamb's Conduit Passage.

The HE bomb wrecked a public house in Red Lion Street and the premises of the National Penny Bank. Owing to the belief in Germany that the Bank of England was damaged, if not destroyed, in this raid, a photograph of this minor banking establishment was confidently claimed in the German press as proof of the belief.

The Zeppelin next threw one HE and one incendiary bomb between Bedford Row, breaking a large number of windows and destroying a house in Jockey's Fields which was used as a cycle club, and injuring four women.

At 10.51 p.m. she dropped seven incendiary bombs, two of them on Raymond Buildings, Gray's Inn, one in Gray's Inn Road and four between Clerkenwell Road and Portpool Lane, followed by one HE off Leather Lane. The incendiary bombs caused several fires, as before, and the HE bomb killed four children (one an infant), and injured six adults and one child in Laney's Buildings, Portpool Lane, besides severely damaging the buildings.

Two incendiary bombs then fell in Cross Street, Kirby Street, and Hatton Garden, doing a certain amount of damage, followed by an HE bomb which burst in Farringdon Road, greatly damaging premises there, and in Great Saffron Hill, but causing no casualties. The headquarters of 6th Battalion, London Regiment were also damaged.

After crossing the boundary of the City at about 10.56 p.m., L-13 passed over Smithfield Market, where three incendiary bombs fell in the roadway south of Central Avenue, and dropped one HE bomb (estimated to have been 300kg) and three incendiary bombs in Bartholomew Close without damaging the ancient and celebrated church, though most of the buildings in the close, and some in Little Britain, were damaged by fire or explosion. Two men were killed, and a boy and two women were injured.

Ten incendiary bombs were next dropped between Noble Street and Aldermanbury, occasioning the heaviest damage inflicted during a Zeppelin air raid on Britain. Several blocks of business premises in Silver Street, Wood Street, Addle Street and Aldermanbury were either entirely burnt out or seriously damaged by fire, the worst sufferers being Messrs Ward, Sturt & Sharpe, wholesale hosiers of Wood Street and Silver Street. The total damage to their premises and stock, chiefly woollen, cotton and silk goods amounted to £207,000.

L-13 dropped an HE and an incendiary bomb between Aldermanbury and the Guildhall. One block of offices was badly damaged, and most of the windows in Aldermanbury were broken. Two incendiary bombs were then dropped in Basinghall Street, beyond the Guildhall, which was not harmed. Several buildings in Love Lane and Basinghall Street were affected by fire, smoke and water. An incendiary bomb fell in Coleman Street.

L-13 bore off in a north-easterly direction towards Liverpool Street Station, on the way dropping three incendiary bombs on the searchlight at Salisbury House, between London Wall and Finsbury Circus; they fell on the roof and were at once extinguished. These were followed by an HE bomb on London Wall Buildings, Bloomfield Street, which wrecked the London Territorial Force Record Offices.

A further HE bomb fell at the corner of Bloomfield Street and Liverpool Street, destroying a motor omnibus, and killing three men and injuring several others.

A HE bomb landed in Sun Street Passage between Broad Street and Liverpool Street Stations, blowing up the pathway, damaging a brick railway arch and severely injuring one man. This was the last bomb thrown in the City during the raid. The next HE bomb fell on the Great Eastern Railway, north of Liverpool Street Station, damaging about 20ft by 10ft of permanent way and some arches used as stores. The last fell in Norton Folgate, destroying another motor omnibus, killing nine people (including the driver) and injuring ten, besides doing a great deal of injury in that street and Shoreditch High Street, where the Electric Light Transmitting Station was also damaged, though the machinery was untouched.

L-13 was seen over Old Street, then went off slowly north, over Dalston and Tottenham to Edmonton where she suddenly rose from about 8,500ft to a height of over 10,000ft and disappeared, emitting a cloud of greenish-grey vapour (water ballast). This was immediately after a shell, probably fired from Parliament

A HE bomb (estimated to have been 300kg) and three incendiary bombs were dropped on Bartholomew Close, London, by L-13, damaging many of the buildings by fire or explosion. Two men were killed, and a boy and two women were injured here on 8 September 1915. (IWM)

Motor omnibus destroyed by the HE bomb dropped by L-13, which fell at the corner of Bloomfield Street and Liverpool Street, killing three men and injuring several others, during the raid of 8 September 1915. (IWM)

Hill, appeared to burst very close to her. It was apparently the only shell that got anywhere near L-13. Firing had begun about the time the Zeppelin dropped its first bomb in Queen's Square and ceased suddenly after her disappearance, which was observed at the same time by all stations. It appears that the raider was fired upon by every gun of the London defences, including those at Woolwich and Erith, 11 miles away from Liverpool Street, the furthest south-easterly point reached by L-13.

A certain amount of damage, none of it serious, was caused by AA shells in Hoxton, Bethnal Green, Mile End Road, East and West Ham, Canning Town and Poplar, chiefly, no doubt, by the Woolwich gun and also at Lambeth, Kentish Town, Highgate, Holloway and Highbury. No fatalities are attributed to the firing, but two women and two children were injured by fragments of shell.

The total number of casualties was 109 – thirteen men, three women and six children killed. Forty-eight men (including three firemen), twenty-nine women and ten children were injured (two women and two children were injured by AA shells). All casualties were civilians.

After the sudden disappearance of the raider, L–13 was momentarily seen near Cheshunt at about 11.20 p.m., east of Hatfield at around 11.25 p.m. and east of Buntingford at 11.40 p.m. She dropped a small balloon to which was attached an electric battery at Elmdon and, at 11.55 p.m., circled round Saffron Walden. She then went off in a north-easterly direction towards Newmarket, west of which town she was seen at approximately 12.20 a.m. Twenty minutes later, she was near Bury St Edmunds and at 1.20 a.m. she was at Wymondham. At 1.30 a.m. she passed Norwich, Martham at 1.45 a.m. and then passed out to sea between Caister and Yarmouth shortly before 2 a.m.

The total number of guns, listed as the Defences of London, engaged in the raid were as follows:

Plumstead Marshes	one 1-pdr QF gun and one 3in QF
Royal Arsenal	two 6-pdr QF guns
Erith	one 6-pdr QF gun
Abbey Wood	one 6-pdr QF gun
Plumstead Common	one 13-pdr gun
Royal Albert Docks	one 13-pdr gun
Nine Elms	one 6-pdr QF gun
Blackheath	one 3in 20cwt gun
Clapton	one 3in 20cwt gun
Gresham Street	one 1-pdr gun
Green Park	one 3in 20cwt gun
Waterloo	one 1½-pdr gun
Temple	one 6-pdr gun
York Road	one 6-pdr gun
Finsbury Park	one 6-pdr gun
West Ham	one 3in 20cwt gun
St Helens	one 1-pdr gun
Honor Oak	one 3in 20cwt gun
Parliament Hill	one 3in 20cwt gun
Foreign Office	one 6-pdr gun
Crown Agents	one 1-pdr gun
Tower	one 3in 20cwt gun
Cannon Street	one 1-pdr gun

An interview with Mathy was published by the German-American correspondent, Karl von Wiegand, shortly afterwards in America in which Mathy boasted of his exploits. L–13 is made to appear to have come much lower down than she did in reality, and even suggests he had considerable difficulty in avoiding a collision with

St Paul's. The real height the Zeppelin maintained during the raid was more like 8,500ft (as opposed to the height of St Paul's dome, which is 365ft tall). Mathy gave no idea of the real height and when questioned on the point discreetly refused to 'give the English his range; they shoot quite well enough already.'

Mathy also took credit to himself for not having bombed St Paul's, although the 'English had established a battery under its shelter'. It is evident from the interview that Mathy knew perfectly well where he was, he steered his course parallel to the river and probably identified St Paul's, the Tower Bridge and Liverpool Street Station. He tried to bomb the two latter structures and the Bank of England, but was never near enough to the bridge to attack it. After bombing Liverpool Street the Zeppelin had passed quite slowly northward, as if her crew were contemplating their work at leisure and without the slightest fear of being hit by the shells that were bursting below. Until suddenly, one large burst had seemed to reach her upon which she threw out her water ballast and shot upwards, thereby eluding the searchlights, and made off at high speed.

It later transpired that a 300kg bomb, the first ever used on England, was dropped by L-13 on London in this raid. It was a '*Liebsgabe*' (love-gift) from the bomb factory to the Zeppelin.

A curious relic of the raid was a scraped ham bone, dropped in a bag attached to a small parachute from L-13 north of London, which landed in Wrotham Park,

The crew members of L-13 wearing the ribbons of the Iron Cross with which they were decorated after the London raid of 8 September 1915, when they dropped the first 300kg bomb on British soil.

Barnet. Round the shank was painted a band of German tricolour, and below on one side was a rude drawing of a Zeppelin dropping a bomb on the head of an elderly civilian in a stiff collar and black tie. He is represented full face and is labelled 'Edwart [*sic*] Grey' – Grey being The Rt Hon. Sir Edward Grey, Secretary of State for Foreign Affairs (1905–16). On the other side of him is written '*was fang ich, armer Teufel an?*' (which translates as 'what shall I, poor devil do?'). On the other side of the bone is inscribed lengthways '*Zum andenken an das ausgehungerte Deutschland*' ('A memento from starved-out Germany').

L-14, under the command of Kapitänleutnant der Reserve Alois Böcker, was first heard off the *Haisborough* lightship at 7.20 p.m., off Overstrand at 7.30 p.m. and was close to Cromer at 7.50 p.m. L-14 skirted the coast westward as far as Blakeney, where at 8.10 p.m. she came overland and developed engine trouble soon after turning inland near Cromer. Passing near Walsingham at 8.20 p.m. and Foulsham at 8.35 p.m., she then passed Bawdeswell at 8.40 p.m.

Five minutes later she dropped her first bomb at Bylaugh Park, where there was a Yeomanry Camp in which lights were showing. Böcker brought L-14 down lower, circled the camp and dropped one HE bomb that fell 150 yards from the camp and a further fourteen incendiary bombs. They did no damage and caused no casualties beyond a cow. They also dropped a German officer's cap (or it was knocked/blown off?), and a parachute was found in a meadow at Scarning; the latter contained German newspapers and a leave pass dated Nordholz, 7 September 1915, signed by Oberleutnant zur See Frankenberg, L-14. Frankenberg was second in command of the airship.

At 8.55 p.m. L-14 was over East Dereham. Böcker was aware that he had flown to the centre of the county and, believing he was over Norwich, he dropped his entire load of twenty-four HE and sixteen incendiary bombs, causing the most severe air raid casualties inflicted upon the county during the war.

L-14 swept over the town from the direction of Scarning. The first four bombs dropped on Church Farm meadows causing little more damage than blasting a gate by a barley stack and blowing the leaves off a hedge, but it did cause the patients in the East Dereham Auxiliary War Hospital to run out in their night attire and investigate what was going on. The next three bombs landed on marshy meadows, blowing out stinking black mud over a large area.

The next bombs were increasingly serious; one landed on the roadway near the Guildhall, leaving a crater 6ft across and 4ft deep and bringing down part of the Guildhall outbuildings, badly damaging the roof of the infant school on the opposite side of the road and smashing some of the glass in the church windows.

The worst damage was caused by the bombs dropped on Church Street. The premises of H.H. Aldiss on the High Street corner were almost completely wrecked; the windows of the King's Arms and Cave's Photographers were blasted in; and the White Lion pub was so badly damaged that it never opened again and

Kapitänleutnant der Reserve Alois Böcker

two patrons, Mr Harry and Mrs Sylvia Johnson, were badly injured. In this upper area of the street, the body of Lance Corporal Alfred Pomeroy was found – his left leg, abdomen and pelvis horribly mangled from the blast. Parts of his body were found on the roof of a building next to the Corn Hall. Mr James Taylor was passing along the top of Church Street to post a letter when the bomb landed. His body was found lying in the road near the National Provincial Bank 'shot in the abdomen by a piece of shell case.'

Cave's Photographers, the White Lion pub and other premises damaged during the air raid on Church Street, East Dereham, on 8 September 1915.

The frontage of Hamerton's grocery shop was blown out and cottages in White Lion Yard were badly damaged, one of them collapsing on top of its occupants, Mr and Mrs Taylor. On the opposite side of the road, many houses were scarred by the flying shrapnel and the orderly rooms of 5th Battalion, Norfolk Regiment TF, had its windows shattered and the roof smashed. The body of Harry Patterson was found in the entrance of the headquarters building of the battalion, a piece of steel shell casing having penetrated his chest.

The Corn Hall had its glass roof smashed and a bomb fell on a nearby house, demolishing it. The occupant, Mr Catton, had heard the commotion outside and ran out to investigate as a soldier ran inside to take shelter; he was extracted alive from the collapsed building but died later from his injuries. The bank also had its windows smashed, along with the Alexander family memorial windows in the Cowper Memorial Church.

The Zeppelin then swept away towards Bayfield Hall leaving a trail of bombs along the way, most of which fell on farm and estate land. But then, ominously, the Zeppelin turned again and made another pass over Dereham.

On this second run an incendiary was dropped on Bradley's Ironmongers in the Market Place, setting fire to the oil store and adjacent cartridge store which began exploding the ammunition in the intense heat. The fire brigade was

The blast-damaged headquarters of 5th Battalion, Norfolk Regiment TF, on Church Street, East Dereham, after the raid of 8 September 1915. The body of Harry Patterson was found in the entrance, with a piece of steel shell casing having penetrated his chest.

Bomb-damaged cottages in White Lion Yard, East Dereham, after the raid of 8 September 1915.

The blast-damaged East Dereham Corn Hall (right) and nearby houses that received a direct hit, which fatally wounded a soldier sheltering inside during the raid of 8 September 1915.

summoned with the firing of maroons, and immediately the fire broke out at Bradley's, Mr Herbert Leech, who had a men's outfitting business at the other end of the Market Place, ran to the King's Arms Hotel where the maroons were kept. He found all the occupants down in the cellar and, as bombs were falling rapidly, nobody answered his request for the maroons to be fired. Eventually he was given the key to the outbuilding where the maroons were kept. Mr Leech enlisted the assistance of a passing soldier and they got the maroons out. He knew nothing about firing them and actually held the match in his hand whilst lighting the fuse! When the first maroon was fired the soldier rolled over amongst the cabbages in the garden and bolted. Mr Leech fired the second maroon as the Zeppelin was hanging directly overhead and it immediately fled. The Zeppelin's departure was attributed by many to Mr Leech's courage, and for this he was complimented by military officers and prominent people of the town.

The casualties were:

Killed:
Harry Patterson (44), watchmaker and jeweller, High Street, Dereham.
Lance Corporal Alfred Edward Pomeroy, 2/1st City of London Yeomanry (Rough
    Riders).
James Taylor (61), an earthenware, china and general dealer, 27 High Street, Dereham.

Died of Wounds:
Pte Leslie Frank McDonald, 2/1st City of London Yeomanry (Rough Riders).
Pte H.G. Parkinson, 2/1st City of London Yeomanry (Rough Riders).

Injured:
Miss Dawson, Scarning Fen – injury to ankle.
Mr and Mrs Johnson, Baxter's Row, Dereham – wounded by shrapnel.
Pte A.W. Quinton, London Mounted Brigade Field Ambulance RAMC – wounded
    in leg.
Mr and Mrs Taylor, White Lion Yard, Dereham – injured by house collapsing on them.

At about 9 p.m. the raiding Zeppelin went off in a north-easterly direction towards Fakenham, was at North Elmham at 9.10 p.m., passed Ryburgh and Pensthorpe, and at 9.30 p.m. was near Walsingham again, where she seems to have turned off eastward towards Holt and back across the sea.

Three planes from RNAS Great Yarmouth were sent to hunt down L-14, but tragically each drew a negative. Two had lucky escapes, but the third pilot was not so fortunate. Squadron Commander C.W.P. Ireland was first in the air at 7.45 p.m. but he was back less than ten minutes later when his patrol was cut short after four of his engine cylinders cracked. Flight Lieutenant J.M.R. Cripps was in the air at 7.50 p.m. and he too had a remarkable escape when his engine spluttered to a stop half way through his patrol. He could see no landing area, so allowed his BE-2c to glide down and he threw himself clear just before it landed. He was unhurt, and his aircraft came down with only minor damage – both man and plane had landed on the Caister Marshes.

After a two-hour sortie patrolling between Cromer and Lowestoft, Flight Sub Lieutenant G.W. Hilliard was attempting to land at the night landing ground at Bacton. Despite the flares being lit on the runway, Hilliard misjudged his approach and touched down heavily in an adjoining field. His BE-2c's undercarriage collapsed, his bombs exploded in their frames and he was killed instantly. This brave 30-year-old pilot was buried with full RNAS honours in Great Yarmouth (Caister) Cemetery.

LZ-77, commanded by Hauptmann Horn, never came overland. The airship appeared at Dunwich at about 8.40 p.m., then went south and dropped eight HE bombs near the *Galloper* lightship at 9.30 p.m. The airship is then thought to have gone to the Straits of Dover, and was last seen off Walmer going east at 10.40 p.m.

The total estimated damage caused by the raid was £534,287.

The full RNAS Honours funeral cortège of Flight Sub Lieutenant G.W. Hilliard passes along Lancaster Road, Great Yarmouth, September 1915.

HIPPISLEY HUT

High on the cliffs at Old Hunstanton in Norfolk, not far from the lighthouse, was a secret installation colloquially known as the 'Hippisley Hut', and therein lies one of the stories of the country's forgotten heroes of the First World War.

Richard John Bayntun Hippisley was born in 1865 to a landed family, blessed with exceptional talents in engineering and science – his grandfather had been a Fellow of the Royal Society.

Bayntun, as he liked to be known, began as an apprentice at the Thorn Engineering Company, where he learned about mechanical and electrical engineering. He was gazetted Second Lieutenant in the North Somerset Yeomanry in July 1888, and Honorary Lieutenant Colonel in 1908, and was awarded the Territorial Decoration.

Bayntun was an early pioneer of radio research, and worked at a wireless station on the Lizard Peninsula in Cornwall, where he had picked up messages from the sinking *Titanic*. In 1913 he was appointed a member of the Parliamentary Commission of Wireless for the War Office Committee on Wireless Telegraphy.

Sir James Alfred Ewing (formerly the Director of Naval Education and Professor of Engineering at Cambridge), was the newly appointed manager of 'Room 40', the Admiralty Intelligence department of cryptanalysis, and was keen to obtain skilled recruits in the comparatively new field of wireless communications – and interception.

In September 1914, Bayntun and barrister Edward Russell Clarke, a prominent amateur wireless expert, called at the Admiralty and informed Ewing that they were receiving messages on a lower wavelength than any being received by existing Marconi stations. The German fleet was using these low wavelengths, and Ewing immediately obtained permission for Bayntun and Clarke to set up a station at Hunstanton in Norfolk.

When they arrived at the coastguard station at Hunstanton, they found a wooden mast with no aerial but, with a little ingenuity, they were soon intercepting signals. The area around the lighthouse and across the immediate farmland was designated a 'prohibited area' and by 1915 a maze of wireless masts stretched across the cliff top. These were controlled from within the lighthouse, which became a miniature wireless monitoring station, listening in to German Navy and airship radio traffic around the clock.

Lieutenant Commander R.J.B. Hippisley RNVR initially lived in the Le Strange Arms Hotel, then took over the old wooden clubhouse of the Old Hunstanton Town golf links and a wooden bungalow adjacent to the Cromer Road, where he installed another wireless device. The bungalow has now been replaced with a brick residence but still retains the First World War nom de plume of the 'Hippisley Hut.'

Lieutenant Commander Hippisley eventually set up a string of listening posts across the British Isles and abroad. The German Navy were confident that they could not be heard and never made any attempt to conceal their wireless traffic. It was these stations that picked up the unusual amount of traffic from Wilhelmshaven, which warned the Admiralty that the Imperial German fleet was putting to sea before the Battle of Jutland.

When the Zeppelin raids intensified in 1916, fears that the wireless stations would become targets led the Admiralty to send a shallow draught monitor, HMS *Cricket,* which was moored off Hunstanton Pier. Locals recalled: 'The ship had an anti-aircraft machine gun, which almost deafened the residents, and awakened the dead when it blasted off at overhead Zeppelins.'

In his article 'Tradition and the Innovate Talent' published in *The Times* on 5 June 1995, William Rees-Mogg pointed out:

> He [Hippisley] was the man who solved the problem of listening to U-boats when they were talking to each on the radio by devising a double-tuning device which simultaneously identified the waveband and precise wavelength. That, it is said, was essential to clearing the Western Approaches in late 1917, when American troops were coming over. Bayntun Hippisley sat in Goonhilly listening to the U-boat captains as they chatted happily to each other in clear German; he told the destroyers where to find them; the food and the Americans got through.

For his services, Bayntun was awarded the OBE on 3 June 1918. He apparently never told his family what he did during the war, and because he was bound as a gentleman and by the Official Secrets Act he did not write it down either, but suffice to say in his obituary which appeared in **The Times** on 11 April 1956, Bayntun was described as 'an almost unique personality' who 'inherited a remarkable mechanical and scientific gift, which put him in the forefront, if not ahead, of most of his contemporaries.' One of the senior officers who had worked with Bayntun remarked that Hippisley 'was one of the men who really won the war'.

11/12 September 1915

LZ-77, commanded by Hauptmann Horn, returned a few days later, passing *Kentish Knock* light vessel at 10.35 p.m. and travelling north–east towards the coast. She made landfall about 11.10 p.m. south of Tillingham coastguard station.

As she passed, rifle fire was opened on her from the coastguard lookout posts at Tillingham and Holiwell Point, without apparent result, except that LZ-77 rose to 5,000ft, and turned slightly south.

Going on westwards, at 11.15 p.m. she passed ½ mile south of Southminster, where eight rounds were fired at her by a 1-pdr pom-pom. The airship carried on westward passing Latchingdon, where she dropped a petrol tank and then veered north-west to Maldon. She passed south of Maldon at 11.20 p.m. and went south-west to Rettendon. Here, she turned west, passing West Hanningfield at 11.30 p.m., Ingatestone at 11.35 p.m. and a point 2 miles north of Chipping Ongar at 11.40 p.m. At 11.45 p.m., on reaching Gaynes Park, 1½ miles south of North Weald Bassett, LZ-77 turned north-west and, after crossing the main road from Epping to Potter Street, steered due north to Thornwood where she turned towards the east and, circling over North Weald Bassett, crossed her original course at Wintry Wood at 11.50 p.m.

Here, she dropped fifty-two incendiary bombs (of which five did not ignite), and eight HE bombs (not one of which exploded because, upon examination, none of them had their safety appliances withdrawn) on the Royal Field Artillery camp situated between Wintry Park and Hayles Farm. The airship commander was clearly uncertain of his location, saw lights at Thornwood and probably thought he was over factories or munitions works in the East End of London, so he dropped his bombs and made off. Amazingly, no damage was done and there were no casualties.

Finding her way back across Essex, LZ-77 turned east along the railway past Thurston towards Elmswell and passed over Great Bardfield at 12.20 a.m. Reaching Bury St Edmunds before 1 a.m. and Wymondham at 1.30 a.m., she was heard approaching Norwich ten minutes later but, owing to fog, could not be observed. The Caister coastguard station next reported the presence of LZ-77 at about 2.05 a.m., travelling very slowly out to sea.

12/13 September 1915

LZ-74, commanded by Hauptmann George, was reported going out to sea over Nieuport in Belgium at 7.50 p.m. After spending a considerable time at sea she passed over the *Kentish Knock* lightship, flying very high, at 10.27 p.m., steering north-west. At 10.45 p.m. she came over the coast near Walton-on-the-Naze and travelled west along the line of the railway over Thorpe-le-Soken, Little Clacton, Weeley and Great Bentley to Wivenhoe, which she passed at 11 p.m. At 11.05 p.m. she reached Colchester and went on in the same north-west direction past Wormingford, where two HE bombs were dropped. These fell in a field, doing no damage. At Mount Bures a third HE bomb fell with the same result.

LZ-74 then circled over Wakes Colne at 11.15 p.m. and moved north-east. She was seen from Boxted and, at Dedham, turned north-west. At approximately 11.30 p.m. she dropped eleven HE and five incendiary bombs at Hills Farm, Stratford St Mary. One HE bomb did not explode, the rest exploded within 15–70 yards of dwelling houses but only broke some glass in a greenhouse and a skylight, and damaged some fences. The bombs had been attracted by the fire from a Maxim gun armoured car of the RNAS. The LZ-74 was originally going north-west, but when fired upon came around towards the east as though searching for the gun.

Going eastward, the airship dropped four HE and four incendiary bombs at East Bergholt, which did no damage. The airship turned north-east, passed Ipswich and was again engaged by a RNAS Maxim car at Rushmere, with the result that, at 11.45 p.m., she dropped four HE bombs, one of which failed to explode. No damage was done. The gunfire caused the airship to take a zigzag course in order to avoid it, as she was flying very low and was described as presenting a 'splendid mark'.

Going on in the same north-easterly direction she passed Woodbridge at 11.50 p.m., where she was fired upon by another RNAS Maxim car without effect. Wickham Market was passed at about 11.55 p.m. and she was at Aldeburgh at 12.05 a.m. Here, the airship was again fired at by a RNAS Maxim car. Passing northward along the coast, she drifted for a short distance with engines cut off so as not to attract fire as she had probably thrown all her bombs.

She went out to sea at Sizewell Gap but came in again, and at 12.15 a.m. was heard south of Blythburgh by a RNAS pom-pom detachment, but was not fired upon as she could not be seen. LZ-74 finally passed out to sea near Southwold at 12.18 a.m. It seems that LZ-74 had difficulty finding any landmarks due to thick ground fog.

LZ-74 was destroyed shortly afterwards when she ran against a hillside in the Eifel. Her gondolas were torn off and her occupants thrown into a wood. The main body of the airship came down near the German General Headquarters, then at Mézières with one or two men on board who had taken refuge in the gangway.

13/14 September 1915

L-14 was commanded by Kapitänleutnant der Reserve Alois Böcker. She was seen off Winterton, Norfolk, at 11.12 p.m., out to sea 25 miles east of Great Yarmouth and Lowestoft at 12.25 a.m. and further south, 18 miles east of Southwold at 12.38 a.m. Her movements were hampered by fog and it is thought that L-14 did not make landfall at all.

L-11, under the command of Oberleutnant zur See von Buttlar, also encountered problems in the fog and clouds on this particular evening. She was first reported off Thorpeness at 11.05 p.m., and midway between Orfordness and the Shipwash at 11.15 p.m. At 11.20 p.m. the Zeppelin was over Hollesley Bay, and off Bawdsey ten minutes later. At 11.37 p.m., when she was between Felixstowe and the *Sunk* lightship, she was fired at by the 6-pdr anti-aircraft guns at Felixstowe and immediately turned back northwards along the coast to Woodbridge Haven, where she came landwards over Bawdsey Marshes.

At 11.48 p.m. L-11 turned westward, and at 11.55 p.m. was at Trimley, where she turned southward to Harwich harbour, arriving there at 11.58 p.m. The low clouds prevented the searchlights from reaching her, so that fire was not opened from the defences. Buttlar was clearly unsure of his location, so he made no attempt to attack Harwich and went away again northwards up the Orwell as far as Pin Mill, where he turned north-east.

The Zeppelin was engaged at 12.10 a.m. by a RNAS Maxim gun mounted on a motor lorry, posted on the Ipswich–Felixstowe road at Levington Heath. In answer to this attack, L-11 dropped five HE and two incendiary bombs on heath and fields near Bucklesham. One of the former, which burst 56 yards away from a house, merely dislodged a few bricks in the chimney and broke some cups and saucers. Going on in a north-easterly direction L-11 dropped two HE and three incendiary bombs on meadow and marsh near Newbourne, doing no damage. Eight HE and four incendiary bombs next fell in a mangold wurzel field in a marsh at Hemley, doing slight damage to the root crops and blowing out a window at Hemley Hall.

Crossing the River Deben, L-11 threw two HE and twenty incendiary bombs on the marshes at Wood Hale Farm, Sutton, to no effect and there were no casualties. Four of the HE bombs, one at Bucklesham, one at Newbourne and two at Hemley appear to have been of 100kg weight, the rest were of the usual 50kg type. After dropping her bombs L-11 went out to sea near Orfordness at 12.12 a.m.

L-15 undertook its first overseas voyage commanded by Kapitänleutnant Joachim Breithaupt. Seen off Shoeburyness at 11.48 p.m., on approaching the coast the raider found an unexpectedly low barometer, thick, foggy weather and heavy clouds. The Zeppelin was off Margate at 12.17 a.m. where she dropped three bombs, probably aimed at a steamer, all of which fell into the sea. L-15 then appears to have turned north, was 25 miles east of Dunwich at 3.15 a.m. and then probably headed home across the North Sea to Germany.

13/14 October 1915

The air raid of 13/14 October 1915 marked a very distinct advance in the attacks of German airships directed against this country. Previously, raids had either been of an experimental character or had been carried out principally by single airships which often exhibited haste in reaching their objective, usually very near the coast, and then darted home.

On this occasion, however, a larger number of airships than hitherto known was employed. The skill exhibited in their navigation was of a very much higher order and there seems to have been a very distinct tactical plan underlying the raid. The damage caused by a single airship when bombing London was of great importance, if only for the reason that the very centre of the city was attacked. The casualties caused during this raid, if considered from the point of view of the total number of airships employed and of bombs expended, are the most severe that have been suffered before or since this date.

On the morning of 13 October 1915, five Zeppelins left their sheds in north Germany with orders to bomb London. The first warning of the attack came from France at 5 p.m., when British wireless direction-finding stations indicated Zeppelins moving towards the east coast. Half an hour later, the Admiralty had news of a Zeppelin sighted from a troll vessel, 45 miles east of the *Haisborough* lightship. Four Zeppelins, L-13, 14, 15 and 16, all arrived in the Bacton area of Norfolk at almost the same time, while L-11 came in too far south and had to retrace her course to find Haisborough and Bacton.

L-13, under Kapitänleutnant Heinrich Mathy, hovered between the *Haisborough* lightship and the coast for nearly an hour, and was first to make landfall when she came in over Haisborough at 6.15 p.m. She went directly west and was next heard at Aylsham at 6.30 p.m. She then turned south-west. The Zeppelin picked up the Midland & Great Northern Joint Railway at Lenwade and followed a train going to Norwich, until the driver, who had sighted the Zeppelin, stopped at Drayton.

After hovering for some time, L-13 went westwards to East Dereham and passed over the town at 7.15 p.m. She then turned south-west and pursued a direct course, passing Griston near Watton at 7.25 p.m., over Brandon at 7.40 p.m., Exning at 8 p.m., Newmarket at 8.03 p.m., Saffron Walden at 8.25 p.m. and was seen north of Harlow at 8.45 p.m. Here, L-13 turned westward to Ware and on to Hertford at 8.58 p.m.

At 9.02 p.m. L-13 was fired at by a 13-pdr AA gun at Birchwood Farm, 1 mile north-north-west of Hatfield. Three rounds had been fired when the main string of the gun broke, and by the time it could be replaced the airship was out of range. On being fired at, the Zeppelin dropped four HE bombs, two in answer to the first round and two after the second. Of the two last, one fell 70 yards from the gun and the other failed to explode. No damage was done beyond the breaking of windows in some cottages adjacent. The third round was supposed at the time to have hit the airship, but no confirmation of this can be derived from her subsequent movements. She was also fired on with rifles by some men of the Army Service Corps. The airship had shut off her engines and was drifting when engaged.

L-13 went off northward, then turned south-west and passed over St Albans at 9.20 p.m., was near Watford about 9.25 p.m. and Rickmansworth at 9.30 p.m. Here she altered her course southward, along the course of the River Colne, past Uxbridge and Staines, and was seen from Windsor. She crossed the River Thames at Weybridge at 9.50 p.m. and was near Effingham five minutes later. She now passed near Gomshall and turned west to Guildford, passing over Newlands Corner at 10.05 p.m.

L-13 was travelling slowly, apparently uncertain of her whereabouts, and dropped a flare at Clandon at 10 p.m. Around 10.10 p.m. she appeared over Guildford, hovered for a few minutes and then went up the valley eastward towards Chilworth, where a second flare was dropped which lit up the surrounding area.

The pom-pom manned by the AA Detachment, Sussex Royal Garrison Artillery, at the Powder Works opened fire with seventy-seven rounds just after 10.15 p.m. The airship at first sheared off westward, but in a minute or two returned to the factory and then definitely went westward as far as Wood Street, and then south-east over the Hog's Back to Artington, St Catherine's and Shalford, dropping a third flare and then twelve HE bombs. No casualties were caused, but a considerable amount of minor material damage was done to houses in St Catherine's, and up the line of the London & South-Western Railway there was damage caused between two tunnels south of Guildford. Two of the bombs fell in the River Wey and one of them killed a swan. The bombs all fell in a direct line from north-west to south-east about ½ mile in length.

The airship subsequently headed off east, and at 10.35 p.m. passed over Chilworth where she was again fired on by the AA gun and rifles, all of which were unable to reach her. She went on without responding and passed Willinghurst, near Cranleigh, going south-east, five minutes later. L-13 kept on this course for some time before turning north-east

Somewhere around 11.05 p.m. she passed over Tandridge Camp, near Oxted, going north-east to Limpsfield and then turned northwards. It is evident that she had seen L-14, which had come up from Shorncliffe and was now close to her to the west and bombing Croydon, and she turned to join her.

When west of Hayes, L-13 waited some minutes for L-14. At 11.30 p.m. both airships were together at Bromley going east, L-14 being to the south of the pair. The two went on, very close to one another, as far as Bickley, where L-13 turned suddenly north and L-14 east. It was no doubt here that the two airships very nearly collided. This fact was later stated by the prisoners of L-33, taken at Little Wigborough on 24 September 1916, many of whom had served under the same commander on L-14 on this occasion. On the return of the airship to Germany, Kapitänleutnant Böcker, the commander of L-14, lodged a very strong complaint against L-13, under Mathy's command, for faulty navigation.

On parting company, L-13 passed over Sidcup at 11.35 p.m., turned over Bexley Heath towards Woolwich, passed south of it and, at 11.45 p.m., approached from the south-west. She was picked up by the Blackheath searchlight and gun, and the latter fired seventeen rounds at her before the gun jammed badly. Honor Oak fired nine rounds at L-13 at the same time.

The Zeppelin sailed right over the whole length of the barracks and Woolwich Arsenal. On the former, she dropped three HE and fourteen incendiary bombs. Extraordinarily little damage was done – the first HE bomb burst in the middle of the front parade and broke the windows of the Royal Artillery mess, the second fell on a barrack dining room in the East Square, completely destroying it and a stable below, and the third blew in the end of a clothing store in the Grand Depot barracks. In the stable, one horse was killed and nine injured and the same bomb slightly injured four men.

The incendiaries which fell in the East Square and the Depot barracks did practically no damage, and the small fires caused by them were put out. An incendiary bomb fell in St John's Churchyard, causing no harm. Eight others fell in streets and gardens close by, slightly damaging St John's Church School and a house in Wellington Street; damaging premises in Cross Street and Beresford Square; and destroying a shop which was burnt out in Thomas Street.

On the Arsenal, the Zeppelin dropped three incendiary bombs near the surgery, another incendiary in the new machine shop, 'Avenue G', another in '5th Street' and one HE bomb in the main machine carriage shop which wrecked a crane and damaged a machine. Nine men were injured in the Arsenal, one of whom died later. Finally an HE bomb was dropped in the magazine area on Plumstead Marshes causing no damage. L-13 then went off north-east.

Fire had been opened from the Woolwich AA guns at 11.47 p.m. and ceased at 11.54 p.m. The shooting was poor and the airship obviously out of range. The roof of a house in Plumstead was damaged by the fuse of an AA shell. West Ham fired fourteen rounds at the airship at 11.52 p.m. and she was also fired on by Clapton.

L-13's commander seems not to have been aware that he had bombed Woolwich Barracks and Arsenal, as he reported on his return that he had attacked the docks. He no doubt saw the Albert Dock plainly. L-13 crossed the river at Barking Reach to Dagenham about 11.55 p.m. and then was engaged by the 13-pdr gun at Sutton's Farm, which fired ten rounds at her. After the fifth, which burst quite near, she turned off at right angles. She then turned again north-east when she was engaged at midnight by Becontree Heath, which also fired ten rounds.

At the fourth road the airship rose to a height estimated at 10,000–12,000ft, and passed on towards Romford. At this time she seems to have been fired upon from Erith and Purfleet, although she must have been far out of range. L-13 was claimed to have dropped 'smoke balls' (actually water ballast that vapourised and probably turned to snow when coming into contact with the cold air) and a light is said to have been shown in her gondola between Woolwich and Romford, which was put out when she was fired on from Becontree Heath.

She went on north-east past Brentwood at 12.10 a.m. and Ingatestone at 12.15 a.m. Here, she was now flying low again and, from observations, it would appear that her propeller was damaged. The light shown at Romford may have been connected with this – repairs were possibly being attempted. At 12.20 a.m. L-13 passed south of Chelmsford, and at 12.30 a.m. south of Maldon. She approached the estuary of the Blackwater at Tollesbury, turning north-east towards Colchester.

At 12.45 a.m., L-13 was fired at by a Maxim from near Mark's Tey. Turning up the railway line and river at Wivenhoe at 12.50 a.m., around five minutes later she was fired upon by pom-pom and rifles. L-13 then picked up the main line of railway and followed it to Ipswich, which she passed at about 1.15 a.m., going erratically. Travelling over Earl Soham at 1.40 a.m. she changed direction, and went off at high speed due east over Framlingham and Saxmundham. She passed north of Leiston around 1.55 a.m. and out to sea at Dunwich at 2 a.m.

L-14, which was under the command of Kapitänleutnant der Reserve Alois Böcker, came in at Bacton at 6.30 p.m., where she was fired upon by a RNAS Maxim armoured car. She passed North Walsham at 6.35 p.m. and was fired on for a second time, in this instance by the machine guns of the 6th Battalion, Norfolk Regiment (Cyclists) TF, posted there.

L-14 traversed the rest of the county via Felthorpe and Thetford unmolested, and proceeded across Suffolk and Essex to Kent, where she dropped her first four bombs at 9.15 p.m., while flying at a height of about 6,000ft, upon the camp of the 8th Howitzer Brigade and 5th Brigade, Canadian Field Artillery at Otterpool Camp, Shorncliffe. The bombs were probably directed at the lights and fires of the camp. A total of fifteen men were killed and eleven wounded, seven horses were also killed. No damage was done to guns or equipment.

L-14 moved north-east towards Westenhanger Camp, and dropped two more HE bombs on the racecourse, which did no damage beyond breaking some windows. She passed on to Sandling and out to sea at Hythe. Here, her commander seems to have realised that he was on the south coast and turned south-west, coming overland again at Littlestone at 9.35 p.m., passing Lydd at about 9.40 p.m. and Winchelsea around 9.55 p.m. He went along the coast as far as Pett, near Hastings, and turned inland having now apparently fully satisfied himself as to his real position and getting his true direction for London, probably by means of the Hastings–Frant railway.

The airship does not seem to have been observed, on account of the mist, until it reached Frant where, at 10.30 p.m., seven incendiary bombs were thrown, one of which did not ignite. No damage was done. The raider then turned north to Tunbridge Wells where, at 10.40 p.m., three HE bombs fell, causing no harm except to windows. The Zeppelin could still not be seen on account of the thick mist.

Going off north-west, L-14 crossed the path of L-13 at Oxted and then passed over Tandridge Camp at 11.05 p.m. L-13 went north over Limpsfield while L-14 pursued a north-west course to Warlingham and Purley at 11.15 p.m. and on to Croydon, which she bombed at 11.20 p.m. Some seventeen HE bombs were thrown, falling along a line from Eldridge Road, over the railway and across Addiscombe Road, immediately east of East Croydon Station, to the corner of Essex and Stretton Roads. A final incendiary bomb was dropped on South Norwood but it failed to explode.

As the line followed is roughly parallel with the railway, it is probable that this was the objective aimed at by the Zeppelin. Since Croydon was an important junction, the railway lights would have been on. The bombs, however, all fell, with one exception, on villa residences, doing considerable damage and causing loss of life to civilians. Three villas were destroyed, eleven seriously damaged and six less seriously damaged. One bomb fell by the railway, but only damaged the permanent way slightly. Thee men, three women and three children were killed, and five men, six women and four children were injured.

The airship passed on immediately east of Crystal Palace and, at 11.30 p.m., was again in company with L-13 over Bromley. L-14, having exhausted her supply of bombs, headed back over Essex and Suffolk and was fired upon by a RNAS Maxim armoured car at Melton. Following the railway to Saxmundham, which she passed at 1.35 a.m., over Leiston and out to sea north of Aldeburgh at 1.45 a.m.

L-15 was under the command of Kapitänleutnant Joachim Breithaupt. She came in at Bacton at 6.25 p.m., closely followed by L-14 and L-16, all of whom were fired on by a RNAS Maxim armoured car there. She went inland on a south-westerly course, reaching Aylsham at 6.35 p.m., and Honingham (6.55 p.m.), and threw a flare west of Thetford, over Elveden, at 7.25 p.m. Changing course southward, L-15 flew over Bury St Edmunds, in Suffolk, and travelled across Essex where she dropped a petrol tank at Matching, east of Harlow at 8.25 p.m. Passing south of Harlow, the Zeppelin was observed signalling by means of flashes to L-13, which was coming up behind her from the north.

Following the line of the river and railway past Roydon and Hoddesdon and Broxbourne, she was fired upon by a mobile 13-pdr AA gun, which fired eight rounds to which the airship replied with four HE bombs. Estimated to be at a height of 5,000–6,000ft, and immediately overhead, the searchlights were of no use because they were too close to the gun. Three of the bombs dropped within 100 yards, blowing down the gun detachment with the force of the explosion, but there were no casualties. A 30cwt lorry belonging to No. 50 Company, Royal Engineers, and the motor car attached to the gun section were wrecked by one of the bombs.

Proceeding to London via Potters Bar, after passing Edgware at 9.03 p.m., L-15 cut her engines and drifted silently with the north-west wind and was not observed until she dropped her first bomb at 9.25 p.m. on Exeter Street, the Strand. The second fell in Wellington Street, between the Lyceum Theatre and the offices of the *Morning Post*. The third landed on Catherine Street, the fourth and fifth in Aldwych and the sixth and seventh between Aldwych and New Inn. Two incendiaries fell on the Royal Courts of Justice, and the eighth and ninth HE bombs in Carey Street. The tenth fell on the roof of a house in New Square, Lincoln's Inn, followed by an eleventh on Old Square. The twelfth HE and a third incendiary bomb fell almost simultaneously on the roadway of Chancery Lane, opposite Stone Buildings.

Three more incendiaries fell immediately afterwards north of Holborn, in South Square, Gray's Inn, followed by the thirteenth HE bomb, which fell in the north-eastern corner of Gray's Inn Square. Four more incendiary bombs were thrown on and near Hatton Garden, and one in Farringdon Road. L-15 then made eastward over the city.

Considerable damage and loss of life were caused by this first fall of bombs. In Exeter Street, houses were damaged, one person was killed and two injured. In front of the Lyceum Theatre seventeen people were killed, twelve seriously injured and nine slightly injured. Panic was fortunately avoided in the theatre, due to the presence of mind of an officer who was present. The theatre itself, the *Morning Post* offices, and other buildings were only slightly damaged by the explosion. In Catherine Street, one person was seriously and eight slightly injured.

In Aldwych, where bombs also fell in the street, the Strand Theatre was slightly damaged, four people were killed, five seriously injured and ten slightly injured. The wooden shed of the Belgian War Refugees Committee clearing house was seriously damaged. The new extension of the Law Courts was damaged by a small fire caused by incendiary bombs. The upper storeys of two of the old houses in New Square, Lincoln's Inn, were wrecked by the HE bomb that fell on them, but the explosion did not penetrate to the lower storeys, thanks to the solidity of the buildings, which date from 1697.

The fragments from the bomb that exploded in Old Square chipped large pieces out of the stone and brickwork of the surrounding buildings, and seriously damaged the sixteenth-century stained glass windows of Lincoln's Inn Chapel.

Entrance of the Strand Theatre, Aldwych, after suffering blast damage during the Zeppelin raid of 13 October 1915.

The Chancery Lane bomb badly damaged the roadway, smashed the water and gas mains, wrecked the fronts of the buildings on the east side and broke all the windows of Stone Buildings, Lincoln's Inn, on the west side. Flying fragments of stone seem to have done as much harm as the splinters of the bombs themselves. In Gray's Inn, the hall and some houses in the square were wrecked, and in Hatton Garden and Farringdon Road several buildings were burned by incendiary bombs.

The next bombs dropped by L-15 fell in the roadway at Finsbury Pavement. Two HE bombs damaged several buildings, including one under construction. One soldier and three other people were killed, and ten injured. Another bomb fell on a small hotel in the Minories, partly demolishing it and damaging a number of houses nearby in the Minories, Aldgate and Houndsditch. One person was killed and eight injured, and a horse was also killed. The next two HE bombs fell on either side of the London, Tilbury & Southend (Midland) Railway between Great Prescot and Royal Mint Streets. Several tenement houses in Great Prescot Street, Chamber Street and Leman Street were damaged and six people injured. Leman Street Station was slightly damaged. The last bomb, also an HE, fell between Wellclose Square and Princes Square. A tarpaulin factory was wrecked and several other buildings were damaged. No human casualties occurred here, but a horse was killed.

Damage to the London & South-Western Bank on the corner of Aldgate High Street
and Minories, East London, after the raid of 13 October 1915. (IWM)

L-15 went off over the London and the West India Docks to Limehouse, turning sharply northwards when fired on by the Woolwich AA guns, and disappeared. During her course over London L-15 had been energetically but fruitlessly bombarded from 9.25 to 9.35 p.m. by the guns of the London AA defences, which seemed quite unable to reach her; the Zeppelin was flying at an estimated 8,000–9,000ft. Considerable damage was done by falling shell fragments, chiefly in Poplar and Limehouse. A house was also hit in Oxford Street and the Westminster Public Library in Great Smith Street had its roof damaged.

On leaving the river at Limehouse, around 9.30 p.m., L-15 went straight north over Hackney and Leyton and turned north-east to Woodford and Barkingside, passing the latter place at 9.45 p.m. She was fired at by the guns at Loughton (until the trigger lever jumped after the fourth round), Hainault Farm, Grange Camp (Waltham) and Kelvedon Hatch between 9.40 p.m. and 9.55 p.m. Ground fog prevented the ascent of most aircraft, but two RFC aircraft went up from Joyce Green and Hainault at 8.10 p.m., patrolling until 9.50 p.m. when they returned and another two went up, but they all returned with nothing to report. The B.E-2c, flown by Second Lieutenant John 'Jack' Slessor R.F.C. from

Sutton's Farm, went up at 9.40 p.m. and had more luck. He spotted L-15 in the searchlights, started off in pursuit and attempted to drop his anti-Zeppelin grenade, but lost sight of his target after passing through cloud. Slessor thus has the distinction of being the pilot of the first plane to challenge a Zeppelin over British soil. This young pilot went on to become Sir John Slessor, Marshal of the Royal Air Force, 1950–52. The flying conditions were indeed difficult that night and two out of the five aircraft that went up crashed on landing. L-15 passed the Kelvedon Hatch gun going east at 10.00 p.m, heading back to the sea over Suffolk. It was fired upon by a pom-pom of the RNAS detachment at Rushmere and dropped four HE bombs in response, which fell harmlessly. Ten minutes later she passed Woodbridge, at a height of 5,000–6,000ft, and was fired on by a Maxim there. Making no reply to the attack, L-15 went on east of Wickham Market at 11.35 p.m., north of Orford at 11.45 p.m. and out to sea near Aldeburgh at 11.55 p.m.

L-16 was commanded by Oberleutnant zur See Werner Peterson, and made landfall at Bacton at 6.40 p.m. She was fired upon by the Maxim gun detachment just as L-14 and L-15 had been. The Zeppelin then travelled west-south-west, and dropped a bomb at Banningham at around 6.50 p.m. No harm was caused. Peterson then altered to a southerly course, passing over Costessey, west of Norwich, at 7.20 p.m., then north-west to Attlebridge and then south-west to Attleborough, where she dropped a petrol tank at 7.50 p.m. and another, a few minutes later, at Eccles. She moved over East Harling at 8 p.m. and to the west of Thetford ten minutes later.

Passing over Suffolk and Essex, L-16 was fired upon by a Maxim at Chelmsford. Visible lights are believed to have attracted the Zeppelin commander to Sawbridgeworth. Approaching over the junction of the River Lea and its tributary the Beane, seven incendiary bombs were thrown, all of which fell in fields close to the swimming bath, doing no damage.

The first HE bomb fell on allotments at the Folly, and the second at Priory New Road, both on the banks of the Lea. Windows and roofs at the Folly were damaged. A house in Frampton Street was demolished. The third HE bomb fell in Bull Plain, where one house was demolished, five others (including the Conservative Club) wrecked, and the museum and eleven other houses slightly damaged. Five men were killed outside the Conservative Club, two of whom were prominent townsmen – the borough surveyor and a well-known organist. A child was killed in the demolished house, a man injured in another house and four men and three women injured in the street.

The fourth and fifth HE bombs fell at the junction of Maidenhead Yard, Mill Bridge and the Wash. A flour mill was partly wrecked; some stables and outhouses badly damaged; a brewery, wine stores and seven shops were also affected; and the windows were broken in about twenty houses. A woman was seriously injured and two men slightly injured in Maidenhead Yard. Two artillerymen were also injured there.

Twelve incendiary bombs fell at the same time on Old Cross, near the Municipal Library and School of Art. The building was slightly damaged, as were also a brewery and several other houses and shops near Old Cross.

L-16 passed over St Andrew's Church, and dropped her sixth and seventh HE bombs, one in the North Road just beyond its junction with the Hertingfordbury Road, and one in a garden, accompanied by seven more incendiaries. An artilleryman was killed in the road and a man injured. A private residence, a motor works and an inn were wrecked; and twenty-nine houses in the North Road, sixteen in St Andrew's Street and two in Hertingfordbury Road were slightly damaged.

The Zeppelin followed the line of the North Road and the River Beane, and dropped an eighth HE bomb which failed to explode, and three incendiaries, in gardens between the road and the river, followed by the ninth HE bomb which exploded in the road near the hospital gate. The hospital suffered slight damage, and also St Andrew's Rectory and five other residences close by. Another artilleryman was wounded and two men killed.

Finally, L-16 dropped five more HE bombs and one incendiary on fields west of the hospital, doing no damage and causing no casualties.

In all, eight men (including one soldier) and one child were killed:

George Cartledge (56), Fore Street
Arthur John Cox (21), 50 Victoria Road, Great Yarmouth (Acting Bombardier in
    2/1st Norfolk Battery, East Anglian Brigade, Royal Field Artillery, stationed
    at Hertford. His body was returned for burial at Great Yarmouth (Caister)
    Cemetery.)
George Stephen Game (4), 37 Bull Plain
James L. Gregory (55), York House, Fore Street
Arthur Hart (51), 61 Port Vale
John Henry Jeavons (67)
Ernest Thomas Jolly (27), Hampden House, Ware Road
Charles Spicer (30), 38 Thornton Street
Charles Waller (43), 34 Hertingfordbury Road

A further eleven men (including three soldiers) and four women were injured.

Peterson took L-16 off over Bengeo in a north-easterly direction, passing Little Hadham at 10.15 p.m. and approaching Newmarket at 10.45 p.m. Here, the Zeppelin was attacked by a Maxim on the heath and turned off northwards, passing Soham at 10.55 p.m. Crossing into Norfolk and moving south of Narford around 11.20 p.m., Swaffham at 11.25 p.m., and Dereham at 11.35 p.m., the Zeppelin was north of Aylsham by 11.55 p.m. and proceeded out to sea at Mundesley at about 1.05 a.m. Ten minutes later, L-16 passed the *Haisborough* lightship, going east.

L-11, under Oberleutnant zur See Horst von Buttlar, crossed between the *Cockle* and *Newarp* lightships at about 7.30 p.m. and then went north. Turning south-west, the Zeppelin passed over the *Haisborough* lightship at 8.15 p.m. and came in over Bacton at 8.25 p.m., where she was fired on by a Maxim armoured car of the RNAS. Then making her way inland at 8.45 p.m., L-11 circled over the parishes of Horstead, Coltishall and Great Hautbois, dropping bombs.

Four HE and three incendiary bombs fell in Horstead parish in open fields, breaking a few panes of glass in the cottages nearby. In Coltishall and Great Hautbois L-11 dropped nine HE bombs, two of which did not explode. All these dropped in fields not far from the Great Eastern Railway station at Coltishall. The only damage done was a shed partly blown down and a few more panes of glass broken. Seven incendiary bombs were also dropped around three cottages. These bombs were quickly put out by pails of water. There were no casualties.

Norfolk National Reserve with one of the unexploded HE bombs dropped by L-11 in fields not far from the Great Eastern Railway station at Coltishall, Norfolk, on 13 October 1915.

'LET FLY THEREFORE!'

Zeppelin commander Horst von Buttlar recorded his personal account of one of the early missions for his book *Zeppelins over England* (1931):

> We started off at 10.00am, made the North Sea coast an hour later, and then, shaping a westerly course, kept the German and Dutch coast in sight. At twilight our position was just north of Terschelling, and at an altitude of about 2500ft we veered west and continued our journey. According to my calculations we ought to have made the English coast between 11 and 11.15 p.m. The ship was darkened; that is to say, all lights were put out in the car and even the use of pocket electric torches was forbidden, our object being to approach the coast as secretly as possible and without being observed by any craft searching out at sea.
>
> It was 11.30 p.m. and still there was no sign that we were approaching land, although the shipping below certainly seemed to be growing more plentiful. Here and there through the pitch-black night we had been able to observe the side-lights of the small steamers and now and again also the lights of fishing smacks. It now occurred to me that the ship was making less headway, for the lights below us took longer to move away. The west wind, which during the day had been only a gentle breeze, had now apparently grown somewhat stronger and was making our progress more difficult; the consequence was that it was taking us longer to reach the English coast than we had reckoned it would. I looked at the clock and was horrified. It was midnight! And still there was no land in sight.
>
> Half past twelve and still no land! Since we had last ascertained our position near Terschelling a good four hours had elapsed. I had our fuel gauged; it was terribly reduced. Should we turn back? We had only enough petrol for another seven hours and still there was no sign of England!
>
> After coming all that distance, however, and on the very point of reaching the English coast, I was determined to push on at all costs. On our return journey we should be able to find some way out of the difficulty. For if the wind from the west really had increased in force, it would be a great help for us on our way back and we should reach home all the quicker. I therefore decided to continue on our westward course until 1 a.m. at the very latest and if by that time I had not made the English coast I should turn

back for otherwise I would most certainly never be able to reach the most westerly airship base at Hage, near Norden, in East Friesland.

At that time a flight of this kind was much more of a strain than it is today [1931]. For instance the cars were open on all sides and this made it extremely cold up in the air, particularly as it was mid-winter. The only car that possessed a small wind-screen was the forward one, but it was just sufficient to protect the steersman and the maps on his chart-table, from the icy blast during the flight. It was 1 a.m. and still no land was in sight! Should I turn back after all? Suddenly a thin shaft of light came into view. The misty arm of a searchlight darted across the sky. It was probably looking for us. We were over England! Then we saw a number of faint lights beneath us, myriads of them. We must be over some town. Let fly therefore!

Schiller hurried aft to drop the hundredweight bombs. I was just able hurriedly to discuss my plan of action with him and told him that I proposed to sweep over the town, which we could just faintly discern, by coming up from the south, turning west and then going off east and arranged with him that when a bell rang he was to begin dropping the bombs.

Steering a south-westerly course, we kept over our objective and climbed to a height of about 3,500ft. It was impossible to take the ship up any higher and I did not wish to make her too heavy, for if an engine broke down or one of our gasbags got hit, I should not have any reserve buoyancy left with which to control her. By this time Schiller must be at his post – now! Then the lights could be seen below. Whitish-green beams came up through the mist again and began searching the sky. They crossed each other. Drew nearer, their light grew brighter and suddenly covered the ship's envelope in a blinding glare. They've got us! Little red spots of fire appear below. They were the anti-aircraft batteries. I gave Schiller the sign. Then I climbed slowly in a spiral to a height of 4,500ft above our objective.

The chief engineer artificer helped the Wachoffizier in the bomb-cabin to drop the bombs, for the incendiary bombs had to be thrown out by hand. A pin had to be taken out to make them 'live', after which they were flung in a gentle curve overboard, to crash and burn below a moment later and burn merrily. Meanwhile the Wachoffizier dropped the three one-hundredweight bombs. By this time the space between us and the ground had become a perfect inferno of gunfire and bursting shells – shrapnel meant for us.

When all the bombs had been thrown, I leant well out of the car and saw a fiery '6' burning gaily below. Now we must get back. At a height of 4,500ft we were soon out of reach of the English anti-aircraft batteries and steered a course south-east by south back toward the North Sea.

The Zeppelin proceeded eastward to Wroxham (8.50 p.m.) and south-west to Thorpe, near Norwich (9 p.m.) where, just north of this latter location, L-11 was fired upon by the 1/1st Leicestershire Royal Horse Artillery TF on Mousehold Heath. From Thorpe she headed towards Acle at 9.10 p.m. and out to sea at Great Yarmouth at 9.15 p.m.

Buttlar described this short and futile trip over Norfolk as a significant attack in West Ham and Woolwich, but did claim that he 'had great difficulty in fixing his position owing to being fired at.'

The four Zeppelins reached the vicinity of the Norfolk coast together, and waited for one another between the *Haisborough* and *Would* lightships. They crossed the coast at the same time, approximately 6.30 p.m. They steered a course in close proximity to one another. After going as far as Norwich, two of the ships seemed to be doubtful of their position. They both followed the course of the River Wensum, then turned south-west again, the formation being three airships ahead and one astern. On reaching Thetford the airships opened out fanwise; two ships (L-15 and L-16) continued on a direct course towards London, two more (L-13 and L-14) laying their course west and east of London respectively. Thus far, the scheme seemed to have worked out according to programme.

Intelligence Summary

On arriving at Braintree, L-15 inclined westwards, then circled and attacked the metropolis from the north-west, being over the city at 9.25 p.m. She was under heavy, but inaccurate gunfire, and her bombing of the city was deliberate. After expending the bombs she returned to Germany, passing over Ipswich. L-13, which meanwhile had laid her course more to the west, must have seen the attack of L-15 with the accompanying reflections of searchlight beams in the sky. L-13, however continued on her course down the valleys of the Colne and Thames and up the River Wey. She them bombed the vicinity of Guildford.

The German communiqué stated that the Hampton Water Works near London had been bombed but it is very difficult to imagine that the commander of L-13 could really have thought that he was actually over Hampton, since the lights of London must have been visible to him some 15 miles away and he had already seen the searchlights at work on L-15. The report was probably mendacious.

It is possible that L-14 on her southward journey may have mistaken the estuary of the Thames for that of Blackwater and consequently came under further south than she intended before heading for London. It was soon after 11 p.m. that L-13 and L-14 met twice, once near Oxted and again at Bromley. After crossing the path of L-13 at Oxted, L-14 headed direct for London but apparently shirked the crossing of the city and bombed Croydon instead. After having thrown all her remaining bombs she turned north-east for home. There is no doubt that the bombing of Croydon by L-14 had been watched by L-13. L-13 then turned to the west again and dropped her remaining bombs on Woolwich before going home.

In this raid the new mobile AA guns, firing from freshly selected positions well clear of London, came into action for the first time, with the result that L-13 was attacked by gunfire near Hatfield and similarly L-15 near Broxbourne. Newly installed mobile guns also came into action north-east of London at Loughton, Romford, Hainault and Sutton's Farm and on the river at the Royal Albert Dock and Plumstead causing considerable trouble to L-13 and L-14 on their departure from London. At Hatfield and Broxbourne the airships resorted to the use of bombs to reply to the attacking guns. The Zeppelin's aim on this day was perhaps better than it has ever been when replying to gunfire. On this occasion the accuracy of the aim was no doubt due to the fact that the airships were flying at the comparatively low altitude which up until this time had been habitual for them. They were surprised by the unexpected position of the guns they encountered and by their increased range. Had the range been estimated the result might have been serious for the raiders. Thenceforward raiding airships tended to maintain a higher altitude when flying over England. The Zeppelin commanders also realised that the danger of being hit by gunfire was greater when acting in such close concert.

TWO

The German Naval Air Service developed plans to bomb the whole of England, and had divided the country into three areas for the purpose of issuing simple attack orders:

England North – Edinburgh and secondarily the Tyne
England Middle – Liverpool and secondarily the Humber
England South – London and secondarily Great Yarmouth and Eastern Counties

31 January/1 February 1916

This raid had been planned as an attack against Midland towns. During its course, the German airships penetrated to the furthest westerly point yet reached by these raiders. It was the most ambitious effort by the German Naval Airship Division to date, as the entire available squadron of their new standardised naval Zeppelins were employed.

The action was prefaced by attacks on the city of Paris by single military airships on 29/30 January and 30/31 January. German airships again crossed the trenches on the three following nights. After the first raid on Paris, the military airship LZ-79 was wrecked in Belgium on her way back.

The events of 31 January/1 February 1916 provide an excellent lesson as to the difficulties of aerial navigation over England by night. There can be little doubt that each of the airships had definite instructions to find a particular target, and it seems equally certain that only three or four ever reached the vicinity of their prescribed objectives.

Nine airships crossed the North Sea that day: L-11, L-13, L-14, L-15, L-16, L-17, L-19, L-20 and L-21. They left the north German sheds on the morning of 31 January, and appeared off our coast in successive groups, the first group of two ships (L-13 and L-21) arriving shortly before 4.50 p.m. off the Norfolk coast. All of the raiders were hampered in their navigation by the mist and fog that were prevalent along the East Anglian coast, where seven of them made landfall on that night, and subsequently they were often wildly inaccurate in identifying in their reports exactly where they had bombed.

L-21, under Max Dietrich, and **L-13** came in together north of Mundesley in Norfolk and passed over Hanworth shortly after 4.50 p.m., but thereafter L-21 showed a greater turn of speed than her consort which lagged behind. Traversing Norfolk, L-21 was over Narborough at 5.20 p.m. and King's Lynn at 5.25 p.m. Passing over Sutton Bridge into Lincolnshire, L-21 proceeded past Nottingham to Derby and approached **Wolverhampton** at 7.45 p.m. She hovered over Netherton for three minutes and then went northwards to Dudley and Tipton where she dropped her first bombs at 8 p.m. An estimated three HE bombs were dropped on Waterloo Street and Union Street, some outbuildings in the rear of the houses were destroyed and one person was killed. In Union Street, two houses were demolished and others damaged, together with the gas main, while thirteen people were killed and ten injured. The canal bank was also damaged. The incendiary bombs fell in gardens and yards and failed to ignite.

Going on from Tipton, the Zeppelin next dropped five HE bombs on the towpath of the canal at Lower Bradley, near Bilston, killing William Fellows and mortally wounded his partner Maud Fellows (they were a courting couple, not related); she died just over a week later at the Wolverhampton & Staffordshire General Hospital. Damage was also caused to a canal bank and the wall of a drainage pumping station. At Bloomfield three incendiary bombs were dropped on some brickworks; two failed to ignite and no harm was done.

The Zeppelin then turned eastward to Wednesbury where, at 8.15 p.m., twenty-three HE bombs and eight incendiary bombs were dropped. In King Street, near the Crown Tube Works, three houses were destroyed and others damaged; thirteen people were killed. Three incendiary bombs landing on the roof of the Crown Tube Works killed one person, damaged the roof and shattered windows. The stable and outbuildings behind Hickman & Pullen's Brewery were damaged. At Mesty Croft Goods Yard slight damage was done to railway wagons and buildings and one person was killed. Slight injury was also caused to the colliery at Old Park.

Walsall was next to be visited. At 8.25 p.m. L-21 flew over the northern part of the town, from west to east, dropping seven HE and four incendiary bombs as she passed. The Wednesbury Road Congregational Church was badly damaged and Thomas Merrylees was killed by a piece of flying rubble as he was walking by. A hole was also blown in the wall of Elijah Jefferies & Sons Ltd. Many windows were broken by concussion. The incendiary bombs dropped in the road and did no damage.

The last of the seven HE bombs landed in Bradford Place outside the Science and Art Institute, shattering some of the windows and showering a chemistry class in glass. Mr A.K. Stephens, who was sitting near one of the windows, was badly cut. In Bradford Place itself the blast killed two men and wrecked the public toilets, and shrapnel struck the passing No.16 tramcar, inflicting severe wounds to the chest and abdomen of Mrs Mary Julia Slater (55), the Mayoress of Walsall. She was removed to hospital where she died from shock and septicaemia on 20 February. A total of seven men and two more women were also injured. The full extent of the damage to buildings in Bradford Place could only be seen in all its horror with the coming of the morning light. A piece of shrapnel can still be seen embedded in the wall of one of the buildings of Bradford Place, and a blue plaque commemorates the death of the Lady Mayoress. There is also a tablet to her memory in the Council House. Walsall's Cenotaph now stands on the spot where the bomb exploded.

Leaving Walsall, L-21 made off eastward at high speed, passing Sutton Coldfield at 8.35 p.m., Nuneaton at 8.45 p.m., travelling near to Market Harborough and Kettering, and dropping six incendiary bombs on the Islip furnaces at Thrapston at 9.15 p.m. The glow of the furnaces no doubt attracted the Zeppelin, which now had nothing but incendiary bombs to throw. Luckily, no damage was done and the bombs fell in fields.

Passing on her way with a very definite direction, L-21 went north of Huntingdon, was over Ely at 10 p.m., Thetford at 10.35 p.m. and traversed south Norfolk to the Suffolk coast, where she went out to sea between Pakefield and Kessingland at 11.35 p.m.

L-13, commanded by Kapitänleutnant Heinrich Mathy, left L-21 at Foulsham and headed south-west to East Dereham at 5.15 p.m. where she altered course north-west and moved up over to the Wash, north of King's Lynn. She crossed Lincolnshire, Nottinghamshire and Derbyshire to south of Stoke-on-Trent, where she dropped six HE bombs on Fenton Colliery, which landed within a radius of 70 yards, two falling in a field, three on shraff heaps and a sixth on an ammonia tank, the top being merely lifted off by the explosion. A few windows were broken by concussion, but there were no casualties.

L-13 passed towards Newcastle-under-Lyme, dropped a flare or illuminating bomb at Madeley at about 8.20 p.m., circled northward towards Alsager and round to Wolstanton, Basford and Stoke again at 8.50 p.m. After leaving Stoke she apparently lost her way, abandoning the direct course which would have brought her into the neighbourhood of Chester, and instead heading south-west along the Trent Valley, following it to Burton. Here at about 9.15 p.m. she joined in a bombardment of the town which had begun at 8.45 p.m. by L-20, dropping an estimated fifteen HE bombs on the town. The number of bombs dropped by L-13 could not be accurately stated, as three Zeppelins bombed Burton that evening and it was never definitely established how many bombs were dropped by each. After an abortive attempt to find Manchester or Sheffield, the Zeppelin disappeared back over Lincolnshire to its base.

L-15, under the command of Kapitänleutnant Joachim Breithaupt, was the third Zeppelin to make landfall on this night, crossing the coast at Mundesley in Norfolk at 5.50 p.m. Passing north of North Walsham, she headed south to pass near Swaffham and reached Mildenhall at 7.10 p.m. The Zeppelin then went to West Row Fen, where three HE and fifteen incendiary bombs were dropped at 7.15 p.m. Only three of the incendiary bombs ignited and no harm was done.

L-15 seemed very uncertain of her whereabouts, circling for some time near Soham, then dropping a flare and, at about 7.35 p.m., she dropped twenty-two HE bombs on the open fen near Isleham. Only fifteen exploded, and resulted in a wrecked fowl house and sixteen dead chickens.

Having fruitlessly ditched most of her bomb load, L-15 hung around the Norfolk and Lincolnshire border, and was seen south of Skegness about 10 p.m. She then turned south-west along the coast to Holland Fen, where an incendiary bomb was dropped at 10.30 p.m. She passed back over Norfolk, by Swaffham and Wymondham, turned east-north-east south of Norwich and reached the Yare valley at midnight. She finally went out to sea at Corton at 12.35 a.m.

L-16, commanded by Oberleutnant zur See Werner Peterson, had not enjoyed a good crossing having developed engine problems. She was left with just two reliable motors and was heavily loaded with snow and ice. Crossing the coast at 6.10 p.m. near Hunstanton, Peterson gave up on his attack on Liverpool as a target too far, and decided to bomb Great Yarmouth.

Believing he was over the coastal town, Peterson reported that he had dropped his 2 tons of bombs from 7,000ft on 'such factories as could be made out.' He had, in fact, been flying over Norfolk and had dropped two HE bombs (one of which did not explode) near Swaffham, causing no damage. Wandering over south Norfolk for about two hours, L-16 found the River Waveney near Pulham and followed it to arrive in the vicinity of Bungay at 8.40 p.m. and went out to sea at Lowestoft at 9.05 p.m.

L-14 was commanded by Kapitänleutnant der Reserve Alois Böcker and came in north of Holkham, Norfolk, about 6.15 p.m. She was over Downham Market at 6.20 p.m., followed by Sandringham at 6.35 p.m. and then pursued a direct south-west course to Wisbech at 7 p.m., where it dropped an incendiary bomb.

Turning north-west at Thorney as far as Knipton, 8 miles south-west of Grantham, it dropped a single HE bomb to no effect.

Travelling over Nottingham and Derby, she appeared at Shrewsbury at 10.05 p.m. This was the extreme western limit of the course of L-14. She then turned, passed south of Wrekin to Ironbridge, circled westward round Wellington, turned east to Oakengates and circled northwards by Gosnall, going directly eastward, south of Cannock, Lichfield and Tamworth. Around 11.35 p.m. she

Photograph of an INCENDIARY BOMB found in Staffordshire after the Air Raid, January 31st, 1916.

The amount realized from the Sale of these Cards to be given to the Fund in aid of the sufferers.

E.B., D.

Postcard of an incendiary bomb recovered after the Zeppelin raid on Staffordshire, 31 January 1916, the proceeds of which were given to those who suffered as a result.

turned north and at 11.50 p.m., attracted by the light of some pipe furnaces at Ashby Woulds, near Ashby-de-la-Zouch, dropped one HE and one incendiary bomb on a cinder heap near the furnaces, which did no damage.

Turning slightly westward at midnight she dropped four HE bombs at Overseal, three of which fell in a field and one in a canal causing no damage, then three further HE bombs were dropped on Swadlincote a few minutes later which broke a few windows.

Ten minutes later, going due north, L-14 reached Derby and dropped twenty-one HE and four incendiary bombs on the town. Nine of these HE bombs were dropped on the Midland Railway Works damaging the engine shed, killing three men and injuring two. Three HE bombs were dropped on the Metalite Lamp Works, doing considerable damage but causing no casualties. At the Rolls-Royce works, two HE bombs fell on the motor track causing no harm beyond some broken glass. Two other bombs were dropped, and three in the yard of a lace factory, without effect. Four incendiary bombs landed in the street, one house was set on fire but there were no casualties.

The airship passed on east, went south and east of Nottingham shortly after 12.30 a.m. and moved north-eastward near Newark and south of Lincoln. She circled for some time over the Wolds, passed Alford at 2.10 a.m. and proceeded out to sea.

The crew of L-33 (many of whom had served in L-14 on the night in question) who were captured at Little Wigborough on 24 September 1916, maintained, under examination, that their Zeppelin had bombed Liverpool on 31 January, as was claimed in the German communiqué published after the raid. The steersman, however, after a prolonged cross-examination abandoned this claim and confessed he knew that they had not been to Liverpool. He claimed however, to have seen the lights of Manchester and Sheffield. But this course lay too far to the south for him to have been able to do this. The crew persisted in refusing to disbelieve the statement made by their commanding officer and the steersman.

L-19, under Kapitänleutnant Odo Loewe, arrived at about 6.20 p.m. near Sheringham, Norfolk, and was observed at Holt at 6.25 p.m. going southwards. Sailing at a moderate rate, she was observed at Swaffham at 7.05 p.m. then up in the Midlands, passing Stamford in Lincolnshire at 8.10 p.m. and on to Exton. She traversed south near Stamford again, then westward once more past Oakham, on an uncertain course and with some circling, until she was in the neighbourhood of Loughborough shortly before 9.30 p.m.

The commander of L-19 then seems to have made up his mind about his course, and made for Burton, attracted there no doubt by the fire caused by the two Zeppelins that had already bombed the town. At about 9.45 p.m. he seems to have added one or two incendiary bombs to the number already having been thrown on Burton since 8.30 p.m. He then turned south-west, passed near Wolverhampton and wandered over the district of Enville, Kinver, Wolverley, Bewdley, Bromsgrove, Redditch and Stourbridge, reaching Wythall at about 11 p.m.

He appeared to be heading towards Birmingham but, as the city was in total blackout, he seems to have been attracted by the lights of Wednesbury. About midnight, a single HE bomb was dropped on the Monway Works at Wednesbury, doing slight damage to the roof and machinery of the axle department.

L-19 then turned towards Dudley, dropping five HE bombs at Ocker Hill Colliery as it went, which broke the windows of the engine house and the dwelling house adjacent. Over Dudley, at about 12.15 a.m., seventeen incendiary bombs were released, all except one landing in fields and in the castle grounds. One fell into the grain shed at the railway station, causing about £5 worth of damage.

Turning northward again, Tipton was bombed next at about 12.20 a.m., with eleven HE bombs landing in the western part of the town and the London & North-Western Railway station at Bloomfield. Either by this Zeppelin or L-21, which had raided four hours earlier, some damage was done to the permanent way of the railway, rails and sleepers being blown some distance away. A large number of windows were broken. L-19 then went on to Walsall where, at 12.25 a.m., three HE bombs were dropped. One landed in a church garden, while another killed a horse and several pigs.

Her load of bombs being apparently expended, the Zeppelin made her way somewhat uncertainly back down towards the Norfolk coast, and was spotted a short distance from Martham at 5.25 a.m. It would appear that the flight of L-19 had been too protracted; the Zeppelin ran low on fuel and her engines were rendered partially useless. It was not until 3 p.m. that she reached the neighbourhood of Borkum, where she broke down entirely and drifted westward. At 7.30 a.m. on 2 February 1916, L-19 was sighted, waterlogged, in the North Sea, 95 miles north-east of Spurn. William Martin, skipper of the fishing trawler *King Stephen*, of Grimsby, was offered money by Kapitänleutnant Loewe to take him and his crew on board. He refused to do so, as he feared that the crew would overpower him and his crew and carry them off to Germany. He therefore returned to the Humber with his report of what he had encountered, but the Zeppelin sank with the loss of all hands before a Royal Navy vessel could get to them.

The weather had been becoming worse, and the crew threw a bottle into the sea containing final messages from the aircrew to their families, and a final report by Loewe which clarifies what actually happened, rather than just the assumptions made in the intelligence report at the time. It translates as follows:

Kapitänleutnant Odo Loewe, lost with his crew when L-19 sank in the North Sea, 2 February 1916.

With fifteen men on the top platform and backbone girder of L-19, floating without gondolas in approximately 3 degrees east longitude, I am attempting to send a last report. Engine trouble three times repeated, a light wind on the return journey delayed our return and, in the mist, carried us over Holland where I was received with heavy rifle fire; the ship became heavy and simultaneously three engines broke down. 2 February 1916, towards one o'clock, will apparently be our last hour. – Loewe.

The bottle and messages were only discovered six months later by Swedish fishermen at Marstrand.

The L-19/*King Stephen* incident received international publicity. British opinion was divided: some, including the Bishop of London, publicly praised Martin for putting the safety of his crew first. Some newspapers promulgated 'retribution' for the 'aerial baby killers.' While, hardly surprisingly, in the German press Martin was vilified for leaving the airmen to die, as was the Bishop of London for his support for Martin's actions.

William Martin himself received many letters of both support and hate at his Grimsby home over the ensuing weeks and months. Being thrust into the media spotlight amid such international controversy undoubtedly took its toll upon him and, just over a year after the incident, he died of heart failure on 24 February 1917, aged 48.

L-17 was under the command of Kapitänleutnant Herbert Ehrlich, who reported that he was having problems determining his location due to dense cloud, when suddenly a searchlight broke through the overcast sky to starboard. He claimed that he saw the glow of 'blast furnaces' nearby and that he came under small arms fire as he steered to attack.

Ehrlich made two runs over the 'industrial area' during which he claimed to have silenced the battery attacking him and extinguished the light. In reality, he had drifted in over Sheringham at 6.40 p.m. and had been caught in the searchlight of the RNAS station at Bayfield, near Holt. He had jettisoned the majority of his bomb load onto it. Ten HE bombs landed 200 yards from the station, five in a field to the south-east, and five more to the south some 400 yards away causing no greater damage than a few craters in open ground. Five more bombs and an incendiary fell on Bayfield Lodge, nearby. One of these wrecked a barn and a greenhouse, blasting out tiles and glass but, fortunately, injuring no one.

L-17 then turned south to Letheringsett, on the way dropping fourteen incendiary bombs in a field and a wood near Bayfield Hall, but causing no harm. A further HE bomb was dropped in Letheringsett. The *History of 25th County of London Cyclist Battalion* records:

In late January 1916 a Zeppelin raided Holt and every telephone line west of it was broken. No lives were lost, but a draft of recruits just arrived from the Depot had a shake-up in their quarters at Letheringsett, which was close to a searchlight station the Zeppelin was apparently anxious to locate and destroy.

L-17 remained in the Holt area for some time, then headed south-west by way of Reepham and north of Norwich, where she was spotted at 8.10 p.m. She was finally recorded out to sea, south of Great Yarmouth at 8.30 p.m.

The two remaining Zeppelins (**L-11** under Oberleutnant zur See von Buttlar, carrying Zeppelin chief Peter Strasser, and **L-20**) came in by the Wash and crossed onto land near Sutton Bridge at about 7.10 p.m., where they parted. L-11 travelled westward and L-20 pursued a south-west course.

L-11 dropped her first bomb (an incendiary), which did no damage at Holbeach at 7.40 p.m. and then turned north-west, setting a course in the direction of the manufacturing districts of south Yorkshire. Shortly after 8 p.m., L-11 dropped one incendiary bomb and three HE bombs on Digby and four HE bombs on Bloxholme Park, two of which did not explode. The Zeppelin kept on her course, passing south of Lincoln, until she reached Retford where she circled for some time around 9.50 p.m. She then altered course to travel north towards Gainsborough and then south-east to Lincoln, dropping an illuminating flare at Hackthorn at about 10.30 p.m.

L-11 then went off at high speed in the direction of Hull and, shortly before 11 p.m., dropped a large number of bombs at Scunthorpe, aiming at Frodingham Iron & Steel works. These works, having received no warning, were in full blast and the untapped furnaces must have offered an easy mark. Sixteen HE and forty-eight incendiary bombs were dropped. All of them missed the Frodingham Works, but several of them fell on the premises of the Redbourne Iron Works, which were closed down and in darkness. Two men were killed there, but the damage done was inconsiderable, involving only slight injury to the engine and boiler house. In the town of Scunthorpe four workmen's houses were practically demolished. One man was killed, two men injured seriously and three men and two women slightly. One bomb dropped on a railway siding damaging the metals. A quantity of window glass was also broken in the town.

Having expended his bombs, the Zeppelin commander made eastwards for the sea and passed near Humberston at 11.35 p.m. He was fired on with thirty-four rounds as he went, by a pom-pom at Waltham wireless station, and finally left the coast north of North Somercotes at 11.45 p.m.

L-20, commanded by Kapitänleutnant Franz Stabbert, held a somewhat erratic course once overland. After parting company with L-11 at Sutton Bridge at 7.10 p.m., she went westward to Peakirk at 7.40 p.m. and on to Uffington, near Stamford, where she dropped an HE bomb five minutes later, with little effect beyond breaking a few windows. L-20 was then on a westerly course, which brought her north of Oakham and Leicester. She was then attracted by the lights of Loughborough, which seem not to have been reduced. She made for the town and, at 8.05 p.m., bombed it. Leicester, it may be noted, was in darkness and escaped a visit, as did Nottingham, where lighting was also restricted.

Four HE bombs were dropped on Loughborough. One fell in the back yard of a small public house, damaging outbuildings and shattering glass. Another fell in a main thoroughfare, and a third in a garden, both smashing a good deal of glass. The fourth dropped in front of the Empress Crane Works, which had not put out its lights until the first bomb dropped. Considerable damage was done to windows and doors but the machinery was unaffected. Four men were killed, two in the street and two in houses, and seven men and five women were injured. One of the injured men was a soldier employed on munition work.

Leaving Loughborough, the Zeppelin dropped an HE bomb, which did not explode, in a field and went north-north-west to Bennerley and Trowell where, at 8.27 p.m., seven HE bombs were dropped, one falling close to a railway viaduct but luckily doing no damage except to a signal box. A cattle shed was wrecked by another bomb.

At 8.30 p.m. the airship appeared south of Ilkeston, where fifteen HE bombs were dropped at the Stanton Ironworks, Hallam Field. The moulding and blacksmith's shops and the stables were damaged and the schoolroom attached to the church at the ironworks wrecked. Two men were killed and two injured.

After Ilkeston the Zeppelin turned south-west in the direction of Burton-on-Trent where, at 8.45 p.m. about a dozen incendiary bombs were thrown. This was the first attack on Burton, which was visited by three airships between 8.45 and 9.45 p.m. Fifteen HE and twenty-four incendiary bombs were dropped in all. It is probable that L-20 had now exhausted her HE bombs and had nothing but perhaps a dozen incendiary bombs to throw on the town, which seems to have been showing a certain amount of light. It had only been possible to reduce gas pressure and dim the public electric lights to some extent half an hour before the first bombs fell. It proved impossible to distinguish between the bombs dropped at Burton by three different ships, or to assign particular damage to a specific Zeppelin.

Much damage was done to the breweries of Messrs Bass, Messrs Allsopp, Messrs Ind Coope & Co. and Messrs Worthington. At Bass's Brewery, an engine house was partially demolished and at Allsopp's, the sawmill was wrecked by HE bombs. At Ind Coope's, the malthouse kiln was destroyed by fire, and at Worthington's the hop room was partially burnt out, both by incendiary bombs. A malthouse was also destroyed at Worthington's by an HE bomb. At Charrington's and Robinson's Breweries HE bombs landed in their yards, doing no damage to buildings. No casualties occurred at the breweries.

The most extensive damage was caused in Wellington and Snobnall Streets, where nine houses were wrecked and several damaged. Five people were killed and many injured. In the High Street, two shops were partially wrecked, and the Grammar Schoolhouse was partly destroyed with two people killed and several injured. At Christ Church mission room and vicarage an HE bomb dropped between the buildings, destroying masonry, blowing out windows and doors and killing four people and injuring others. Further bombs dropped near a billiard hall, damaging it and killing one person, and on the Midland Railway Company's goods depot, where a warehouse, siding and trucks were partially destroyed and several people injured.

The railways suffered considerably. At Moor Street Crossing and Leicester Junction sidings the permanent way was damaged, and the signal and an empty train destroyed. A large amount of minor damage was also caused. Eight incendiary bombs failed to ignite.

In all, three men, six women and six children were killed at Burton, and twenty men, thirty-five women and fifteen children injured.

The dead, their home addresses and the locations of where they were killed were:

Margaret Anderson (60), 195 Scalpcliffe Road, at Christ Church Mission Room.

Ada Brittain (15), Waterside Road, at Christ Church Mission Room.

John Lees Finney (53), 5 Slater's Yard, near the Midland Railway Good Shed.

Bertie Geary (13), 89 Blackwood Street, near Peel Croft.

Charles Gibson (52), 34 Wellington Street.

Edith Measham (10), 32 Wellington Street.

Mary Rose Morris (32), 32 Easton Place, Brighton, at Christ Church Mission Room.

Lucy Simnett (15), 150, Branstone Road, at Litchfield.

Elizabeth Smith (45), 73 Park Street, at Christ Church Mission Room.

George Stephens (16), 332 Blackpool Street.

Florence Warden (16), 206 Uxbridge Street, at Christ Church Mission Room.

George Warrington (6), 108 Snobnall Street, at 109 Snobnall Street.

Mary Warrington (11), 108 Snobnall Street, at 109 Snobnall Street.

Rachel Wayte (78), 72 New Street, at Christ Church Mission Room.

Florence Jane Wilson (23), 8 Casey Lane, at 109 Snobnall Street.

Leaving Burton-on-Trent, the Zeppelin turned away north-east, going north of Derby and Nottingham and then bearing south-east to the neighbourhood of Stamford. She then passed Thorney going east at 10.30 p.m., was spotted near Swaffham at 11.15 p.m., crossed out to sea at Blakeney at about 11.45 p.m. and passed Cromer while out at sea at 11.52 p.m.

5–6 March 1916

Three Zeppelins, L-11, L-13 and L-14, left the western airship shed, situated near the Bight of Heligoland, before noon on 5 March. The Zeppelins moved together westwards, steering for Flamborough Head.

L-14, under Kapitänleutnant Alois Böcker, crossed Flamborough Head from south to north, dropping some flares on the shore in order to locate the land. Finally, she re-crossed the headland going southwards.

Bridlington was passed at about 10.50 p.m., and the Zeppelin kept southwards along the coast at some distance inland as far as Aldborough. Here, she suddenly turned sharp north-east at 11.05 p.m. and out to sea. About 11.25 p.m. she came over the coast again, near Barmston, and went west dropping an incendiary bomb in a field near Gembling, halfway between the coast and Driffield.

She now turned southwards after verifying her bearings and, at 11.40 p.m., was seen at Cranswick going south. Later, she passed Lockington, where she was going slowly southwards along the railway to Hull. She seems to have gone off south-east towards Hedon at this point and at midnight was over Beverley, where she dropped three HE and three incendiary bombs in open fields in the parish of Woodmansey, 1 mile south-east of the town. None of them caused any damage.

The Zeppelin must have then perceived Hull, and bore directly southwards to the city. The attack began at 12.05 a.m., when L-14 passed slowly over the western part of the city, dropping seven HE bombs as she went, the first of which fell in the grounds of Hymer College. The second fell on the railway embankment south of the college and east of Anlaby Road junction, doing no damage beyond breaking windows in adjacent streets. The four following bombs destroyed houses in Regent Street, Linnaeus Street and Day Street and damaged a grocer's shop in Bean Street. The last bomb to drop on the quay of Albert Dock wrecked a crane. Nine incendiary bombs were dropped on the town and four in the river.

The bombing of the defenceless city was slowly and deliberately carried out by Kapitänleutnant Böcker. The ship remained above the city for ten minutes at a height estimated at no more than 5,000–7,000ft. The instant release of the bombs was visible when light in the interior of the ship appeared as the trap door opened to drop the bombs. The sky was clear at that moment, but snow had fallen intermittently before the raid, so that the area of the docks and the Rivers Humber and Hull would have shown up like ink stains on white paper, thus giving the Zeppelin a well-defined target in between the drifts of the snow cloud above which she was slowly passing.

This was further borne out by the testaments of the prisoners of L-33, captured at Wigborough on 24 September 1916, who served on L-14 on the occasion of the raid. Böcker was also taken prisoner and he, along with the steersman, Jensen, must have known perfectly well where they were and what they were doing when they bombed Hull. The men stated, under examination, that the visibility of the city was perfect. The steersman, at least, showed himself fully conscious of the doubtful morality of his commander's action.

Having been successful over Hull, Böcker took his Zeppelin off, passing over Paull at 12.25 a.m. and moving down the river to Killingholme where, at 12.30 a.m., he was fired on by the AA guns, without result. He immediately turned off eastward and, over Birstwick, dropped a single HE bomb, which had no effect, followed by four HE and two incendiary bombs, which were equally innocuous, at Owstwick about 12.40 a.m. The airship then went out to sea, apparently at Tunstall, five minutes later.

L-11, commanded by Korvettenkapitän Viktor Schütze, crossed the Yorkshire coastline between Tunstall and Withernsea at about 9.45 p.m. After circling for some time, she seems to have made south, passing over Grimsby at 10.10 p.m. Continuing her southern course, she was heard at Ludborough at about 10.20 p.m., turned south-west and was heard again near Market Rasen at about 10.30 p.m., and both seen and heard at Waddington, 2 miles south of Lincoln, at 10.40 p.m.

Schütze now seems to have abandoned the idea of penetrating any further in a south-westerly direction, but did not propose to run for Belgium so instead he beat back in the direction of the Humber. At around 10.50 p.m. L-11 was seen between Wragby and Hainton, going west. Schütze seems to have discovered near Lincoln that he was going too far south and east, and so turned and made off north-west towards south Yorkshire.

L-11 was next seen at Carcroft, going towards Selby. He then appears to have given up the idea of dropping bombs where he was and turned eastwards to the coast. This was due to the weather, which was becoming more unfavourable. Snow clouds prevented him from seeing where he was and also prevented him from being seen, which accounts for the paucity of reports regarding his movements. The Zeppelin accordingly returned to the Humber below Hull before 1 a.m. and, having definitely fixed her whereabouts, moved up the river to Hull which she bombed exactly one hour after the attack by L-14.

The city must have been as perfectly visible to von Buttlar and the crew of L-11 as it had been to Böcker and his crew. The bombing was equally deliberate. The Zeppelin came down to a height of 3,000–4,000ft and, for a time, was stationary over the city. As before, the light shown by the opening of the bomb-dropping trap was clearly visible. A HE bomb was dropped in the river opposite Earle's Shipyard, and this caused the partial collapse the next day of a half-finished 3,000-ton steamer. An incendiary bomb also fell into the river.

The Zeppelin went north-west over the city, passing from the mouth of the Hull, where another incendiary bomb landed, across Prince's Dock and over the Paragon Station, dropping eight more HE bombs and ten incendiaries as she went. One of these incendiary bombs fell in the dock. The CRA's office in

Queen Street was seriously damaged by HE bombs. After hovering above the station for two or three minutes, L-11 suddenly turned about and went off down the river.

The damage done in Hull by this Zeppelin consisted of several houses destroyed and a water mains broken in Queen Street and Collier Street; a fire at the Mariners' Almshouses in Carr Lane; the breakage of some painted glass windows in the south aisle of Holy Trinity Church and the roof of the Paragon Station; and the previously mentioned damage to the CRA's office. A small fire, caused by an incendiary bomb which fell on the quay of Prince's Dock, was soon extinguished.

The total casualties from both raids at Hull were: eight men, four women and five children killed, and twenty-two men, twenty-two women and eight children injured. One of the men was an old seaman, who was burnt to death in the fire caused by an incendiary bomb at the Mariners' Almshouses.

Those killed were:

Frank Cattle (8), Little Humber Street.
Robert Cattle, Little Humber Street.
James William Collinson (63), 14 Johns Place, Regent Street.
Edward Cook (38), 33 Lukes Street.
Ethel Mary Ingamells (33), 8 The Avenue, Linnaeus Street.
Martha Rebecca Ingamells (35), 8 The Avenue, Linnaeus Street.
Mira Lottie Ingamells (28), 8 The Avenue, Linnaeus Street.
Edward Ledner (89), Mariner's Almshouses, Carr Lane.
John Longstaff (71), 6, William's Place, Upper Union Street.
Annie Naylor (6), 32 Collier Street.
Charlotte Naylor (30), 32 Collier Street.
Edward Naylor (4), 32 Collier Street.
Jeffery Naylor (2), 32 Collier Street.
Ruby Naylor (8), 32 Collier Street.
James Pattison (68), 33 Regent Street.
Edward Slip (45), 23 Queen Street.
John Smith (30), 2 Queens Alley, Blackfriargate.
George Henry Youell (40), 4 Post Office Entry, High Street.

Passing down the river, at 1.15 a.m. L-11 was over Killingholme and was fired at by the AA guns there, without result. She retaliated with four HE bombs. One of these dropped on wasteland, another struck a railway siding, damaging the rails, and the others fell harmlessly. One man was killed.

The airship kept on her course down the river, and at 1.25 a.m. was near Grimsby, until at 1.40 a.m. she passed the Spurn on her way out to sea.

L-13, under Kapitänleutnant Heinrich Mathy, struck the Lincolnshire coast at North Coates Fitties at 9.14 p.m. and, passing south-west, went over Tetney, was heard over Binbrook soon after 9.30 p.m. and at East Barkwith around 9.50 p.m. At 10.10 p.m. she dropped a flare at Branston, south-east of Lincoln. Proceeding east of Newark, more flares were dropped at 10.25 p.m., at a point north-east of Claypole Station.

L-13 then turned north-west, circled in the direction of Southwell and thence round to the south-east again, being heard between Whatton and Bingham, east of Nottingham, at about 10.40 p.m. Near Whatton another flare was thrown, lighting up the whole country. The commander of the Zeppelin was evidently very uncertain as to his whereabouts. It is not impossible that he imagined himself to be much further to the north-west than he really was. He seems to have decided, in the face of the strong northerly wind and snow, to run direct for Belgium, instead of attempting to return to Germany across the North Sea.

In order to lighten his ship, he determined to get rid of most of his bombs at once on any objective which might present itself, and so used his flares. Obtaining no result from them, he nevertheless began to unload at once. Going slowly south-east, at 11.15 p.m. he dropped thirty-two incendiary bombs which fell harmlessly in open fields between the village and church of Sproxton. Next, at 11.30 p.m., fourteen HE bombs were dropped in grass and arable fields in the angle formed by the Thistleton, Market Overton and Greetham–Market Overton roads in the parish of Thistleton, 8 miles north-east of Oakham. No damage whatever was done. These bombs were heard at Norwich, 85 miles away!

L-13 then circled back north-west and, at 11.45 p.m., dropped a single HE bomb at Edmundthorpe, which also had no harmful effect. She then resumed her south-easterly course, which she now pursued to the Kent coast. Gathering speed, she was heard at Yaxley at midnight. She passed over Cambridge at 12.20 a.m. and Saffron Walden at 12.33 a.m., flying now very fast with the northerly breeze behind her. She passed some miles east of Dunmow at 12.45 a.m., south of Witham at 12.55 a.m., south of Maldon at 1.03 a.m. and then was over Burnham at 1.13 a.m., apparently going out to sea. She turned south, however, and was off Shoeburyness at 1.25 a.m., making a wide circle just inland of the coast. Here, two rounds were fired at her by a 3in AA gun, until the traversing gear jammed, no more rounds could be fired and L-13 disappeared behind a cloud. The searchlights were neither needed nor used.

L-13 crossed the Thames estuary, and appeared off Sheerness at 1.30 a.m. She hovered by Sheerness for about ten minutes at a height of about 9,000ft, dropping four explosive bombs at Ripley Hill Marshes and Danley Farm near Minster-in-Sheppey, without result. She was fired at by AA guns at Sheerness, standing bows to the wind almost motionless in the searchlight beams for over five minutes, under heavy, but useless, fire from the 6-pdr guns, which at that time constituted the only AA armament of the fortress.

The Zeppelin then made off over Harty Ferry, where she was fired at by a heavier piece, a 3in 20cwt gun; one of its rounds probably hit her. She headed at once south-east, and disappeared from sight. At Faversham, at 1.55 a.m., three rounds from a 6-pdr Hotchkiss were fired at her. One of these also was thought to have hit. She passed slowly over Canterbury, apparently drifting with engines cut off, at 2 a.m. and, when over Adisham, dropped a small fan-driven oil pump.

She was heard at Sibertswold at 2.10 a.m. and went out to sea in the neighbourhood of Deal at 2.25 a.m. At 3.10 a.m. she was seen going east in mid-straits by the SS *James Fletcher*. Her behaviour when crossing the Channel appeared somewhat erratic, and she drifted for about twenty minutes after passing offshore, but eventually found her bearings and finally made it to Belgium.

31 March/1 April 1916

Seven Zeppelins left their sheds in north Germany around noon on 31 March. Two of the seven (L-9 and L-11) developed defects and had to turn back when they were north-west of Terschelling. The remaining five continued on towards their objective of London.

Two of the Zeppelins were sighted by a Lowestoft minesweeper at 6.50 p.m., and at 8 p.m. the Admiralty ordered two destroyer divisions out from Harwich, one off Cromer and the other off Lowestoft to deal with any Zeppelins that may have been brought down.

L-15, under Kapitänleutnant Joachim Breithaupt, made the Suffolk coast in company with L-13 shortly before 8 p.m., one south of Southwold and the other north of Aldeburgh. L-15 was leading, and came in over Dunwich at 7.45 p.m., dropping two HE bombs in the sea as she came. She dropped a petrol tank and an incendiary bomb at Yoxford, to no effect, at 7.50 p.m., and passed Sweffling at 7.55 p.m.

At 8 p.m. she showed a flare at Framlingham and, flying fast, passed Wickham Market at 8.05 p.m. She proceeded between Bealings and Grundisburgh, east of Woodbridge, at 8.15 p.m., and on to Ipswich where, at 8.20 p.m., she dropped two HE bombs and one incendiary near the docks. One fell into the water, others badly damaged two small houses, killing three people and injuring two.

L-15 kept on at a speed of 35–40mph east of Manningtree, passing Wherstead at 8.25 p.m. and Stratford St Mary at 8.35 p.m. At Colchester, at 8.45 p.m., a single HE bomb was dropped, causing no casualties and only slightly damaging the glass roof of a printing works in Hawkins Road, besides breaking a few cottage windows nearby.

L-15 passed on east of Witham at 9 p.m. and across Essex to Pitsea, where the Zeppelin commander was able to see the course of the Thames and turned his vessel south-south-west, passing over Orsett and keeping the river on his port beam. The Zeppelin was picked up by the Perry Street searchlight, and the Dartford AA guns opened fire immediately, at a range of 6,000 yards and an estimated height of 8,000ft. She was also observed by an aeroplane, piloted by Lieutenant C.G. Ridley, which fired about twenty rounds and gave chase, but lost the Zeppelin in the darkness.

Other searchlights at Erith, Dartford, Abbey Wood, Purfleet and Plumstead now illuminated L-15 and, at 9.40 p.m., the guns at Purfleet, Abbey Wood, Erith, Erith Marsh, Southern Outfall, Plumstead Common and Plumstead Marsh began to deliver heavy fire. L-15, in attempting to elude this bombardment, turned northwards and, at 9.43 p.m., dropped twenty explosive and twenty-four incendiary bombs on Rainham. These fell on open fields, and did no damage at all.

Just as the Zeppelin was crossing the Rainham–Wennington road at 9.45 p.m. she was hit by a shrapnel shell from Purfleet which made a large rent in her side, damaging three gas cells. As he admitted in his examination after capture, this injury determined Kapitänleutnant Breithaupt to make off at once, and he went promptly north-east over Upminster.

As she passed to the west of Brentwood at a height of 9,000ft, L-15 was overtaken at 9.55 p.m. by a BE-2c aeroplane from Hainault Farm, piloted by Second Lieutenant Alfred de Bath Brandon. He climbed above the airship, which could no longer rise owing to her loss of gas, and dropped some explosive darts without apparent result. The pilot then boldly passed along the side of the Zeppelin in order to get to the rear of it, at the same time coming under rapid machine gun fire from the crew. After an unsuccessful attempt to drop an incendiary bomb and more darts on the Zeppelin, he lost sight of his quarry and, being unable to locate her again, turned back to his aerodrome. The attack took place as L-15 was passing over Ingatestone at about 10.05 p.m., and at the time the airman lost her she had altered her course and was heading due east for the coast.

If the commander of L-15 had at first entertained the idea of returning to Germany, he had now ascertained the serious nature of the injury which his craft had suffered, and consequently altered his course with the intention of making for Belgium. In order to lighten the ship as much as possible and enable her to rise, he threw overboard two machine guns and all other movable objects, all of which landed in fields at Stock, South Hanningfield and Woodham Ferrers.

Having picked up the River Crouch, he followed the left bank and passed Althorne at 10.25 p.m. At this point his wireless operator sent the first of two

Kapitänleutnant Joachim Breithaupt, commander of Zeppelin L-15.

messages to their fellow raiding airships as to the Zeppelin's injuries. Then, between Burnham and Southminster and after circling twice over Foulness as if undecided whether to risk leaving the land, or perhaps in order to try his ship's steering powers, he attempted to cross the sea.

A second message repeating the state and location of the Zeppelin was sent at 10.50 p.m. The ship's framework, however, was so severely tried by the collapse of the central balloons that it finally broke her back and she fell into the sea from a height of 2,000ft at Knock Deep at 11 p.m. The crew were rescued by two

The downed L-15 shortly before she sunk off Kentish Knock, 1 April 1916.

trawlers at about midnight. Only one man drowned. An attempt was made to tow the wrecked Zeppelin to ground, but this foundered off Westgate.

L-13, the second ship of this group, under Kapitänleutnant Heinrich Mathy, came in at Sizewell and passed over Leiston at 8 p.m., going south. Her first objective was Stowmarket, which she approached from the south-west at 8.45 p.m., dropping two flares, one of which landed at Badley. It had been claimed the Zeppelin had followed the 8.15 p.m. train from Ipswich. The searchlight and AA guns tried to locate her while she dropped twelve HE bombs very close to both the guns and works. Twenty-six rounds were fired by the 6-pdr AA guns without effect. The action lasted ten minutes, during which time the Zeppelin only came for a moment into the full beam of the searchlight. Her height during the attack varied from 6,000–7,500ft.

The Zeppelin then sheared off westward, passed south of Haughley and round southwards to Finborough, returning to Stowmarket at about 9.15 p.m. She was again attacked, and was hit by a tracer shell. This was recorded at the time by the battery, and is confirmed by a copy of a message intended to be sent by wireless and written by the commander of the Zeppelin, which was picked up near Stowmarket next morning. It had evidently blown overboard before it could be given to the wireless operator. The message is translated:

To Chief of the Naval Staff, High Seas Fleet, 10 p.m. Have bombarded battery near Stowmarket with success. Am hurt; have turned back. Will land at Hage about 4 a.m. L-13.

The only damage done at Stowmarket was to the railway, the permanent way being torn up. No casualties were reported, except for a soldier who was slightly injured by a fragment of AA shell.

The injury to L-13, whilst it did not disable her, was evidently sufficient to render a speedy return advisable. She dropped no bombs when hit, but passed on her course east-north-east and was over Stonham at about 9.30 p.m. She then dropped three large petrol tanks (which indicates that she was losing gas), and was not seen again until she was near the coast, flying very low.

Sighting the headlights of the armoured cars of the RNAS Machine Gun Section, L-13 dropped eleven HE bombs and five incendiary bombs upon them at Wangford at 10.10 p.m., and a further seven HE and twenty incendiaries were dropped on the aerodrome at Covehithe at 10.20 p.m. No casualties were reported, and no damage done.

L-13 was flying at the unusually low altitude of 2,000ft. Rising as the result of the loss of the weight of her bombs, she made full speed for Germany, reaching her shed at 2 a.m., after hugging the Dutch coast all the way back to base in consequence of her injury.

L-14, commanded by Kapitänleutnant der Reserve Alois Böcker, and carrying Korvettenkapitän Strasser, approached the Norfolk coast shortly before 8 p.m., followed two hours later by L-16. L-14 passed the *Would* lightship at 8 p.m. and

Obverse and reverse of one of the solid gold Wakefield Medals, presented to the gun crews involved in the shooting down of Zeppelin L-15 on 1 April 1916.

moved inland over Palling at 8.15 p.m., steering west and then south-west in a wide circle round Norwich. She found the Great Eastern Railway line at Tivetshall, which she followed past Diss and into Suffolk, and was over Sudbury at 10.30 p.m. Passing over the town from east to west, L-14 dropped her first bomb just after the church clock had chimed the half hour. A total of eight HE and nineteen incendiary bombs were dropped on the town, killing three men (one of them a soldier) and two women. One man was also wounded. An inn and two dwelling houses were demolished and glass was broken in a large number of houses.

It was noted that Sudbury was bombed almost immediately after Böcker had received the first wireless calls for help sent out by L-15 at 10.25 p.m., and it was by no means improbable that Sudbury suffered in consequence. The only other reason was that Sudbury was perhaps mistaken for Stowmarket.

The Zeppelin proceeded south-west past Bulmer and Halstead to Braintree where, at 11.05 p.m., three HE bombs were dropped, causing seven casualties, four of which were fatal. One small dwelling house was demolished and four partially wrecked.

Pursuing her course towards London, past Great Waltham at 11.15 p.m. and over Highwood Quarter at 11.25 p.m., at 11.35 p.m. the Zeppelin was fired at over Doddinghurst by a pom-pom at Kelvedon Hatch. She travelled for about a mile or two further and then, when searchlights were turned upon her, she circled to the south and turned right about, passing directly over the Kelvedon Hatch gun, which again fired at her at about 11.40 p.m. She dropped two HE bombs north of Doddinghurst in answer to the gun, followed by nine more at Blackmore at 11.45 p.m. No damage was caused beyond a few windows broken at Blackmore.

Whilst over Braintree at 12.55 a.m., L-14 dropped one HE bomb at Springfield, to no effect, and passed over Chelmsford, appearing to be making for Thames Haven. At 1.25 a.m. she dropped an incendiary bomb at Stanford-le-Hope, causing no damage, and at 1.30 a.m. five more HE and twelve incendiary bombs were dropped over Thames Haven. These struck two empty tanks at the Asiatic Oil Company's works and caused a small fire on the pier, with no casualties.

The raider was fired at by the AA guns at Thames Haven, Kynoch Town and Pitsea, without any apparent result, and then made off north-east after dropping a flare on Canvey Island, which did not ignite. She passed Rochford at 1.40 a.m., Southminster at 1.50 a.m., Bradwell at 1.55 a.m. and Mersea at 2 a.m. By 2.05 a.m. she was east of Colchester, and then went between Mistley and Stratford St Mary at 2.18 a.m., over Ipswich at 2.27 a.m., near Saxmundham at 2.25 a.m., to Theberton at 2.50 a.m. and out to sea via Dunwich at about 3 a.m.

L-16, under Oberleutnant zur See Peterson, passed the *Would* lightship at 9.20 p.m., and crossed the coast at 10.10 p.m. at Winterton Ness. She passed inland over Potter Heigham to Wroxham at 10.25 p.m., turning south-south-west over Thorpe, and over Stoke Holy Cross to Long Stratton. Turning west at 10.50 p.m. to New Buckenham, she headed south-west to East Harling and North Lopham, where she dropped a petrol tank at 11.15 p.m.

Passing Ixworth at 11.35 p.m. and reaching Bury St Edmunds by 11.45 p.m., the Zeppelin was fired upon by a mobile pom-pom section of the RNAS, without result. She then aimed her bombs at the guns, and over the town. Twenty-one HE and five incendiary bombs were dropped, killing six people (one of them a soldier) and injuring six others. Two cottages were wrecked, nine badly damaged and several other houses slightly damaged. The incendiary bombs fell in a field and did no damage.

Traversing Suffolk past Otley, Framlingham and Laxfield, she passed out to sea at Lowestoft at 1.05 a.m., dropping a heavy HE bomb (estimated to have been 100kg) on a tram shed and causing considerable damage, but no casualties. She was fired upon by the naval 12-pdr guns at Pakefield and by 6-pdr guns at Lowestoft, without result, other than causing her to rise higher.

Three military airships, **LZ-88**, **LZ-90** and **LZ-93** also set out to attack England on 31 March, but their actions are less easy to trace. Two Zeppelins were reported over England but not positively identified. One of them approached the Suffolk coast and returned to her north German shed at Namir at full speed early next morning. The other came overland, but dropped no bombs. Evidence points to this latter airship having come inland near Alderton. She was sighted at Wickham Market at 12.55 a.m., at Woodbridge at 1.05 a.m. and was seen circling over Ipswich after 1.18 a.m. She was fired at by a RNAS pom-pom and then disappeared. The purpose of the journey of this airship is difficult to fathom, unless it had been a reconnaissance or a practice flight. In any event, two nights later an army airship, probably LZ-88, steered almost the same course.

L-22, under Kapitänleutnant Martin Dietrich, was sighted by the *Inner Dowsing* lightship at 12.15 a.m., and reached the Lincolnshire coast at Mablethorpe at 1 a.m. on 1 April. Passing northwards, she was sighted at Donna Nook, where she dropped an incendiary bomb at 1.20 a.m. It was thought that either Grimsby or the important wireless station at Cleethorpes were her intended points of attack.

L-22 then went out over the Humber and came inland again, dropping fourteen HE and twelve incendiary bombs on Humberston at 1.35 a.m.; five of these did not explode and the only result was some windows broken at a farmhouse; there were no casualties.

Passing on over Cleethorpes, she dropped six HE bombs at 1.48 a.m. The Alexandra Road Baptist Chapel, which was being used as a billet for seventy men of E Company, 3rd (Special Reserve) Battalion The Manchester Regiment (who had only arrived the day before), received a direct hit and was completely destroyed, killing twenty-nine members of the battalion and leaving fifty-three wounded. Two of the wounded later died of their injuries, taking the death toll to thirty-one – the greatest loss of British soldiers as a result of enemy action in Britain during the First World War. A wing of the Town Hall was also damaged, and windows were broken in many houses and shops.

The majority of the soldiers of 3rd Battalion, The Manchester Regiment killed during of the air raid 1 April were laid to rest with full military honours, at Cleethorpes Cemetery on 4 April 1916, in a ceremony attended by many of their surviving comrades, the massed bands of the 3rd Battalion The Manchester Regiment, 4th Battalion The Manchester Regiment and 3rd Battalion The Lincolnshire Regiment. The coffins were carried on eight motor lorries, draped with the Union Jack and covered with wreaths and other floral tributes received from the general officer commanding and staff, officers and ladies of the regiment, sergeant's messes, all companies of the battalion, other regiments, batteries and schools of instruction in the command, members of St Peter's Church and Baptist churches, the Hebrew congregation of Grimsby and the Sisters of Brighowgate Hospital. The mourners also included a voluntary aid detachment, a naval detachment, representatives of Cleethorpes District Council, local tradesmen and hundreds of local people.

The following list is based on information kindly supplied by The Commonwealth War Graves Commission:

Cleethorpes Cemetery:
Private Wilfred Ernest Ball, 30358, 1 April 1916.
Private Joseph Beardsley, 32238, Died of Wounds 3 April 1916.
Private Louis Archie Beaumont, 34637, 1 April 1916. Age 34. Son of Harry B. Beaumont. Born at Oldham.
Private Samuel Bell, 9779, 1 April 1916. Age 28. Son of Joseph and Hannah Bell of Street Bridge, Chadderton; husband of Sarah Jane Bell of 45 Park Avenue, Broad Way, Chadderton, Oldham.
Private William Robert Bodsworth, 2425, Died of Wounds 4 April 1916.
Private William Henry Brown, 34621, 1 April 1916. Age 19. Son of Emily Schofield (formerly Brown) of 20 Elder Grove, Northfield Road, New Moston, Failsworth, Manchester, and the late William Milner Brown.
Private Ernest Budding, 34619, 1 April 1916. Age 19. Son of the late William and Mary Ann Budding, born at Hollinwood, Oldham.
Private Jospeh Chandler, 32997, 1 April 1916.
Private Job Clowes, 34618, 1 April 1916. Age 19.
Private Harry Cuthbert, 34633, 1 April 1916.
Private Frederick Dimelow, 30241, 1 April 1916.
Private Thomas Diviney, 30401, 1 April 1916. Age 37. Son of Patrick and Catherine Diveney.

Private Albert Edward Downs, 32215, 1 April 1916. Age 20. Son of the late Albert
Edward and Hannah Maria Downs of Haughton Green, Denton, Manchester.

Private Robert Fox, 34620, 1 April 1916.

Private William Francis, 32034, 1 April 1916. Age 19. Son of Joseph and Mary Ellen
Francis of 35, Worsley Street, Hulme, Manchester.

Private Thomas Hannon, 32263, 1 April 1916.

Lance Corporal Alfred Haynes, 32323, 1 April 1916.

Private William Hetherington, 30126, 1 April 1916.

Private Tom Pierce, 32278 1 April 1916. Age 35. Son of Mr and Mrs Hartley Pierce
of 31 Mawson Street, Ardwick, Manchester; husband of Mary Ann Pierce of
22 Durham Place, Chorlton-on-Medlock, Manchester.

Private Henry Ramsden, 27902, 1 April 1916.

Private James Russell, 30179, 1 April 1916. Age 21. Son of John and Maria Russell of
13 Keats Street, Collyhurst Road, Manchester.

Private Thomas, Tomkinson, 32275, 1 April 1916.

Private John Wheeler, 27591, 1 April 1916.

Private William Wild, 34639, 1 April 1916. Son of Mrs. Mary Hannah Wild of 38
Burnley Lane, Oldham.

Private Robert Wood, 27537, 1 April 1916. Age 32. Son of John Wood of Royton,
Oldham; husband of Mary Elizabeth Wood, of Devon Street, Oldham.

A large stone memorial with three panels of York Stone bearing the names of
all thirty-one men who died in the zeppelin raid was erected in Cleethorpes
Cemetery and unveiled on 9 March 1918.

Six casualties were also buried, by family request, at:

Oldham (Greenacres) Cemetery, Lancashire:

Private Thomas Brierley, 30117, 1 April 1916. Age 24. Son of the late John James and
Elizabeth Brierley; husband of Annie Power (formerly Brierley) of 35 Middleton
Road, Oldham. Grave Ref. L9. CE. 169.

Tonge (St Michael) Churchyard, Lancashire:

Private Frank Chandler, 2914, 1 April 1916. Grave Ref. Lower ground. M. 381.

Oldham (Hollinwood) Cemetery, Lancashire:

Private John Henry Corfield, 34625, 1 April 1916. Age 19. Son of Edward and
Elizabeth Corfield of 73 Hawksley Street, Hollinwood, Oldham.

Ashton-Under-Lyne (Hurst) Cemetery, Lancashire:

Private Percy Harrison, 33107, 1 April 1916. Age 21. Son of William and Mary Ann
Harrison of Hurst. Grave Ref. M. 324.

Stalybridge (St Paul) Churchyard, Cheshire:

Private Joseph Radford, 27724, 1 April 1916. Age 19. Son of George and Matilda
Radford of Manor House, Grosvenor Street, Stalybridge. Grave Ref. Spec.
Memorial.

Ashton-Under-Lyne (St Michael) Churchyard Extension, Lancashire:

Lance Corporal Jack Swift, 33055, 1 April 1916. He was cremated and is
commemorated on the Screen Wall.

1/2 April 1916

At noon on 1 April, Zeppelins L-11 and L-17 set out under orders to raid London.
On the way, the winds proved so unfavourable that Zeppelin commanders
received instructions by wireless to bomb the Midlands or north of England
instead. On the evening of Saturday 1 April 1916, two 'hostiles' were reported
heading for the Yorkshire coast.

L-11, commanded by Korvettenkapitän Victor Schütze, arrived over Seaham at
11.05 p.m. At 11.10 p.m. she passed over Eppleton Colliery, where the first of two
HE bombs were dropped on colliery refuse heaps, to no effect. At 11.12 p.m. the
Zeppelin was over Hetton Downs, where two more HE bombs were dropped,
breaking the windows of three houses and destroying several wooden outhouses.

She now turned due north, passing over Houghton-le-Spring, and when she
arrived over Philadelphia at 11.15 p.m. she dropped two HE bombs, the blast from
which smashed the glass in thirty-five houses but caused no other disruption. Still
keeping on her course due north, she reached the River Wear and followed it to
Sunderland where, at 11.20 p.m., she dropped fourteen HE and seven incendiary
bombs. Considerable damage was caused. Four shops and fifteen dwelling houses
were entirely destroyed and many partially destroyed. A workman's hall was partly
demolished, and three stables destroyed. A railway goods yard was damaged,
a tramcar was wrecked in a shed and one small fire broke out. The casualties
amounted to twenty-two people killed, twenty-five seriously injured (including
two soldiers) and 103 very slightly injured.

The Zeppelin was picked up by the Hylton searchlight, and engaged by a 6in
AA gun at Fulwell Quarry, Sunderland, which fired one round of common shell.
The Zeppelin then evaded the light and the gun did not fire again.

Leaving Sunderland, L-11 went out to sea, following the coast southwards, and
passed over the mouth of the River Tees near Grantham at 12.05 a.m. She then
went over Port Clarence five minutes later, where she dropped an HE Bomb which
fell on a heap of steel billets at Messrs Sell's works, doing no damage. Crossing the
river, she dropped two HE bombs in the mud. At Middlesbrough, three minutes

later, two HE bombs were dropped on vacant land adjoining the Cargo Fleet Steelworks. Windows were broken in a large number of houses in the vicinity and a hotel in Middlesbrough had all its windows blown in. Two men were injured.

L-11 followed the railway to Eston Nab, departing for Saltburn at 12.25 a.m. and circling there. Flying low again she followed the railway line past Brotton, and on to Skinningrove, dropping two HE bombs on a grass field at Cattesty Farm, 250 yards from the Skinningrove slag tips. No damage was done. At 12.30 a.m. the Zeppelin passed out to sea over Skinningrove.

L-17, under the command of Kapitänleutnant Herbert Ehrlich, was east of Flamborough Head at about 9 p.m. but did not come inland. Instead, she wandered along the coast for about an hour, emptying her bombs into the sea and finally turning back. Seventeen explosions were heard at about 9.23 p.m. from Tunstall, East Riding, and the last 'sounds of aircraft' reported at 10.20 p.m. off Hornsea.

2/3 April 1916

Four Zeppelins rose under orders to bomb the Firth of Forth on 2 April, most especially the naval base at Rosyth and the Forth Bridge, while two military airships set off with the objective of bombing London.

L-13, under Kapitänleutnant Heinrich Mathy, developed engine trouble shortly after departure and had to turn back.

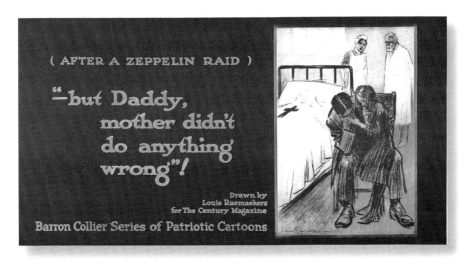

British propaganda poster produced in 1916 in the wake of mounting casualties suffered during Zeppelin air raids.

L-22, commanded by Kapitänleutnant Martin Dietrich, was claimed to have been heard off Bamburgh at 8 p.m. but only crossed the coast just north of Berwick-upon-Tweed at about 9 p.m. Nine HE and five incendiary bombs were dropped near Lamberton Farm, nearly 3 miles north of Berwick. The Zeppelin then proceeded some 6 miles inland as far as Chirnside, where two HE and four incendiary bombs were dropped at about 9.15 p.m. Three more HE and five incendiary bombs were also dropped in an open field at East Blanerne Farm. No harm was done by any of the bombs.

The Zeppelin then turned east, following the River Whiteadder, dropped a flare over Clarabad Mill and passed out to sea south of Lamberton at about 9.45 p.m. She dropped five more HE bombs in the sea, presumably at a ship. She next steered northwards past St Abb's Head, and came inland again over North Berwick at about 10.30 p.m. She then went north-west along the coast, past Dirleton where an incendiary bomb was dropped on the links near Archerfield House, and on to Gullane and Aberlady.

She flew across Portobello Bay, passing Tranent at 10.50 p.m., to Leith, and appeared over Edinburgh while L-14 was bombing the city, but at a much greater height during this attack, in which she took no part. She finally passed slowly round the south of the city from west to east and dropped three HE bombs in succession, the first two falling in the parish of Colinton – one at 11.50 p.m. at Slateford, the second at 12.15 a.m. at Comiston – and the third at 12.20 a.m. at Bridgend, near Craigmillar, in the parish of Liberton. No harm was done by these bombs beyond broken windows. The Zeppelin was going very slowly in contrast with L-14, which was circling the city at higher speed.

At around 12.30 a.m. L-14 was seen leaving Edinburgh eastward, and L-22 must have passed over her, probably north of Craigmillar Junction. At about 12.40 a.m., L-22 went out to sea near Portobello and flew away down the Firth of Forth.

L-14 was commanded by Kapitänleutnant der Reserve Alois Böcker and was first sighted off St Abb's Head at about 9.30 p.m. She appears to have followed the coast as far as Dunbar and then to have passed north across the Firth of Forth. She picked up her bearings by means of the Isle of May, where she was spotted turning west. After making Elie, on the north shore of the Forth, at about 11.05 p.m., she headed south-west and, after passing over Inchkeith, came in over Leith at 11.30 p.m.

L-14 attacked the docks at once, dropping her first HE bomb in the Edinburgh Dock, sinking two small boats and breaking the skylights of a couple of Danish steamers. The second and third bombs were incendiaries and they fell on the Albert Dock warehouse, damaging it considerably. The fourth bomb was another HE, which fell on the roof of a grain warehouse, and the fifth, also HE, fell on the edge of the harbour quay breaking the quay wall. The sixth, another HE bomb, fell on Commercial Street, destroying a building and killing a man who was sleeping in it.

The next four bombs were incendiaries landing on Commercial Street and Sandport Street. One fell through a roof into a room at 14 Commercial Street, which was occupied by an old woman. It fell through her floor into the room underneath, where it burst into flames. The old woman calmly got out of bed, poured water through the hole made by the bomb and extinguished the fire. The eleventh bomb was HE, and did serious damage after it fell on the roof of Messrs Innes & Grieve's whisky bond warehouse, destroying it and with it the whole of the stock of whisky in bond, causing £44,000 worth of damage. The premises, like several others in Edinburgh at the time, were unfortunately not insured against damage from aircraft.

The twelfth and thirteenth incendiary bombs fell through the roofs of two houses, the second being the manse of St Thomas's Kirk. Both houses were set on fire and the manse practically destroyed. Those inside had a miraculous escape; eleven people were in the first house and three in the manse. The next three bombs were all incendiaries and the seventeenth, an HE, did no harm. The eighteenth, another HE, fell in a court off Bonnington Road and killed a child. The nineteenth HE bomb fell on the railway line close to Bonnington Tannery and did not explode. The twentieth, again HE, fell through the roof of the tannery and damaged it considerably.

The course of the Zeppelin from the docks had accurately followed the course of the Water of Leith, which was evidently visible to her steersman, Jensen.

L-14 came over Edinburgh and dropped her first HE bomb on the city at 11.50 p.m. It landed in a vacant piece of ground at Bellvue Terrace and windows were broken in the adjoining streets. The second bomb, an incendiary, landed on the roadway at the Mound. The third bomb, an HE, fell into the grounds of George Watson's College, 150 yards due south of Lauriston Place. No damage was done, and there were no casualties.

An incendiary bomb fell harmlessly in the meadows south of the college, and at 11.55 p.m. the sixth bomb, an HE, fell on tenement houses in Marchmont Crescent, going through one house to the basement but injuring nobody. The Zeppelin then circled to the east and dropped an HE bomb on a tenement on Causewayside, wrecking the building and injuring six people therein. At the same moment an incendiary bomb fell harmlessly in a garden at Blacket Place, followed by another that did no damage in a garden at Hatton Place. The Zeppelin had, by then, turned north-west and was making for the centre of the city again.

The tenth bomb was an incendiary which fell on the grounds of the Royal Infirmary. It was followed by an HE, which fell on the pavement in the Grassmarket immediately in front of the White Hart Hotel, injuring four people, one of whom subsequently died. Another HE bomb fell on the south-west corner of the Castle Rock, followed by another, which badly damaged the County Hotel in Lothian Road and slightly damaged the Prince's Street terminus of the Caledonian Railway, injuring one person.

The Zeppelin then circled to the west, passed over Haymarket Station and dropped a group of three HE bombs as she was passing round north-eastward over the Water of Leith, near the Mill Lade and Donaldson's Hospital School. They did no damage, beyond breaking glass and damaging the embankment, but the next group of bombs proved more destructive.

The Zeppelin again passed over the centre of the city, going from west to east and dropped her seventeenth bomb, an HE, on the pavement in Marshall Street, a working-class quarter. A house was destroyed, six people were killed and seven injured. The eighteenth bomb, also HE, now fell on a spirit store in Haddon's Court, Nicholson Street, wrecking it and injuring three people. The nineteenth bomb fell on St Leonard's Hill, wrecking a tenement, killing a child and injuring two people.

It was now 12.15 a.m. and the Zeppelin was passing off eastward over the southern slopes of Arthur's Seat, from which she was being fruitlessly fired at by machine guns. As she went, she dropped three bombs in answer, in the southern portion of the King's Park, which were harmless, apart from breaking the glass in one building and throwing down part of a boundary wall. The last bomb, an incendiary, fell on the grounds of Prestonfield House, doing no damage.

L-14 then flew eastwards, passing Tranent at 12.40 a.m., Haddington at 12.45 a.m., moving south of Dunbar, and then north-east of Innerwick at 12.55 a.m. and finally out to sea at Cockburnspath about 1 a.m.

The steersman of L-14, captured with his commander when L-33 came down at Little Wigborough on 24 September 1916, admitted under examination that he had known perfectly well where he was, as the configuration of the Firth of Forth was unmistakable, and confessed that, since it was evident that Edinburgh was unprotected by guns he had purposely brought the Zeppelin down to a height of 700m (2,200ft) in order to make the bombardment effective and be certain of results. He said that he had tried to find Rosyth but, failing to make the dockyard, had steered for Edinburgh.

It was impossible to acquit the commander and his steersman of having, on purpose, avoided the more dangerous task of bombing the military objective Rosyth, in order, as at Hull, to terrorise and damage an unfortified city with no possibility of danger to themselves or their ship. Kapitänleutnant Böcker, who in civilian life was an officer of the Hamburg–American line, had been to Edinburgh in peace time as a ship's captain, and so knew the appearance of the coast and the position of Edinburgh in regard to Leith. The ordinary members of the crew seem to have been under the impression that they had attacked Rosyth.

L-16, under the command of Oberleutnant zur See Werner Peterson, made the coast of Northumberland between Creswell and Druridge at 11 p.m. L-16 then steered south-west, crossing the River Wansbeck north-west of Guide Post. The North Eastern Railway main line was picked up south of Morpeth and followed very slowly in a south-east direction as far as Plessey Station. The Gosforth and Bedlington colliery searchlights from Newcastle were then turned on, which seemed to disconcert the raider for she turned south-west.

The attention of the Zeppelin was attracted by the flares of the aircraft landing station at High West Houses, near Ponteland, which may have been mistaken for the lights of Newcastle (Elswick). Twelve HE and eleven incendiary bombs were then dropped in that area at 11.35 p.m., doing no damage.

The Zeppelin circled and made for the north-west, directly onto the flares of the aerodrome at Cramlington where, at 11.50 p.m., five more HE and six incendiary bombs were dropped. The only damage was a little burnt woodwork. She then steered north-west along the railway to the vicinity of Morpeth, and afterwards north-east to the neighbourhood of Broomhill Colliery where, at 12.15 a.m., more HE bombs were dropped in the open fields, three on Hadston and four on Togston Barns Farm. L-16 passed out to sea at the mouth of the River Amble at 12.20 a.m.

LZ-90, commanded by Oberleutnant Lehmann, was reported by the *Kentish Knock* light vessel as approaching east-north-east. The airship made the coast at the mouth of the River Colne at 10.40 p.m. and Tiptree at 11 p.m., striking the Great Eastern Railway at Witham a few minutes later and then steering south-west towards London.

At Great Baddow she was fired on by an 18-pdr of the Kirkcudbrightshire Territorial Artillery and bore off to the north-west, before resuming her course towards London. Passing between Blackmore and Ingatestone at 11.30 p.m., and on approaching Kelvedon Hatch at about 11.35 p.m., she was engaged by a pom-pom. Seventy-five rounds were fired, and the bursts of three shells were said to have been observed. It is doubtful whether the airship was hit.

She passed over Stanford Rivers at 11.40 p.m. holding course for the factories at Waltham and Enfield. She was now slowing down. At 11.50 p.m. she was passing Epping and Theydon Bois when she was picked up by the Chingford searchlight. Other lights now caught her, and at 11.54 p.m. she was heavily engaged by the guns of the Waltham Abbey Control. She immediately replied by dropping bombs as she approached the Woodridden Farm and Windmill Mill near Waltham Abbey. Four houses were slightly damaged.

LZ-90 flew over the Farm Hill gun at 11.57 p.m. and suddenly disappeared, having risen sharply after dropping her load of bombs, and so escaping the searchlight beams. Making off at high speed, she was seen passing between Nazing and Epping

Upland at 12.05 a.m., High Laver at 12.10 a.m., Barnston, south of Dunmow, at 12.20 a.m., Braintree at 12.25 a.m., Coggeshall at 12.30 a.m., Colchester at 12.40, Brightlingsea at 12.48 and finally out to sea north of Clacton at 1 a.m.

LZ-88 and her commander, Hauptmann Falck, reached the coast of Suffolk in the neighbourhood of Orfordness soon after 11.30 p.m. She appears to have been uncertain of her bearings, but it may be more than a coincidence that she followed much the same track as that supposed to have been taken by the military airship in the district on the night of 31 March/1 April.

Proceeding inland, she appeared over Wickham Market at 11.47 p.m. She then made a circle, and steered in the direction of Woodbridge, where she was reported at 12.07 a.m. The airship was next engaged at Rushmere Heath by four machine guns of the RNAS Anti-Aircraft Corps, and by a Lewis gun of the 58th Division.

At 12.20 a.m. she was over Ipswich, and made a complete circle south-west of the town without dropping any bombs, proceeding south-west to Copdock and then south-east round to Wherstead at 12.30 a.m. Turning north towards Ipswich, she went off due east to the coast, and was engaged by RNAS AA Corps guns at Levington Heath. After crossing the River Deben, the airship unloaded the bulk of her bombs, namely ten HE and fifty-three incendiary bombs, in the vicinity of Ramsholt and Alderton, where she hovered from 1 a.m. to 1.08 a.m. One HE bomb was also dropped at 1.15 a.m. near Hollesley, and the airship then passed out to sea at Orfordness at about 1.20 a.m. The bombs were harmless apart from a few windows broken at Ramsholt.

The airships that raided East Anglia were carrying an unusual number of incendiary bombs – at least sixty-five. This suggests that this was the method proposed for the attack of explosive factories in particular, since the incendiary bomb would be a poor weapon against guns or personnel.

3/4 April 1916

Zeppelins L-11 and L-17 set off from their north German sheds with London as their objective, but ran into strong winds on their way over. Although L-17 got near the coast off Haisborough, she made such little headway that her commander abandoned the raid and returned home.

L-11, with Korvettenkapitän Viktor Schütze, was heard 2 miles off the coast of Weybourne, Norfolk, at 1.30 a.m. on 4 April, going east. Visibility was poor, but she seems to have crossed the coast at Sheringham and was heard at Gunton at 2 a.m., having dropped an incendiary bomb 300 yards from Hanworth Hall.

She was later heard between North Walsham and Aylsham, and afterwards drifted slowly southwards, probably with her engines cut off, for 25 minutes. At 2.30 a.m. she was heard from Drayton and the Norwich observation post and, at the time, dropped two incendiary bombs at Buxton Lamas. She then flew direct along the line of the River Bure to Salhouse, where one incendiary bomb was dropped in a field at 2.45 a.m.

The Zeppelin was next heard 2–3 miles north of Acle at 2.50 a.m., and south of Winterton at 2.55 a.m., having followed the lines of the Rivers Bure and Thurne. She went to sea at Caister at about 3.05 a.m. and, 1 mile out, dropped nine HE bombs into the sea (probably at a ship) at 3.15 a.m. She appears to have done no damage whatever.

Owing to the peculiar atmospheric conditions of the night, the noise of the Zeppelin seems to have been deadened, and at many places such as Cromer, Wroxham, Mousehold and Martham, it was not heard at all. The local opinion, that the Zeppelin possessed an efficient silencer, is probably not justified by the facts. The mist, perhaps, and the fact that the Zeppelin was probably travelling very high and very slowly may account for its comparative silence. The late hour at which it arrived may also have contributed to the fact that it was not generally heard. The raid was probably intended for Norwich which, as usual, eluded observation.

5/6 April 1916

L-11, again commanded by Korvettenkapitän Viktor Schütze, made the coast at Hornsea Mere at 9.10 p.m. on 5 April. She then steered straight in the direction of Hull, following the railway. Arriving over Sutton at 9.17 p.m., she dropped one HE bomb, which fell in a field. The concussion slightly damaged a farmhouse nearby. At the same moment, AA guns came into action around Hull, and the raider sheared off south-east to East Park where she dropped three more HE bombs. Some windows were broken in Holderness Road and the side streets. Jesse Matthews of 11 Cotton Terrace, Barnsley Street, died of shock.

The Zeppelin then went on to Marfleet. The shooting was now becoming more accurate, and she accordingly retreated northward. After an interval during which she circled round over Swine and Burton Constable, she then returned towards Hedon and Hull at about 9.35 p.m. She turned east again without hesitation and passed over Burstwick at 9.46 p.m. and moved out to sea south of Aldbrough at 9.50 p.m.

Unluckily, a house in Park Avenue, Hull, was hit by gunfire and two men and two children were wounded by shrapnel. The AA gunners at Sutton and Marfleet both claimed to have made hits, and numerous observers confirmed their opinion, which is borne out by the fact that the Zeppelin spent a number of hours off the Yorkshire coast, probably engaged in repairs (indeed, she was reported as being 10 miles north of Flamborough Head at about 11.30 p.m.). Soon after, she was reported 8 miles east of Scarborough.

Three hours later, at 2.29 a.m. on 6 April, L-11 came back inland over Skinningrove. It is reported that the local ironworks attracted the raider by tipping slag. After circling the ironworks at a height of less than 3,000ft and dropping nine HE and fourteen incendiary bombs on them, she ran south-east along the coast and passed out to sea over Hinderwell at about 2.50 a.m. The laboratory at the ironworks was wrecked, but otherwise no damage was done there. The Council School, Co-operative Society shop and several dwellings were damaged at Carlin How. There were no casualties. The 6in gun at Skinningrove, not yet having received its searchlights, could not see the target and hence did not come into action.

L-16, under Oberleutnant zur See Werner Peterson, passed Hartlepool coming up from the south, hovered for some time and then crossed the coast over the Black Hall Rocks at 11.30 p.m. Steering south-west, she passed over Sedgefield at 11.45 p.m. and then south of Bishop Auckland, going west at great speed. At midnight she passed Evenwood, and at 12.03 a.m. she suddenly turned in her tracks when over Railey Fell Colliery, probably attracted by the fiery waste heaps. She returned at slower speed, dropped thirteen HE and ten incendiary bombs at Evenwood and Randolph Colliery. Fifteen miners' cottages were seriously damaged and about seventy slightly affected. A man and a child were both injured.

The Zeppelin carried on her way eastwards and appeared over the Auckland Park Collieries at Close House, Eldon, south-east of Bishop Auckland at 12.15 a.m. A single HE bomb was dropped on Coronation Colliery, followed by sixteen more and ten incendiaries at Close House and Eldon. Two miners' cottages were destroyed and eleven seriously damaged. Windows were broken in twenty-eight cottages and eighteen shops. A child was killed, and one woman and two children injured. No damage was done to the colliery works and eleven of the HE bombs fell on a cycle track and waste heaps, to no effect. The glow from the fiery heaps no doubt attracted the Zeppelin, as at Evenwood.

Proceeding east L-16 passed again near Sedgefield, and at 12.45 a.m. she was sighted over West Hartlepool. Turning north-east, she travelled back over Black Halls Rocks. Running parallel to the coast, she was next sighted over Seaham at 1.15 a.m. after which she went off north-east, passing Sunderland at 1.35 a.m. She was over the mouth of the Tyne at 1.50 a.m. and from there, steered for home.

A number of RFC aeroplanes went up from Beverley and Scarborough, but failed to locate the Zeppelin. Unfortunately, the aircraft from Beverley crashed due to engine trouble. Aircraft also went up from Cramlington, and one of them was damaged after colliding with a hedge, while another was completely wrecked against a house and the pilot, Lieutenant J. Nichol RFC, was killed.

Action of Anti-Aircraft Guns

Place	Gun	Range (ft)	Rounds	Time in Action (minutes)
Sutton	12-pdr	4,500	8S	11
Dairycoates Crossing	6-pdr	5,400	3 NT	2
Hornsea	12-pdr	4,500	8S and NT	5
Hull Cricket Ground	12-pdr	4,500	7S	6
Cottingham	13-pdr (mobile)	7,500	19 HE	2¼
Marfleet	13-pdr (mobile)	3,000	13S	7
Immingham Halt	12-pdr	8,000	6S	6

24/25 April 1916

The raid of 24/25 April 1916 was closely bound up with the bombardment of Lowestoft and Yarmouth by a German cruiser squadron early on 25 April. British Intelligence suggested:

> In view of the climatic conditions prevailing on that day the raiding airships would not in all probability have attempted to raid if the naval operations had not demanded an accompanying aerial demonstration. The naval attack was in turn connected with the political crisis in Ireland. The raid is unlike earlier attacks in that it had no definite objective on land.

Intelligence was correct only in parts. The German government had sought to support the Easter Rising in Ireland, and had taken the Irish nationalist leader, Roger Casement, from his exile in Germany by submarine to Banna Strand in Tralee Bay. The bombardment of Lowestoft had been planned by Admiral Scheer as a diversion, and the Zeppelin raid was independent to the naval operation. Eight Zeppelins had set out to 'attack England south, London if possible.' Their mission, however, was badly hampered by the wind over the central North Sea (40–50mph at 1,500ft), and the mist and cloud that hung over much of the coastline.

L-16, under Oberleutnant zur See Werner Peterson, crossed the Norfolk coast at Trimingham at 10.15 p.m. and pursued a winding course in a south-westerly direction, by way of Hanworth, Colby, Aylsham, Cawston, Ringland and Kimberley, where a small parachute was dropped with a bundle of German illustrated newspapers attached to it. Then, on it droned to Hardingham, Attleborough and west of East Harling to the neighbourhood of Thetford, which it reached about 11.30 p.m. It circled north of the town for about twenty minutes, then passed over Brandon at 11.55 p.m., and Mildenhall at 12.10 a.m.

She then circled to the west, south of Cheveley, and turned back north-east passing over Newmarket Heath at 12.30 p.m. where she was fired upon by two machine guns She retaliated at about 12.35 a.m. by dropping eighteen HE bombs, which landed along a line across the middle of the town of Newmarket, from the Heath to Warren Hill Station on the Bury road. One incendiary bomb was also thrown at the junction of the Bury and Norwich roads, beyond Warren Hill Station. Several houses, especially on St Mary's Square in the centre of the town, were damaged, and a racing stable on the Bury road was badly affected. It was here that the racehorse Coup-de-Main was killed. The only human casualty was inflicted by the last HE bomb, which damaged a house close to Warren Hill Station and seriously injured the owner.

Outside the limits of Newmarket, beyond the junction of the two roads, one HE and one incendiary bomb were dropped on the training ground known as 'The Limekilns' in the parish of Snailwell. No appreciable damage was caused and there were no casualties.

L-16 then made off at high speed north-east, passing Bury St Edmunds at 12.45 a.m. and west of Thetford at 12.55 a.m. She dropped five incendiary bombs at Honingham at about 1.15 a.m. and they landed in fields of wheat near the hall, with no real effect, but the fifth fell within yards of a large oat straw stack; the flames were fanned by a good breeze and the stack soon caught fire and damaged farm sheds, known as 'Read's Sheds', owned by farmer Walter Bartram. It was reported that the occupants of a cottage adjoining these buildings were unaware of anything untoward happening until they were roused by the police. The only other damage done in this neighbourhood was the destruction of an old straw stack ladder that had also caught fire and a hen turkey was roasted alive on her nest.

L-16 made for the coast via Ringland and North Walsham, where she was spotted at 1.30 a.m., and went back out to sea near Mundesley at 1.35 a.m.

As the German Admiralty reported that Cambridge had been bombed, it is probable that Peterson, judging by the position he had obtained at 11.35 p.m., when north of Thetford, was pleased to imagine that Newmarket was Cambridge!

L-21, under the command of Kapitänleutnant der Reserve Max Dietrich, struck the coast of Kessingland, south of Lowestoft, at about 11.10 p.m., after having dropped two HE bombs in the sea. She pursued a straight course for Stowmarket, passing over Halesworth at 11.30 p.m. and Eye at 11.50 p.m. On approaching Stowmarket she was once again engaged by AA guns of the RGA at Badley Park and Stowupland (Thorney Green). Eleven rounds were fired, one of which is claimed to have been a direct hit. This claim was, however, hardly substantiated by the subsequent movements of the airship. The fixed guns (two 6-pdrs) at Stowmarket also fired eleven rounds at her.

Shell-scarred and gutted premises on London Road South, Lowestoft, after the bombardment of 25 April 1916.

One of the houses destroyed by the naval bombardment on Lowestoft Esplanade, 25 April 1916.

Some of the naval shells and bombs that had been dropped or fired at Lowestoft, 25 April 1916.

It is alleged that she dived, stern-first, for 1,000–1,500ft in order to avoid the shells and then, emitting a cloud of smoke, beat a retreat. The statement that she dived stern first was improbable, and perhaps due to an optical illusion. The 'cloud of smoke', as on several other occasions, was likely to be a discharge of water ballast to enable the Zeppelin to rise quickly away from the fire. In any case, she was prevented by the guns from reaching her objective, the munitions works at Stowmarket. When fired upon, she dropped nine HE bombs at 12.16 a.m. on two farms at Old Newton, 2 miles north of Stowmarket before moving off. No damage was caused beyond a few holes in the fields and a broken window. There were no casualties.

L-21 proceeded across Norfolk, passing over Diss at 12.25 a.m., New Buckenham at 12.40 a.m., Wymondham at 12.45 a.m., Horsford at 1.15 a.m. and Worstead at 1.25 a.m.. Five minutes later, she dropped an HE bomb at Witton, which did no damage, and then went out to sea near Bacton at 1.35 a.m. L-21 subsequently took part in the operations of the fleet.

L-23, commanded by Kapitänleutnant Otto von Schubert, came in at Caister at 11.50 p.m. and immediately dropped three incendiary bombs, of which only one exploded. Next seen over Filby at midnight, L-23 seemed to follow the line of the Broads, north-west at low speed to Stalham, which she passed at 12.10 p.m. Arriving over Ridlington at 12.30 a.m., the Zeppelin dropped nine HE bombs, breaking windows and killing a bullock at Church Farm, and partially wrecking a cottage, but fortunately there were no human casualties. Windows were smashed and damage was also caused to the parish church of St Peter.

The Zeppelin then dropped a further six HE bombs near the RNAS station at Bacton, damaging glass and woodwork of the windows in two houses known as 'Beech Bough' and 'The Croft'. The blast also displaced the searchlight. L-23 then turned seaward and departed.

L-17 was commanded by Kapitänleutnant Herbert Ehrlich and came in to the Lincolnshire coast at Chapel St Leonards. She proceeded only a short distance inland, passing over Anderby at 1.40 a.m., to Alford where she dropped three HE bombs between 1.50 a.m. and 2 a.m. from a height of about 10,000ft. The only result was one broken window. The Zeppelin then immediately went out to sea again at Sutton-on-Sea at 2.05 a.m. The raid was represented in the German Admiralty communiqué as an attack near Lincoln.

The activities of **L-13**, under the command of Kapitänleutnant Eduard Prölss, remain a little uncertain. She was only overland for a short time, was not observed by the military and dropped no bombs. She seems to have made landfall at Cromer at 10.20 p.m., was at Hanworth in company with L-16 five minutes later, and then vanished into the night. At 11 p.m. she was at Weybourne and at 11.08 p.m. was heard at Bodham, near Holt, going south-east. She probably went out to sea near Sheringham shortly afterwards.

L-9 and **L-21** were active with the Imperial German Fleet. The squadron destined to attack the English coast on the early morning of 25 April was accompanied by L-9, commanded by Hauptmann Stelling. She was seen between 3.25 a.m. and 3.50 a.m. north-east of the *Cross Sand*, *Newarp* and *Cockle* lightships, going south-east. She was off the *St Nicholas* lightship at 4.10 a.m., about the time of the bombardment of Great Yarmouth and Lowestoft, and then disappeared eastward. L-21, when well on her way homeward after her expedition to Stowmarket, turned at about 2 a.m. to join the fleet, and accompanied it westward towards the coast, where she was seen off the *Haisborough* lightship at 4.30 a.m.

L-20, commanded by Kapitänleutnant Franz Stabbert, scouted for the fleet to the southward and did not approach the coast.

L-6 and **L-7**, commanded by Hauptmann Manger and Kapitänleutnant Hempel respectively, went out with the rest, and reconnoitred to the north of the advancing naval squadron. They did not go west of 4° E *long.*, and returned to Germany around 4 a.m. when the objective of the naval attack was attained. L-7, coming from her station in Schleswig, made for the sheds near Emden at about midnight. There she lay for some two hours whilst being refilled with gas, petrol and oil. She then resumed her patrol work and returned to her base next morning.

L-11, under Korvettenkapitän Viktor Schütze, made landfall near Bacton shortly after 12.30 a.m. and went over Honing Hall, and shortly after emptied what was probably her entire cargo of bombs – nineteen HE and twenty-six incendiaries over Honing and Dilham. Only one of the HE did not explode, and the blasts from the others stripped many tiles from farm premises at Dairy House Farm and four nearby cottages, while another four cottages had numerous panes of glass broken. Mrs Fanny Gaze, a 79-year-old widow, who was living with her son at the farmhouse of Hall Farm near to where the bombs fell, died of shock.

L-11 then turned east, passed Stalham at 1.05 a.m. and was sighted at Palling, heading out to sea at 1.18 a.m. where she turned north-west and skirted the coast as far as Bacton. At 1.25 a.m. she was fired on by the 3-pdr guns of the Eastern Mobile Section of The Royal Naval Air Service Anti-Aircraft Corps.

L-11 then turned out across the sea and flew back to base, but not without injury. Independent observations claimed that the Zeppelin appeared to have been hit by one round, an achievement borne out in the statement of the prisoners of L-7, captured in the North Sea on 4 May, who confirmed that L-11 went home with one gas balloon collapsed, after it had been pierced by a shell.

Aftermath

The inquest into the death of Mrs Fanny Gaze was held on Wednesday 26 April 1915, before the Norwich District Coroner, Mr H. Culley. Her son, William Bowyer Gaze, with whom she was living, confirmed that she had previously enjoyed fairly good health, though she had suffered with rheumatism 'but was able to get about.' She had retired to bed about 10 p.m.

William Gaze stated that they had been awakened between 1 and 2 a.m. on Tuesday morning by '… terrific explosions. Four within the period of a minute.' He went outside, and saw a Zeppelin. Not more than five minutes elapsed between the time he first heard the explosions and his seeing the aircraft. The bombs had dropped in a field about a quarter to half a mile from the house. On returning indoors, he went to check on his mother and found her in a state of collapse and unconscious. He called up their servants to help him, but Mrs Gaze did not revive. He then ran to get help. Seeing Robert Earle, the gamekeeper, he sent him to summon the district nurse but, by the time she arrived, Mrs Gaze was already dead.

The findings of the inquest were recorded on her death certificate – 'Heart Failure from shock endured by the terrifying effect of explosions produced by bombs unlawfully dropped from a Zeppelin aircraft.'

A personal memoir of this Zeppelin raid was written by Edith Everitt Owen (1873–1946) who, at the time, was living with her sister, Alice, and their widowed father, the Reverend Canon John Smith Owen, at Witton Vicarage:

April 24th 1916, Easter Bank Holiday
The night was dark but starlit with a strongish westerly breeze which had been more evident during the day. Father retired as usual about 10 o'clock and Alice soon after but returned to tell me that the aerodrome was illuminated and the search lights were being used. On going into the garden I heard one of our biplanes but though I could trace his course by the whine of the engine I could not see the machine. I came indoors about 11 p.m. and went upstairs soon after. When I heard a shell explosion which I put down (rightly as we afterwards heard) to our airman letting off his red flare to show he was descending. The subsequent intermittent purring of his engine told me he was landing. I would have dropped off to sleep but was awakened at 11.45 by a deafening report, quickly followed by several others, all sounding very near.

The household was active by this time, the maids in the kitchen and Alden, Alice and myself in hastily dressed clothes in father's room. Bombs fell at greater intervals and apparently at a greater distance and finally ceased altogether the last one sounding a long way off. Of us all Father took it all in, I think, with the greatest composure but the noise of the explosion of the bombs was ear splitting and the whole place shook and the windows rattled so that we expected every moment to hear the crack of shattered glass. Going out into the garden I saw the beam of the search light still lighting the sky but the raider had apparently made off and presently the search light went out and we agreed to go to bed again though some determined to keep some of their clothes on. Personally I undressed and consequently had the trouble of searching for raiment in the dark again when a second Zeppelin began dropping bombs about a quarter of an hour afterwards. Three fell in quick succession, the flashes of the explosions lighting up the entrance of the hall through the skylight and roar of the exploding bombs telling of their proximity. This time father, somewhat reluctantly, consented to get up and dress and we all retired to the cellars where we drank hot coffee and sat on uncomfortable chairs. All the time bombs were dropping at frequent intervals but seemingly further away. After a lull we decided that the dining room was preferable to the cellars and we accordingly adjourned thither and the maids going into the kitchen. Someone remarked that the wind was rising and we listened, I going to the front door. It was no rising of the wind but a third Zeppelin for as I stepped out there was a flash and a fearful bang. I turned to go indoors and as I did so flash after flash lit the garden tree tops and several bombs (five in reality) dropped a distance of 200 to 300 yards away their reports succeeding one another so rapidly as to merge into one stupefying roar. The mighty 'swish' that the five made, falling in such rapid succession, was distinctly audible, we sat up in the dining room until 2.45 when everything being quiet we all turned in, the sun having risen by that time.

I must have slept for over an hour when the shaking of the windows awoke me again. It was daylight, the sun was rising – a glorious morning. I lay and listened. A distant deep throated growl followed at an interval of a few seconds, by a rattling crash that made the house quiver awakened father's inquiry, from the next room, as to whether the Zeppelin had returned, but eventually suggested a naval bombardment, probably Yarmouth. For about 20 minutes we lay listening to the reports of the guns and the crash of the exploding shells as they fell (as we afterwards learnt) on Yarmouth and Lowestoft.

The sun was nearly above the horizon and between the distant muttering of the guns and the succeeding explosions I more than once (and father also) heard the cry of an imperturbable cuckoo for the first time this year. The firing gradually lessened, grew more distant, faded to mutterings and ceased. We went to sleep.

To describe one's feelings during the actual bomb dropping is almost impossible. Fear was there without a doubt but no unreasoning panic. Anger at our utter helplessness coupled with the desire to keep a 'stiff upper lip' as a help to each other predominated I think. It was perhaps curious but easily to be understood that we gave little thought to anyone outside our immediate neighbourhood where we afterwards found that people living some little distance from us outside the range of the bombs were almost, if not quite, as disturbed as we were and told us that they thought the bombs dropped were surely meant for them. But some of them have heard them really close. Let's hope they weren't.

The damage done was mercifully small. One bomb dropped in Mrs Cozen's garden, breaking the palings and glass in her house and the adjoining one (Mrs Baine's). The five bombs above referred to dropped in Brady's field between us and Bacton making holes 10–12ft across and 5–6ft deep. Mrs Randall Cubitt's house had all windows broken and doors wrenched off. She, bare footed as she was, carried her baby to Ridlington vicarage. The bomb that wrecked her house dropped within 20 yards of her door, close by the wall of a barn, carrying away a large piece of the wall but injuring none of the 6 horses in it. The south windows of Ridlington Church were all broken and the lock at the south door freed. One old lady in Honing died of shock. Mrs Cubitt of the Grange, Bacton, aged 93, slept soundly though it all. Forty-three bombs dropped within a radius of 4 miles from us some of them fell on the shore but failed to explode.

25/26 April 1916

Five army airships set out to raid England on 25 April 1916; one of them turned back before it reached the coast.

LZ-87, probably commanded by Oberleutnant zur See Barth, was reported off the Downs at 9.50 p.m. At 9.55 p.m. she dropped eight HE bombs round the steamer *Argus* in Deal harbour, without doing any damage. She was next sighted from Walmer, and on being fired at by AA guns there, turned seawards. She afterwards passed north along the coastline and was off Ramsgate at 10.24 p.m. She then appears to have turned eastward and to have returned to Belgium.

The second airship, most probably to be identified as **LZ-97**, commanded by Hauptmann Linnarz, reached the coast near West Mersea at 10 p.m. and flew across Essex, following the course of the River Roding and dropping forty-seven incendiary bombs in a line between Fyfield and Ongar at about 10.50 p.m. The only damage caused was a partially destroyed shed in Ongar.

She continued to follow the course of the River Roding, past Stapleford Abbotts, and went on south-west over the gun at Dog Kennel Hill, which opened fire at 11.08 p.m., and on to Barkingside. There, between 11.10 p.m. and 11.15 p.m. the airship dropped twelve HE bombs along a line curving from south-west to south-east between Forest Farm and Aldborough, west of Hainault Farm Aerodrome.

Four bombs fell at Forest Farm doing slight damage, one at Fairlop Station and one in the roadway at Fairlop, damaging several cottages. One then landed in a field at Monk's Well, one at Barkingside Station Bridge and one in a field east of the station. No damage was done at Barkingside. Three fell near Aldborough, one in a well, which was blown to pieces, and two, one of which did not explode, in a field.

At Aldborough the airship turned south, and dropped a thirteenth HE bomb at Newbury Park, which did no damage. The airship, which was then under heavy gunfire, next turned sharply south-west towards Ilford but, when over Seven Kings, finding the gunfire increasing in accuracy and being simultaneously attacked by aeroplanes, she turned about abruptly and made off to the north-east, dropping two HE bombs which fell between Goodmayes and Chadwell Heath. One, which dropped in a field, did no damage save the breaking of some windows, but the other fell on a house and destroyed it. The occupants were luckily absent and there were no casualties, either there or at Barkingside.

The airship was then seen going off as fast as possible past Romford, along the parallel lines of road and railway, but soon had to twist about in her course in order to avoid the searchlights turned on her from the London defences, and the guns at Brentwood, Kelvedon Hatch and Billericay, of which the airship had to run the gauntlet on her retreat. Around 11.35 p.m. she passed near the Brentwood gun and approached Billericay about 11.45 p.m. Headed off by the searchlights and gunfire to the southward, she now finally made off north-east.

A considerable amount of damage was caused by fragments of AA shell. Sixteen houses at Ilford, Seven Kings, Barking, East Ham and Wanstead were slightly damaged in this way, by the fire of friendly guns, and one man at Ilford was injured by a fall of debris caused by an unexploded shell which fell on the roof of his house. This was the only casualty, directly or indirectly, caused by the raid.

A number of pilots also contributed to the departure of the airship. Eight pilots rose, three from Hounslow, three from Sutton's Farm and two from Hainault Farm. Lieutenant Robinson, from Sutton's Farm, engaged the airship with machine gun when she was over Seven Kings but Robinson, flying at 8,000ft, was still some 2,000ft below her and his fire did not prove effective. Captain Harris from Hounslow also got within 2,000ft of the airship, the height of which he estimated at well over 15,000ft, but his Lewis gun jammed and his own aircraft was hit by machine gun fire and shrapnel bullets. The airship then eluded the searchlights and disappeared eastwards.

In her retreat she passed north of Chelmsford at 11.55 p.m., over Maldon at 12.04 a.m., Tollesbury at 12.15 a.m. and Brightlingsea at 12.25 a.m. She went out over Clacton at 12.34 a.m. and was last heard going along the coast off Harwich at about 12.45 a.m.

LZ-93 was first reported at the mouth of the Orwell at 10.30 p.m., approaching from the direction of the *Cork* lightship. She passed over Landguard Fort, dropping two water flares in the sea before coming overland, one HE bomb

A Zeppelin caught in the searchlights, with anti-aircraft fire bursting around it.

landed on the common north of the Fort, to no effect, and three incendiary bombs fell in the mud on the other side of the spit.

The airship went across to Harwich, where she dropped two HE bombs close to Government House, neither of which exploded. She then made for Shotley, where three HE and four incendiary bombs were dropped in and around the Royal Naval Training Barracks, none doing any damage beyond the breaking of a little glass, even though one of the incendiary bombs fell between two dormitories. A fifth incendiary bomb was dropped in the mud west of the barracks.

The airship turned south over the River Stour, appearing over Parkstone Quay, where a single HE bomb was dropped on reclaimed land between the station and the village. It was probably a large and heavy bomb of the 240lb type and it sank through the soil to the mud beneath, where it could not be found, even at a depth of 20ft. The airship then passed again over Harwich and Landguard, dropping four water flares in the harbour west of the fort. The airship was fired at by the AA guns of the garrison, without effect.

The airship went out to sea about 10.45 p.m., in the direction of the *Sunk* lightship, near which she dropped an uncertain number of HE bombs, before finally departing.

LZ-88 approached from off Herne Bay at about 12.20 a.m. and came in near Whitstable about 12.30 a.m. From Whitstable, she went south-east to Sturry. She had stopped her engines at 12.45 a.m. and was drifting.

At 1 a.m. she turned south-west, went over Canterbury at 1.05 a.m. and circled round to the south-east and north-east, passing over Bridge to Wingham, where she was spotted at 1.15 a.m. She then turned north and, at 1.20 a.m., dropped nine incendiary bombs in a straight line parallel to and 500ft west of the main road north and south of Preston. They merely burnt some turf.

Further on, thirteen incendiary bombs, two of which did not ignite, were dropped at Sarre and Chislet Marshes at about 1.25 a.m. with no results, and then at 1.30 a.m. an HE bomb was dropped at St Nicholas-at-Wade, falling into the vicarage gardens, and uprooting two trees that fell against the house breaking several windows.

The airship then proceeded to Birchington, and dropped four HE bombs in fields between St Nicholas and Shuart Farm and another north of the farm, followed by four more in the marshes, all doing no damage. Two incendiary bombs were thrown, one near the railway line and the other on the sea wall at Minnis Bay, also doing no damage. The airship then went out to sea at 1.35 a.m., going north-east, and dropped three HE bombs in the bay as she went. No casualties occurred during the raid.

26/27 April 1916

LZ-93 (or **LZ-95**) was sighted off Nieuport at 9.05 p.m., going along the coast towards Calais. She was fired at from La Panne, passed Dunkirk and was 5 miles north of Calais at 9.25 p.m. She then approached the Kent coast, and at about 10.30 p.m. came inland over Kingsdown after having dropped three HE bombs in the sea off Deal, where she was heavily fired on.

The airship then passed north-east inland, at a distance of 2 miles from the coast, went over Sandwich at about 10.40 p.m. and then over Minster and Westgate, where she was bombarded by two AA guns about 10.48 p.m. and was reported to have been hit. The noise of traffic in Margate and passing trains at this point, is said to have made the locating of the airship difficult.

The airship then turned east over Margate at 10.50 p.m. and went out to sea in the direction of the *Tongue* lightship, from which she was reported at 11.05 p.m.

as going east with an unusual noise of engines as if damaged. No bombs were dropped on land, and no damage or casualties resulted.

2/3 May 1916

Eight Zeppelins left their sheds on 2 May 1916, with Rosyth and the Forth Bridge as their main objective. When they were approximately 100 miles off the Firth of Forth they encountered adverse winds and, with the exception of L-14 and L-20, they turned off to attack alternative objectives in the Midlands.

LZ-98, an army airship under Kapitänleutnant Lehmann, seems to have hovered off the coast of Lincolnshire for an hour and a half without ever coming over the land. She approached the Humber and passed Spurn at about 7.55 p.m., and was engaged by the gun as she passed, without result. On being fired at she turned south, passing near Donna Nook, and then headed south-east discharging water ballast freely in order to rise out of range. The gun was in action for ten minutes. After lingering off the Spurn from 7.50 p.m. to 8.15 p.m., and having verified her position, she made off direct across the North Sea back to her north German shed.

L-23, commanded by Kapitänleutnant Otto von Schubert, appeared at Robin Hood's Bay at about 9.15 p.m. She passed directly inland past Goathland and over the moors as far as Danby High Moor, where at about 9.40 p.m. she dropped a single incendiary bomb near Danbury Head, which set fire to a large expanse of heather.

She then turned northward, and was heard shortly afterwards at Castleton, appearing over Skinningrove at 10.05 p.m. and dropping seven HE and four incendiary bombs there. A storehouse in the ironworks was partly wrecked and set on fire. No lights were shown at the works, and all outside work that would occasion lights had been suspended. When bombs began to drop, the searchlight on Huntcliffe located the airship, on which fire was at once opened by the 6in gun at Brotton. She was believed by some observers to have been hit.

She passed on her course along the coast eastwards, dropping six incendiary bombs at Easington at 10.10 p.m., one of which damaged a dwelling house and injured a child. At about 10.25 p.m. she appeared at Whitby, where she looked to be damaged; the sound indicated that all her propellers were not working. She then left the coast, and was last seen dropping flares over the sea in the direction of Scarborough at 10.35 p.m.

Her return journey to Germany was made very slowly, and it is not impossible that she had received some damage from the Brotton Gun. The damage could not have been too severe, as ten days later L-23 was reported as being seen cruising over the North Sea.

L-21, with Kapitänleutnant der Reserve Max Dietrich in command, came inland at Cloughton, north of Scarborough at 9.40 p.m. and then pursued a very clearly marked course in a direct line for the city of York. It may well have been that this was purely by chance.

The Zeppelin commander steered south-east across Wykeham High Moor and via Wrelton, near Pickering, at 10.05 p.m., Hovingham at 10.10 p.m., Terrington at 10.15 p.m. and Stamford Bridge at 10.30 p.m., then on to Naburn, where Dietrich seems to have perceived the railway running north to York. He immediately turned to follow it, going over Copmanthorpe and Dringhouses, where he dropped five HE bombs and thirteen incendiary bombs. Glass was shattered in houses from the concussion caused by the exploding bombs, and an officer and private on duty were slightly injured.

L-21 then went on to York where, between 10.40 p.m. and 10.50 p.m., twelve HE and four incendiary bombs were dropped along a line south-west to north-north-east across the southern and eastern part of the city, from Knavesmire to Skeldergate Bridge to St Saviourgate. Taking her route in detail, on leaving Dringhouses L-21 passed across the Knavesmire, in a north-easterly direction to Nunthorpe Hall. One incendiary bomb fell on Knavesmire, doing no damage, and three HE and two incendiary bombs landed at Nunthorpe Hall. One of the incendiary bombs fell on a building in which the Voluntary Aid Detachment Hospital was housed and destroyed the roof. One other bomb fell in the grounds and did no damage.

A HE bomb landed in Nunthorpe Avenue, damaging three or four houses badly. This was followed by one in Upper Price Street, which blew one house to pieces, badly damaged another and considerably injured several people. Two people were killed there. Two HE bombs fell in Nunthorpe Road, one in Victoria Street and one in Price's Lane, smashing the doors and windows and damaging the roofs of a large number of houses. Another HE bomb then hit Newton Terrace, wrecking four houses and injuring three people, followed by one in Kyne Street, where windows and doors were blown in and one woman was injured. Finally, the last bomb, also HE, fell in St Saviourgate, blowing out the fronts of several houses, killing six people and slightly injuring several others.

In all, eighteen houses were destroyed or wrecked, and considerable damage was done to a very large number of others, mostly workmen's dwellings. Happily, none of the old and historic buildings of York were damaged as the Zeppelin passed well east of the centre of the city, and the nearest bomb to the Minster, which fell in St Saviourgate, burst more than ¼ mile away.

Three people were killed inside houses, and six in the streets. Six men were killed (including two soldiers) and three women. Nineteen men (including three soldiers) and eight women were injured, making a total casualty list of thirty-six.

L-21 went off in a north-easterly direction, passing near Stamford Bridge at 10.55 p.m. to a point north of Bishop's Wilton, where she seems to have veered off in the direction of Settrington, where she was seen at 11.10 p.m., afterwards circling westward to south of Norton at 11.15 p.m. She then went off east by south, dropping a single incendiary bomb at Kirby Grindalythe, and was seen at Sledmere at 11.30 p.m. going south.

The Zeppelin went out to sea at Bridlington at 11.52 p.m. She apparently circled when out at sea, came inland again and finally went to sea at Ulrome, near Skipsea, where she was seen and heard at 12.10 a.m.

L-16, under Oberleutnant zur See Werner Peterson, appears to have made landfall near Scarborough before 10 p.m. but was not observed until she was over Rosedale Abbey at 10.30 p.m., turning north, and making for the heath fire on Danby Moor, set alight by the incendiary bomb of L-23 an hour before. By this time, the fire had got a good hold on the moor and Peterson evidently took it to be an important conflagration as he steered directly for it, and dropped an uncertain number of HE bombs and probably six incendiary bombs on it, passing rapidly northward.

At 10.50 p.m. the airship was over Castleton and then turned east along the valley of the Esk. Around 10.55 p.m., she dropped five HE bombs at Lealholm, causing slight damage to a farm building and breaking the windows of a few cottages. Then, after wandering about for some time she went off north-west and, at 11.15 p.m., dropped five incendiary bombs at Moorsholm, doing no damage. Going north, at 11.25 p.m. she was seen at Skinningrove and then went out to sea.

L-17, commanded by Kapitänleutnant Herbert Ehrlich, came in near Saltburn at about 10.50 p.m. and made straight for Skinningrove. Shortly before 11 p.m. she dropped thirteen HE and four incendiary bombs on Carlin How, which wrecked six dwelling houses and damaged others, but caused no casualties. L-17 proceeded inland over the moors and was seen west of Whitby at 11.10 p.m., before turning west to the scene of the fire at Danby and, at about 11.30 p.m., added several HE bombs to the number already thrown there and went off, passing out to sea 3 miles north of Whitby.

L-13, commanded by Kapitänleutnant Eduard Prölss, appears to have come overland near Whitby at about 10.30 p.m. She was seen making westwards inland, and then turning south in the vicinity of Goathland, after which all trace of L-13 was lost until around 11.40 p.m. when she was located over Market Weighton, going west. The Zeppelin dropped two flares near Pocklington at 11.50 p.m., was heard at Bishop's Wilton five minutes later, having turned north and, just before midnight dropped an incendiary bomb at Fridaythorpe, which did no damage. She next passed over Rillington at 12.25 a.m., turned east and was observed shortly afterwards from Wykeham and, at 12.35 a.m., was over Ganton. Here she turned abruptly north-east, following the railway, and dropped another single incendiary bomb at Seamer, to no effect. She then passed out to sea north of Scarborough at 12.50 a.m.

L-11, under Korvettenkapitän Viktor Schütze, was engaged at 8.40 p.m. 10 miles east of St Abb's Head by HMS *Portia* and *Semiramis*. Thirteen rounds were fired but, although the Zeppelin was flying low and was broadside to the attack, it does not appear that she was hit. She escaped eastward, but returned to the coast of Northumberland near Beal at 10.20 p.m., dropping an incendiary bomb at Goswick as she came in. Another was dropped further south, on the sands between Holy Island and the mainland. No damage was done by either.

L-11 continued on her way southward along the coast and, at 11.15 p.m., went out to sea again at the mouth of the River Coquet without having dropped any more bombs. She was observed en route to Belford, Seahouses, Alnwick (where she dropped a flare) and Amble. On leaving the coast she seems to have gone north towards the Firth of Forth again.

L-14, under Kapitänleutnant der Reserve Alois Böcker, came overland near Eyemouth, north of Berwick, at about 8.25 p.m., dropped no bombs and immediately went out to sea again. She then made northwards, intending to attack the Forth. She was, however, driven too far north by the strong south-east wind and, at 9.30 p.m., was 15 miles east of the mouth of the Tay.

She did not come overland until she had reached Lunan Bay, north of Arbroath. After passing Arbroath at 11 p.m., Böcker seemed very uncertain of their position and circled about for half an hour near the town (probably due to bad visibility caused by low rain clouds). He then went off down the coast at 11.40 p.m., and dropped three HE bombs in a potato field on the farm of Panlathy, in the parish of Arbirlot. A horse was slightly injured, and a single pane of glass was broken.

L-14 went on slowly in the direction of Dundee, passed over the burgh of Carnoustie and crossed the Tay east of Monifieth. At 12.07 a.m. she was seen from Tayport going towards St Andrew's Bay, where some bombs were dropped in the sea. She must have crossed Fife Ness and was last seen going east from the Isle of May after 12.15 a.m.

L-20, commanded by Kapitänleutnant Franz Stabbert, struck land at Redcastle, Lunan Bay, at about 9.55 p.m. and appeared to be heading for the Cromarty Firth. She pursued her course as far as the Caledonian Canal, where she was heard at Errogie, near the north end of Loch Mhor at midnight, and at Balmacaan, Milton and Lennie on the west shore of Loch Ness, half an hour later.

L-20 was then seen turning south and east, and she began her return journey. Her commander no doubt realised that, with the weather changing to drizzle and rainclouds rendering all landmarks invisible, his object was unattainable. At 1.10 a.m. he verified his position, and L-20 was then over Aviemore, going east, and about 40 minutes later was at Rhynie, north of Lumsden, in Aberdeenshire.

Here, Craig Castle, being beyond the restricted lighting area, was lit up and thus attracted the Zeppelin commander's attention. Six HE bombs were dropped, falling within 40ft of the house, the windows and roof of which were damaged. No casualties were caused.

Going on her way after passing Kennethmont at 2 a.m., L-20 dropped four HE bombs and one incendiary at Knockenbaird and Scotstown, north of Insch and 10 miles east of Lumsden. These bombs landed in fields doing no damage. Three more HE bombs were dropped in a field near Freefield House in the parish of Rayne shortly afterwards, again to no effect.

L-20 was seen between Rothienorman and Oyne, passed near Old Meldrum and went out to sea at Peterhead at about 2.40 a.m.

At 5 a.m. L-20 was seen by a trawler about 95 miles due east of Aberdeen and half an hour later by another further east.

Stabbert found that his supply of petrol was insufficient to carry him back to Germany against the wind, so he steered for the nearest land, the coast of Norway, where at about 10 a.m. he came ashore, low down, over the coast of Jaeren, south of Stavanger. Some of the crew, including the second in command, Leutnant zur See Schirlitz, saved themselves by jumping into the water before the land was actually reached, but the rest remained in the Zeppelin, which flew on over the land until she collided with a hill near Sandnaes.

The impact wrenched the whole afterpart of L-20 up almost at a right angle.

Zeppelin L-20 in the water of the Hafrsfjord, after colliding with the Jaataaberg, a hill near Sandnaes, Norway, while returning from a raid on Britain, 3 May 1916.

She rose, in this terribly damaged condition, and almost immediately fell into the water of the Hafrsfjord, where she broke in two. The rest of the crew were now rescued, and the Zeppelin was subsequently destroyed by the Norwegian military authorities since it was impossible to either repair or intern her. Those of the crew who swam to land were treated as shipwrecked sailors and sent back to Germany. Schirlitz was later posted to L-33 when the Zeppelin was completed, and was captured along with the rest of the crew and the captain when she came down at Little Wigborough on 25 September 1916.

Those who were taken prisoner in the Zeppelin after she had crossed Norwegian territory, including Kapitänleutnant Stabbert, were interned. Stabbert, it was later reported, escaped captivity and returned through Sweden to Germany in November 1916.

28/29 July 1916

On 28 July 1916, the weather conditions were most favourable for Zeppelin navigation on the European side of the North Sea, and so ten naval Zeppelins set off to bomb England. The problem for them when they arrived was unexpected sea fog on the east coast.

Four Zeppelins turned back before making landfall. L-24, L-17 and L-13 made landfall over the Humber and Lincolnshire, and then conducted their missions over the Midlands. L-31 came in at Corton, circled over Suffolk, and left without dropping a bomb, while L-16 and L-11 droned over Norfolk.

L-24 made for land north of the mouth of the Humber. She was observed by the SS *Montebello* flying low at 12.40 a.m., east of Withernsea. She was seen from Kilnsea, and passed over Sunk Island at 12.50 a.m. then across the Humber to Immingham where, at 1.10 a.m., six HE bombs were dropped on Stallingborough Marsh close to the Immingham Electric Railway. No damage was done, other than slight injury to the electric wires. The Zeppelin was engaged by a fixed 12-pdr gun at Immingham Halt, but the gun was at a disadvantage owing to fog. Two rounds were fired, apparently unsuccessfully, as the raider went off very swiftly eastwards across the Humber and on to Withernsea.

She went out over sea at 1.25 a.m. and while crossing the coast, flying high, she was fired at by a 3in 20cwt AA gun and a 6in mobile gun without result; this was still due to thick fog which the searchlights could not penetrate. She held on her easterly course without making any reply, and at some distance from land she moved north. She was 15 miles east of Hornsea at 1.45 a.m. and at about 1.55 a.m. turned west in Bridlington Bay, came overland and went over Carnaby at 2 a.m., to Burton Agnes at 2.15 a.m. She was reported from Driffield at 2.25 a.m. (though she probably passed some distance east of the town) and then turned south-east to Hornsea, where at 3 a.m. or shortly before, she dropped two bombs aimed at the Swedish SS *Thor* from Kalmar, at that point lying at anchor off the town. There was a thick fog which obscured the bombers' aim and the bombs fell about 150ft ahead of the ship to no effect. L-24 finally made off eastwards at an estimated speed of 60–70mph.

L-17 appeared near Grimsby at about 12.10 a.m. and passed north-west in search of her objectives on the south shore of the estuary, but was unable to place her bombs with accuracy due to the density of the fog. The first batch, comprising eight HE and four incendiary bombs, was dropped at about 12.45 a.m. on Killingholme, doing no damage. Continuing northwards in the direction of Hull, at 12.48 a.m. two HE and two incendiary bombs were dropped by the raider at East Halton, resulting in the destruction of a straw stack, damage to some outbuildings and the death of a calf.

She turned north, then east over the Humber, was heard from Hedon at 1.04 a.m. and then seems to have gone straight out to sea in a north-easterly direction between Hornsea and Withernsea at about 1.10 a.m. When she was a few miles south-west of Flamborough Head, at or shortly after 1.15 a.m., she dropped three bombs about 100ft astern of the SS *Frodingham* to no effect. The fog was dense at the time, but very low so that the ship's masthead lights were visible above it, while the Zeppelin itself was not visible from the deck of the ship.

Both L-17 and L-24 dropped a good number of their bombs while over the sea. After L-24 passed out over Withernsea, between 1.40 a.m. and 1.55 a.m., a number of explosions to the eastward were heard at several places as far inland as Crowle and Goole.

L-13, under Kapitänleutnant Eduard Prölss, came in at North Somercotes near Donna Nook on the Lincolnshire coast at 12.37 a.m., and steered a direct course for Lincoln, passing over Louth at 12.45 a.m., over Hainton at 12.55 a.m. and reaching Fiskerton, 5 miles east of Lincoln, at 1.10 a.m. At Fiskerton, which the Zeppelin commander seems to have mistaken for Lincoln (both places being roughly the same distance from his landfall at North Somercotes), one HE and one incendiary bomb were dropped, doing no damage beyond the breaking of a few windows.

L-13 passed on in the direction of Newark, going south of Lincoln over Bracebridge Heath at 1.20 a.m., and on to Bassingham where a single incendiary bomb was dropped a few minutes later, doing no damage. The raider then turned south, passing east of Newark over Balderton to Long Bennington, where at about 1.30 a.m. another incendiary bomb was dropped. At 1.45 a.m. the raider dropped two HE bombs at Stubton and, at 1.52 a.m., circling south-west over the railway again, dropped four more HE bombs at Dry Doddington. All of these bombs fell within 50–500 yards of the railway. Turning east and again re-crossing the line, she dropped four HE and seventeen incendiary bombs at 2 a.m. within 200 yards of the railway station at Hougham. The only result was a few broken panes of glass at Stibton and Doddington.

The raider made off east by Ancaster and Rauceby, passed south of Sleaford at 2.10 a.m., between Hubbert's Bridge and Langrick at 2.25 a.m., and finally south of Boston and out to sea in the direction of Old Leake at about 2.30 a.m. She then crossed to the Norfolk coast and was heard, by Cromer Lighthouse, to drop a bomb out at sea at 3.10 a.m.

L-16, which was under the command of Kapitänleutnant Erich Sommerfeldt, made for the Norfolk coast at Brancaster Bay at 12.50 a.m. She passed inland over Thornham at 12.55 a.m., dropped a 'water-indicating bomb' at Ringstead and went over Hunstanton at 1.05 a.m. to Heacham at 1.15 a.m. She circled first south-east, and then south-west, dropping two incendiary bombs at Snettisham, moved out over the Wash, returned to Heacham at 1.40 a.m. and headed out to sea over Thornham at 1.45 a.m.

L-11, with Leutnant zur See Otto Mieth, first appeared off the Norfolk coast at Sheringham at 2.35 a.m., where she dropped an HE bomb in the sea. The raider then came inland at Weybourne at 2.40 a.m., dropping an HE bomb on the cliff edge which killed a cow and damaged some tiles on a farm building. Then, manoeuvring south-west, L-11 dropped an illuminating flare at Holt and, while passing Sharrington at 2.45 a.m., dropped a third HE bomb and, a few minutes later, a fourth on Gunthorpe, neither caused any harm. From Gunthorpe she pursued her way south to the neighbourhood of Melton Constable where, at about 2.50 a.m., she suddenly turned and flew at high speed eastwards, passing over Hanworth at 3 a.m., North Walsham at 3.10 a.m. and Paston at 3.20 a.m. where an HE bomb was dropped, but again, caused no damage.

The Zeppelin then went out to sea at Mundesley, flying very high, at about 3.25 a.m., dropping one HE and one incendiary bomb. L-11 was last seen off Haisborough by the Danish steamer *Rai*, evidently heading out to sea, and was finally heard from the *Haisborough* and *Would* lightships going north-east

L-31 came over the coast at Corton, going from north-east to south-west at about 1.15 a.m. At 1.18 a.m. she was heard at Blundeston, going south-west, at Carlton Colville and then at Beccles at 1.27 a.m. She turned eastwards near Bungay at 1.37 a.m., circled over the coastline and went out to sea at Kessingland in a north-east direction, having dropped no bombs. There was a heavy ground mist at the time, which probably accounts for the inactivity of L-31.

30/31 July 1916

Reports were received from units of the North Army Home Defence that two Zeppelins were seen north-east of Wells-next-the-Sea, between 7.50 p.m. and 8.20 p.m., on 30 July. Intelligence indicated that at least one Zeppelin was far out in the North Sea on the afternoon of 30 July, and she may have reached the British coast in company with another on a reconnoitring cruise at about the time given. But at the same time, there was much mist at sea and it was suggested that two of the funnel-less motor patrol boats showing beyond the mist in a sort of mirage may have accounted for the report.

The observers, eleven in number, were, however, very confident about the accuracy of their reports. Two aircraft went up from Yarmouth, one from Bacton and one from Holt in response to these reports and others were sent up from Covehithe, Felixstowe and Killingholme.

One of the Covehithe aircraft, a Blackburn 100hp BE-26 (No. 8612) flown by Flight Sub Lieutenant J.C. Northrop, engaged a Zeppelin 30 miles east of Covehithe at 5.15 a.m. on 31 July. His seaplane was armed with four 16lb bombs, forty-eight Ranken darts and a Lewis gun. Northrop was firing the Lewis gun vertically upwards, and fired two trays of explosive and tracer bullets and four rounds from the third tray, when the magazine came away and hit him in the face, stunning him. By the time he had recovered control the airship had disappeared and Northrop returned to his base.

31 July/1 August 1916

On the afternoon of 31 July 1915, ten naval Zeppelins left their sheds on the north German coast, and flew over the North Sea with orders to attack London and southern England. Four of these raiders were sighted by the trawler *Adelaide* at 8.30 p.m., about 50 miles south-east by south of the *Humber* light vessel. The raiders were flying west-north-west. One of them came close to the trawler but did not molest her.

A similar report was received from the trawler *Lilac*, which sighted four Zeppelins at 9 p.m. about 45 miles south-east of the *Humber* light vessel. L-13 came close to this trawler and manoeuvred over the vessel. The trawler *Exeter* also reported four Zeppelins in the same area. According to the report from this latter vessel, the Zeppelins remained almost motionless for about an hour and a half (until about 10 p.m.) and then headed west.

L-16 came in over the Lincolnshire coast in the neighbourhood of Skegness (L-14 at 11.15 p.m. and L-16 about 20 minutes later). They both pursued the same course as far as Wainfleet. L-16 then went, first northward towards Spilsby, where she was heard, and then turned west across country. She passed north of Sleaford at 12.50 a.m., Caythorpe at 1.15 a.m., Claypole at 1.28 a.m. and Caythorpe again at 1.35 a.m. where two incendiary bombs were dropped.

She continued northward to Skinnard, dropping one incendiary bomb there at 1.40 a.m., and then eastward again to the neighbourhood of Newark. The raider seems to have passed round to the west of the town, and to have hovered in its neighbourhood for some minutes without discovering it, and then, turning north-east, dropped an incendiary bomb at Langford Common at 1.55 a.m.

Going on seawards in the same direction, she appeared south of Bracebridge Heath (near Lincoln) at 2.10 a.m., bore south-east towards Blankney, dropping a fifth incendiary bomb at Metheringham at 2.15 a.m. She proceeded east-north-east to West Ashby where, at about 2.25 a.m., the sixth and last incendiary bomb was dropped. Nothing further was heard of this Zeppelin, which then passed out to sea, presumably near Mablethorpe at about 2.45 a.m. No harm was caused by any of the bombs dropped by L-16.

L-14, after coming in over the Lincolnshire coast to Wainfleet and on to Old Leake, passed out southward over the Wash and came overland again at Sutton Bridge at 12.10 a.m. She went on to Wisbech at 12.15 a.m. and, at 12.33 a.m., dropped her first bombs, two HE and two incendiaries, at March. The bombs struck close to the railway at Whitemoor Junction, but did no damage beyond cutting three telegraph wires.

After circling north-east, the Zeppelin passed March again, going south-east shortly after 1 a.m. She flew north of Littleport, dropped an incendiary bomb at Hockwold, passed over Mundford and Lynford Hall and, at about 1.30 a.m., dropped seven HE and one incendiary bomb on Croxton Heath near Stanford. Five minutes earlier L-22, passing north from Thetford, had dropped a flare which set fire to the heath. The commander of L-14, seeing the fire, no doubt concluded that something important lay beneath him and promptly bombed the place.

Continuing on the same course, the raider passed north of Thetford, and reached East Harling at 1.40 a.m. and Bunwell at 1.45a.m. Here, L-14 dropped two incendiary bombs.

At 1.50 a.m. she was over Long Stratton, and at 2 a.m. she dropped an incendiary bomb ½ mile east of Buckenham Station, near Acle, and then changed course south-east. She passed Reedham, where four HE bombs were dropped at 2.05 a.m., and went to sea south of Yarmouth at 2.15 a.m., passing over the *Cross Sand* light vessel, going east-north-east, at 2.25 a.m. Her speed was, at times, very high – approaching 70mph.

L-**13**, under the command of Kapitänleutnant Eduard Prölss, came in over the Wash and was first spotted near Sutton Bridge at 11.55 p.m. At midnight, she dropped her first bomb (an incendiary) at Walpole St Peter. At 12.05 a.m. she was seen north-east of Wisbech, where she seems to have turned abruptly north-east and, passing King's Lynn, reached West Newton near Sandringham, where she dropped another incendiary bomb and changed course due east. On the way, she dropped a third incendiary at West Rudham at 12.25 a.m. and then on over Guist where L-13 dropped her first HE bomb of the raid.

At 12.40 a.m. she was near Cawston, when she dropped one incendiary and two HE bombs. Moving northward, she passed over Blickling at 12.50 a.m., Roughton at 12.57 a.m. and then back over the coast at Cromer at 1 a.m., where she turned south-west approaching the coast again at Sheringham and finally out to sea in a north-easterly direction at 1.05 a.m.

L-11, with Korvettenkapitän Viktor Schütze in command, was the last Zeppelin to emerge during this raid, and she cruised slowly off the Norfolk coast for some time before coming overland. At midnight the Zeppelin was heard in the neighbourhood of the *Haisborough* lightship on the one side, and from the *Newarp* light vessel on the other. From other indications she is known to have been immediately north of Cromer at 1.18 a.m.

She came in west of Cley at 2.04 a.m., dropping an incendiary bomb on the sands at Warham Hole, and proceeded south to Binham, where she dropped an incendiary bomb. Changing course south-east, the Zeppelin steadily pursued this direction, showing a flare at Field Dalling. She dropped an incendiary bomb at Galthorpe, an HE bomb at Briningham, a fourth incendiary bomb at Briston, a fifth incendiary and another HE at Thurning (which did some damage and injured two bullocks), a sixth incendiary bomb at Wood Dalling and a seventh at Cawston. Here, at 2.30 a.m., her course was slightly altered to the eastward. She dropped an incendiary bomb at Wroxham at 2.45 a.m., another at Hoveton St John and another at Neatishead. The Zeppelin went out to sea between Horsey and Winterton at 3 a.m. No harm was done by any of the bombs, except at Thurning.

L-17, under Kapitänleutnant Herbert Erlich, passed over the *Cross Sand* light vessel at 11.25 p.m., made landfall at Ormesby, north of Great Yarmouth at 11.45 p.m. and followed the river Yare to Loddon where, at 12.15 a.m., she turned north-west at moderate speed in the direction of Norwich, which may have been her objective.

She dropped an incendiary bomb at Bixley, passed south of the city over Stoke Holy Cross, and headed towards Tuddenham at 12.30 a.m., apparently drawn to the aeroplane searchlight at Honingham and the flares at Tuddenham aerodrome. The searchlight may have induced the belief that a battery was close at hand, and the Zeppelin unloaded the majority of its HE bombs. Ten were dropped on Tuddenham, and the raider then turned east, dropping seven more HE and five incendiary bombs between Tuddenham and Honingham. No damage was done by any of them. The Zeppelin passed over Costessey at

1 a.m., then Wroxham (1.15 a.m.) and finally, out to sea between Mundesley and Bacton at 1.40 a.m.

L-22, commanded by Kapitänleutnant Martin Dietrich, came in early and covered a large area of ground. She passed near the *Cross Sand* light vessel at 10.15 p.m., and crossed the coast near Lowestoft at 10.30 p.m. She pursued a direct south-west course for Lowestoft, passing Pakefield at 10.40 p.m., Wrentham at 10.45 p.m., Halesworth at 11 p.m., Earl Soham at 11.20 p.m. and north of Needham Market at 11.25 p.m.

Reaching Stowmarket at 11.35 p.m., she hovered for about five minutes, apparently keeping out of the range of the guns and went to Postlingford (north of Clare) where, at 11.45 p.m., she dropped her first bomb, an incendiary, which did no damage. She then changed her course southward, and was in the neighbourhood of Yeldham and Castle Hedingham from 11.55 p.m. to 12.05 a.m. Turning north, she passed near Haverhill and was seen at Dullingham in Cambridgeshire at 12.15 a.m.

Turning south-south-west at 12.20 a.m., she dropped four HE bombs on the aerodrome at West Wickham, having been attracted by a hurricane lamp which was burning on the ground in order to enable the ground guard to find their flare positions. One bomb fell in the aerodrome, the other within 300 yards of it.

Going east, six incendiary bombs were dropped in the fields near the gasworks at Haverhill at 12.25 a.m. L-22 harmlessly dropped five HE bombs at Withersfield at 12.30 a.m. and one at Great Wratting. The raider then made off in a north-westerly direction, passing near Newmarket at 12.45 a.m. She altered course to the north-east, passing near Chippenham at 12.55 a.m., Mildenhall at 1.05 a.m. and, reaching Snarehill near Thetford, fruitlessly dropped three HE bombs aimed at the aerodrome.

Her course was altered northward over Croxton Heath, where a flare was shown and this started a small fire on the heath, and also attracted L-14, who probably thought a target had been found; she also dropped an incendiary. She then made off at very high speed north-east, passed near Attleborough at 1.30 a.m. then over Honingham to Hevingham where, about 2 a.m., she dropped six HE and four incendiary bombs, with the result that one horse was slightly injured. Further on, at Burgh-next-Aylsham, an incendiary bomb was dropped. The Zeppelin then passed by North Walsham and out to sea between Mundesley and Haisborough at approximately 2.10 a.m.

L-23 came in at Benacre around 11.15 p.m., went over Covehithe and Wrentham, reached Brampton at 11.27 p.m. and then turned seaward, passing over Wingford at 11.33 p.m. At Southwold, at 11.35 p.m., a single incendiary bomb was dropped on the common, 20 yards from the military ammunition store, doing no damage. The raider went out to sea, then came back again to

Walberswick at 11.45 p.m. and then out to sea in a south-south-east direction, being observed from Aldeburgh at about 12.10 a.m.

L-31, under Kapitänleutnant Heinrich Mathy, approached the North Foreland from a north-east direction at 11.10 p.m. and at 11.20 p.m. began to drop bombs as she neared the coast. At least a dozen were dropped in the sea, in groups of three, in the vicinity of Kingsgate. The Zeppelin came in over the North Foreland lighthouse, going in a westerly direction along the coast.

At 11.25 p.m. she came overland between Westgate and Margate, passed over the western extremity of the latter borough and proceeded south-east towards Ramsgate. She was seen at 11.30 p.m. from Broadstairs, and was fired at by the AA guns at Manstone without result. At 11.35 p.m. she dropped three HE bombs in a field near Ramsgate, which did no damage beyond breaking windows in ten houses and a greenhouse. The concussion of the bombs broke glass in a house at Westwood 1½ miles away. The Zeppelin was fired at by the AA guns at Ramsgate, also without result.

The raider then moved off out to sea. Several bombs were dropped in the sea on the way. She was heading in a south-easterly direction, which she soon changed to south-west, coming in close to land at Sandwich, off which several more bombs were dropped. The Zeppelin then approached Dover, but when fire was opened upon her, she at once turned tail and retired north-north-east. She made no further attempt to come inland, but returned to Germany forthwith.

L-30 never came overland in Britain, but may have come across the North Sea in conjunction with L-31. She was located at about 10.50 p.m. midway between Zeebrugge and the North Foreland, and went back by the sea route.

2/3 August 1916

On the afternoon of 2 August, six Zeppelins left their shed and flew westwards over the North Sea, and by 7.30 p.m. British Intelligence had located all of them.

L-11 and L-16 raided Suffolk and Essex, L- 31 south-east Kent, while L-21, L-13, and L-17 raided Norfolk and Suffolk.

Before coming inland **L-21**, under the command of Hauptmann August Stelling, cruised out to sea north of the coast of Norfolk, between Cromer and Wells-next-the-Sea. She was heard at 9.30 p.m. from Wells, and between 10.40 and 10.55 p.m. from the Cromer Lighthouse, flying from east to west. She reappeared and made landfall at Wells-next-the-Sea at 11.55 p.m., flying south-west. She momentarily changed her course at 11.59 p.m. to the north-west, but very soon turned south again.

At 12.05 a.m. she passed over Little Walsingham, flying fast in a southerly direction. Passing Fakenham at 12.09 a.m., she seemed to reduce speed but carried on in the same direction. She was next seen from Wendling going south-west at 12.20 a.m. and arrived over Swaffham by about 12.27 a.m. She hung over this place for about three minutes, and then went due south, making for Thetford.

She was heard at Ickburgh and Mundford at 12.37 a.m. L-21 was attracted by the flares of Thetford Aerodrome and dropped five HE bombs, then altering course eastward, flew between Attleborough and East Harling and passed New Buckenham at 1 a.m. Twenty-five minutes later she was over Suffolk, passing across the county and out to sea over Covehithe and dropping two incendiary bombs as she did so. Eight more HE bombs fell into the sea close by, but no damage was caused and there were no casualties.

L-13, commanded by Kapitänleutnant Eduard Prölss, was heard over the *Haisborough* lightship going south-west at 11.37 p.m. She made landfall over Bacton at 11.45 p.m., where fire was at once opened by the mobile anti-aircraft guns at Bacton coastguard station and, at 11.47 p.m., the airship was driven off along the coast in the direction of Haisborough. Here, L-13 changed course at 11.52 p.m. to the south-west, and passed over Stalham at about 11.44 p.m.

She now changed course, and passed over Horning at midnight, Wroxham at 12.05 a.m., and turned due south. She dropped an incendiary bomb at Panxworth and went over Blofield at 12.10 p.m. She passed Mundham at about 12.25 a.m., where the Zeppelin dropped three incendiary bombs that damaged five panes of glass at Grange Farm.

Progressing towards the Norfolk/Suffolk border, L-13 dropped five HE and three incendiary bombs on Ditchingham, breaking some seventy panes of glass at Ditchingham Hall, although fortunately there were no casualties.

Flying over Suffolk, the Zeppelin dropped four incendiary bombs on Earsham soon after. Manoeuvring west and slightly north, she dropped three HE bombs over Shelton, which broke a few windows, then turned north-west, arriving over Tacolneston at 12.55 a.m., where she dropped one incendiary and six HE bombs. Still bearing north-west, she passed Silfield at 1.10 a.m. and dropped three HE bombs, damaging two farmhouses and two cottages and smashing twenty windows.

L-13 then passed to Wymondham, where she dropped one incendiary bomb. North of Wymondham, she turned north and then sharply to the east at 1.12 a.m. Between 1.15 a.m. and 1.25 a.m. she circled round Stoke Holy Cross, Shotesham All Saints, Saxlingham, Thorpe, Shotesham St Mary and back to Stoke Holy Cross. It was thought by military observers that L-13 was looking for Norwich.

At about 1.25 a.m., when the Zeppelin arrived over Stoke Holy Cross for a second time, she seems to have abandoned the attempt, as she turned north-west and arrived at Costessey at 1.35 a.m. She then steered off in the direction of

The RNAS Mobile Anti-Aircraft Brigade under Lieutenant Mackenzie Ashton (far right) photographed on Yarmouth Road, around the corner from their headquarters on Grammar School Road, North Walsham, 1915.

the coast, passing over Horsford at 1.45 a.m., hovered over North Walsham at 1.55 a.m., before flying in the direction of Mundesley, turning south-east towards Bacton as she approached and was then observed flying out to sea between these villages by the Mundesley coastguard at 2.15 a.m.

L-16, under Kapitänleutnant Erich Sommerfeldt, was cruising between the *Haisborough* lightship and Mundesley off the Norfolk coast 9.35 p.m.–10.15 p.m. She was sighted over the *Cockle* lightship at 12.15 a.m., flying south-west, and made landfall over Hemsby at 12.25 a.m.

She was seen out to sea at 12.21 a.m., and after passing Hemsby at 12.24 a.m. she flew south-west. After Ormesby St Margaret she turned towards Acle, which she reached at 12.40 a.m. Going due south, she passed over Cantley at 12.45 a.m., then headed west to Rockland St Mary, before turning south again, passing over Brooke at about 12.55 a.m., then south-west to Hempnall at about 1 a.m.

Setting a westerly course, L-16 was over Long Stratton at 1.10 a.m. where she dropped three incendiary bombs. The Zeppelin was next spotted at Bunwell at 1.20 a.m., but her course after that was lost until she reached Ashby St Mary, where she dropped four HE and three incendiary bombs, breaking all the windows in Ashby Lodge and two nearby cottages. On leaving Ashby the Zeppelin seems to have reduced her speed and was next seen over Limpenhoe before 1.55 a.m. and over Acle again at 2 a.m. L-16 was last seen flying out to sea over Yarmouth.

L-17, commanded by Kapitänleutnant Herbert Ehrlich, passed over the *Cockle* lightship at midnight, and made landfall between Caister and Great Yarmouth at 12.20 a.m. At 12.30 a.m. she was seen in the neighbourhood of Ormesby St Margaret, in company with L-16.

She was next spotted over Halvergate at 12.30 a.m., turning south and passing over Reedham at 12.35 a.m., Loddon at 12.45 a.m., before manoeuvring south-west to Pulham, where the Zeppelin dropped three HE bombs.

She followed the valley of the River Waveney to Eye, in Suffolk. Turning due west, she dropped an incendiary bomb on the Great Eastern Railway line at Mellis at 1.03 a.m. Returning to Eye at 1.10 a.m., she went down the Waveney, passed over Thorpe Abbots at 1.15 a.m. and, while over Billingford, she dropped three HE bombs, killing six horses. Between 1.15 a.m. and 1.20 a.m. she dropped a further three HE and three incendiary bombs at Brockdish, and these damaged a farmhouse.

Turning north, L-17 arrived at Hardwick at 1.30 a.m. where she harmlessly dropped five HE bombs. At Long Stratton she dropped two HE bombs at 1.37 a.m., then a single incendiary bomb on Fornsett St Mary (1.40 a.m.) which landed near a signal on the Great Eastern Railway, doing no damage to the permanent way. She then turned south-east again to the valley of the Waveney, dropping two HE bombs at Starston before 1.45 a.m., which killed three horses and injured one. From Starston she next went to Redenhall, where she harmlessly dropped one HE bomb at about 1.45 a.m., then on to Denton, at about 1.50 a.m., where she dropped six incendiary bombs.

The Zeppelin turned north-east to Broome, where she dropped three incendiary bombs between 1.50 and 1.55 a.m. She was then thought to have headed direct for Southwold, no further bombs were dropped, and she was last seen over Southwold lighthouse going out to sea at 2.15 a.m.

As the Zeppelin appeared to have followed the course of the River Waveney for the greater part of her journey, both coming in and going out, it was suggested in the Intelligence summary that the chief objective for L-17 was Pulham Air Station.

L-11 was identified about 10 miles north-east of Aldeburgh at 12.16 a.m., and again at 12.40 a.m. about 10 miles out to sea off Orfordness. The Zeppelin had previously had an encounter with a RNAS seaplane over Burgh Castle, but does not seem to have sustained any damage, although two trays of Lewis gun ammunition were expended and the pilot claimed to have hit repeatedly.

At 12.48 a.m., Hollesley heard her approaching the coast from the north-east. At 1.10 a.m. the Zeppelin was seen over Bawdsey Manor Camp. Here, she turned north-west and went up the River Deben in the direction of Woodbridge. Two bombs were dropped in the water at Bawdsey Ferry and another near Bawdsey. She then passed Ramsholt and, when close to Waldringfield, caught sight of the electric lights at Playford Heath, near Woodbridge, and turned south to Kirton. Passing over Kirton at 1.20 a.m. she dropped a flare on Kirton Marshes, and one large and one small HE bomb on Kirton village. Six cottages were seriously damaged and a great deal of glass was broken in twelve others. One boy was slightly injured.

After bombing Kirton, she seems to have shut off her engines and drifted down the River Deben without being seen or heard, and to have left the coast at about 1.25 a.m. Between 1.25 a.m. and 1.30 a.m. she was fired at by the naval guns at Shingle Street, while flying at a height of about 6,000ft.

At 1.30 a.m. she came inland again near Felixstowe, but went out to sea almost at once at 1.33 a.m. At 1.45 a.m. she was heard from Felixstowe, approaching Landguard from the north-east. At 1.54 a.m., she was seen from Landguard and was observed to throw out a starlight while slowly descending to a height estimated at 2,000–3,000ft. She was watched from the Naval Air Station and Fire Command Post, but was not found by the searchlight.

One gun opened over the starlight she had dropped, and the Zeppelin went out to sea again without dropping bombs. Between 1.55 a.m. and 1.58 a.m. she was observed hovering off Landguard and, at 2 a.m., she came over the entrance to Harwich harbour. Being heavily fired at by both naval and military guns at an estimated height of 5,000–6,000ft, she turned to the north and was over Landguard at 2.05 a.m. Here, she dropped two HE bombs and four incendiaries in the sea, and one HE bomb on the parade ground at Landguard, damaging a few tents.

Having apparently been hit by the guns of the Harwich defences, L–11 went out to sea over Landguard at 2.10 a.m. Returning towards Dovercourt and being engaged by guns, she made a nosedive as soon as fire was opened and again turned out to sea. At 2.17 a.m. she was seen going due east away from the land.

At about 2.25 a.m. she once more came over Harwich, possibly assisted in steering her course by the play of the searchlights. She repeated the manoeuvre of going north to Landguard, but dropped no further bombs. At 2.35 a.m. she finally went out to sea over Landguard, having again been fired at by the naval guns at Shingle Street. No damage was done at Harwich or Landguard beyond a few broken windows in huts at the latter place, and a few tents perforated by splinters.

L-31, under Kapitänleutnant Heinrich Mathy, was reported by Dover seaplanes 35 miles north-west of Dunkirk at 10.15 p.m. Between 10.45 p.m. and 12.50 p.m. she was cruising the Straits of Dover, before turning westward for the east coast of Kent. The coastguard at Kingsdown reported having seen a Zeppelin to the east, going north at 12.50 a.m., making for Deal. L-31, having previously dropped twenty HE bombs in the sea as a result of three rounds fired at her by the steamer *Duchess of Devonshire*, turned south from Deal and she left the land between Kingsdown and St Margaret's at 1.08 a.m., following the outline of the coast until she was just out at sea due east of Dover at about 1.10 a.m.

Having been picked up by the searchlight at St Margaret's at 1.08 a.m., and on approaching Dover, she was heavily fired upon at 1.10 a.m. She turned eastwards, going out to sea over Cornhill coastguard station. She flew steadily eastwards, illuminated by the lights until out of range.

Aftermath

Scaremongering during the Zeppelin offensive was taken very seriously, and some of those who repeated rumours even ended up being brought before the magistrates. This case was reported in the *Norfolk Chronicle*:

> At Wymondham Petty Sessions, John Quantrill of Silfield, Wymondham, was charged with unlawfully spreading a false report, viz. that twenty-two Zeppelins were crossing the channel. Florence Ellen Chilvers, a single woman, said that on 3 August she saw Quantrill at work in his garden. He said there were twenty-two Zeppelins crossing the channel. This was at 8 p.m., she was with two other women at the time. She then went to Bunwell and told some people there and the consequence was the report soon spread all over the parish. Quantrill claimed his boy had been to Wymondham and when he came back he said he had been told of the report and that the news had come from Norwich and had been sent on to Wymondham to let the people in the town and surrounding villages know. The Chairman (Mr W.B. Fryer) said it was the first case of this kind to come before the Bench at Wymondham and they would deal leniently with the defendant but added it was, however, quite time that the public should know reports of this nature must not be spread around without taking steps to ascertain if they were true. The defendant had rendered himself liable to a fine of £100 and they would deal severely with any future case of this kind. Quantrill was fined five shillings.

8/9 August 1916

Eleven Zeppelins departed Germany on 8 August, but only nine of them reached the British coastline and raided parts of the Midlands, the north and Scotland.

L-14, under Hauptmann Manger, passed northward along the coast past Goswick, came overland at Berwick at 12.25 a.m. and went west over Chirnside and Duns. At 12.37 a.m. the Zeppelin was over Fallside Hill, near Gordon, where one incendiary bomb was dropped on a farm without doing any damage.

Her course was then changed south-east and, at 1.08 a.m., three HE bombs were dropped at Grahamslaw, in the parish of Eckford, south of Kelso. A minute later, an incendiary bomb fell on Kersknowe, and three minutes afterwards, two more incendiary bombs fell at Clifton, in the parish of Morebattle. No harm was done, the bombs having dropped in fields and merely set fire to some thistles.

When close to Kelso the Zeppelin threw a flare. The raider then made off over the Cheviots, dropping an incendiary bomb on the hills ½ mile south of Southern Knowe, in Northumberland, as she went. She passed Alnwick at about 1.43 a.m., Alnmouth at 2 a.m. and headed out to sea, passing Amble, going south-east

L-11 came in at Whitley Bay, near Cullercoats, north of the Tyne, and bombed this small watering place. After a flare had been dropped, which fell in a field south of the cemetery, eight HE and five incendiary bombs were dropped. The first seven HE bombs fell in a line from Whitely Road, across Albany Gardens and Clarence Crescent to the railway. The last HE bomb, and the five incendiary bombs landed further on, between the corner of Alma Place and Burnfoot Terrace, across Lish Avenue, and the corner of Carlton Terrace and Marden Crescent. One man, one woman and three children were injured. One dwelling house was totally wrecked, another and three shops extensively damaged, many others slightly damaged and a considerable quantity of glass broken. Slight damage was done to signals, telephone and telegraph wires at the station.

It was suggested that the 'arcing' of electric trains on the railway may have attracted the Zeppelin, but in fact, no trains were running on the line after 11.12 p.m. The bombs fell practically simultaneously with the opening fire from the Whitley Bay guns. As the searchlights were greatly hampered by the mist, the gun opened fire without waiting for the light. The second shell burst just under the tail of the Zeppelin, which immediately threw out a 'bouquet' of bombs which fell at sea. The Zeppelin then rose 3,000ft, afterward descending again. As she went out to sea she threw out what was described as a burning petrol tank, which fell into the sea and continued to burn for some time. It was believed to have been an acetylene flare.

L-31, after dropping a number of bombs in the sea off Seaham and Sunderland between 12.15 a.m. and 1.15 a.m., came inland near Whitburn, passed over Boldon and Cleadon and, at about 1.45 a.m., went north-east to Marsden. Here, at Salmons Hall, six HE bombs were dropped on miners' dwellings at 2 a.m. Very little damage was done, with only a few windows and roofs suffering. A horse was killed.

The raider then went on northwards, and circled over the mouth of the Tyne two or three times. As the whole district was in absolute blackness and the valley was hidden by mist, the Zeppelin commander probably failed to realise that he was over the river. He passed over South Shields and Tynemouth and at least one military camp. The Zeppelin was sighted without the aid of a searchlight by the 1-pdr mobile guns at Monkwearmouth, but was not within range and was not fired at. When the searchlights lit she appeared to vanish in a cloud. Unable to find any definite objective, the Zeppelin seems to have gone off to sea, probably at approximately 2 a.m.

L-13 crossed the coast at Denemouth, north of Hartlepool, at about 1.30 a.m. She passed over Castle Eden at 1.40 a.m., thence following the Hartlepool–Ferryhill Railway to Wingate where, at 1.45 a.m., an HE bomb was dropped in a field, doing no damage beyond breaking a little glass. Then, probably attracted to the Coxhoe district by a burning waste heap at Kelloe Colliery or, according to another account, by a limekiln at Quarrington which is said to have shown a 'strong flame', she made in the direction of Kelloe. Here she dropped twelve HE and fourteen incendiary bombs on the colliery at Quarrington Hill and Bowburn. At Kelloe, 30ft of colliery railway was demolished and a dozen panes of glass broken in a weigh cabin. At Quarrington, glass was broken in about forty houses and a shop.

The Zeppelin went off north-east, passing Thornley at 1.50 a.m., Haswell at 1.55 a.m. and out to sea over Easington at 2.05 a.m.

L-30 was heard approaching Skinningrove from the north-east between 12.20 a.m. and 12.28 a.m. The Zeppelin was picked out by the No. 2 light at Skinningrove and, after being in a beam for a minute and a half, turned seawards, dropping bombs in the sea as she went.

She was next heard over Hartlepool at 12.50 a.m. going north, and at 12.55 a.m. having turned half a circle west she was over the Central Marine Engine Works and the docks, without dropping bombs. The town, being completely in darkness, was probably invisible to the Zeppelin.

At 12.58 a.m. L-30 turned south and dropped six HE bombs in a field west of Seaton Carew Ironworks and three HE bombs on a slag tip at the ironworks. According to local opinion, the bombs were attracted by the slight glow from the chimneys and furnaces. The same would apply to the zinc works at Seaton

Snook, where one incendiary bomb was dropped a minute or so later. No damage was done at the zinc works, but many windows were smashed by concussion at Bellevue and Long Hill, near Seaton Carew.

The No. 9 searchlight at Port Clarence now found the Zeppelin, at 1.05 a.m., and she immediately turned east, crossing the River Tees and thence out to sea, where she was next located 60 miles out on her way homeward at 2.02 a.m.

L–22 approached Redcar from the direction of Hartlepool at 12.55 a.m. and (possibly attracted by three flares that had been used to guide a seaplane taking off 12.30 a.m.) dropped five HE bombs on grass fields forming part of the Redcar aerodrome landing ground at about 1.05 a.m. No damage was done.

L–22 then proceeded south-east and dropped four HE bombs on a camp east of Wheatlands Farm, again doing no damage. She proceeded to Saltburn, where she was spotted by an AA gun at Huntley Hall, which opened fire without the aid of searchlight at 1.12 a.m. Considerable difficulty appears to have been experienced by the gun commander at Huntley Hall in dealing with the Zeppelins, owing to his having received very little help from the searchlight, which was not connected to him by telephone. One round was fired, which fell short, and the Zeppelin continued her course to Huntcliff at 1.15 a.m., going towards Skinningrove.

At 1.17 a.m., hearing the bombs dropped by L–21 on the cliffs immediately east of Skinningrove, L–22 turned at Carlin How, apparently with the object of spotting the light and gun at Huntley Hall, which had again fired upon the Zeppelin at 1.19 a.m. she swept round to the north of Warsett Hill, passed over the gun and turned southward inland, pursued by a parting shot from Huntley Hall.

Her southerly course was continued to Houlsyke, near Lealholm, where at 1.35 a.m. she dropped a single incendiary bomb, which fell on the moor and did no damage. She now turned east following the River Esk and the railway to Whitby, passing south of it at 1.45 a.m. The Zeppelin was last seen from Robin Hood's Bay going out to sea at 1.50 a.m.

L–21 was in company with **L–23** (which never came overland) 12 miles north-east of Scarborough at 11 p.m., when both were attacked by the armed trawler *Itonian*, who reported that one of the Zeppelins appeared to drop by 25°. This was probably L–23, which no doubt received some slight injury, necessitating repairs and this prevented her coming overland. She went slowly northwards, and at 1.18 a.m. was returning back to base.

L–21 continued on her course and, at 12.45 a.m., she was heard by HMY *Miranda*, stationed 2 miles north-east of Skinningrove. At 12.58 a.m. the Zeppelin attacked the *Miranda*, dropping three HE bombs, one of which exploded 400 yards from the ship. At 1 a.m. the Zeppelin was visible from the ship, 6,000ft overhead, and five 3-pdr shells were fired at her. The third was claimed to have struck the Zeppelin, bursting on her side but, in view of her subsequent action this seems improbable.

On fire being opened, L-21 immediately went south-west over the cliffs and scars east of Skinningrove. At 1.17 a.m. three HE bombs dropped in a quarry, and the raider went over Skinningrove itself at 1.20 a.m., dropping seven HE bombs which destroyed a small wooden office and damaged some pipes and tanks, in total worth about £25. She then went out to sea, where more bombs were dropped and was 15 miles out on her way home at 1.35 a.m.

A RNAS BE-2c, flown by Fight Lieutenant de Roeper, went up from Redcar Aerodrome and followed an airship (either L-21 or L-22, or possibly both), for over an hour but he could not get sufficiently near or high enough to bomb it.

L-24, under Kapitänleutnant Robert Koch, came in south of Flamborough Head at 12.15 a.m. and passed Bridlington, going inland at 12.20 a.m. At 12.25 a.m. she was heard from Burton Agnes, and at 12.30 a.m. reported between Hulton-Crenswick and North Frodingham, going south, and following the line of the railway from Driffield to Beverley. Shortly afterwards, she passed between Lockington and Leven, and by 12.37 a.m. was at Beverley, where she turned west. At 12.45 a.m. she passed over Market Weighton and went on 1 mile westward of the town, turned south-east over Holme-on-Spalding Moor and then north to Kiplingcotes, which she reached at 12.56 a.m.

She then turned south again, possibly attracted by the flares of Bellasize Landing Ground, which had been lit up. She reached the Hull–Selby Railway at 1.10 a.m., passed Elloughton and thence followed the river eastward, passing North Ferriby a few minutes later. At 1.18 a.m. she went over Hessle, dropping her first bombs: eight HE and two incendiary, in open fields near the town. No damage and no casualties were caused.

Hull lay immediately on her track and could scarcely be missed, and at 1.20 a.m. the Zeppelin reached the north-western portion of the city. She steered over the golf course, where three HE bombs were dropped, and arrived at the Hull & Barnsley Railway, by Spring Bank Junction. Here, one HE and four incendiary bombs were dropped; one of the latter buried itself in the permanent way without exploding, while the others fell in fields.

L-24 went on by Anlaby Road, near Sandringham Street and the North Eastern Railway line at the bottom of Walliker Street, and then turned north-east circling round West Park, over Hymers Collage, to the neighbourhood of Newland.

The Zeppelin was at a great height, much higher than those on previous raids. She dropped eight HE and twenty-eight incendiary bombs on the city itself, doing considerable damage on: Walliker Street, where a shop was demolished, a house wrecked and two houses damaged; Selby Street, where two houses were badly damaged and another slightly; Granville Street and Sandringham Street both had two houses wrecked and others damaged. Lesser damage was done in Park Avenue, with four fires started in houses, and in other streets nearby there were a number of windows shattered. A haystack was partially burnt on Arnold Street.

The neighbourhood was close to a number of railway lines, and the course of the bombs seems to show that the Zeppelin followed them for some distance. It was stated, on authority of the police, that a train was said to have been passing on the line shortly before the raid. This, however, was denied by the North Eastern Railway. Accusations regarding the railways being responsible for attracting bombs were rather freely made in respect of this raid.

The railway lines close to Walliker Street were brought down.

There were several casualties, with those killed being hit in the street. Others died later from injuries or shock:

- Emmie Bearpark (14), 35 Selby Street (Died 10 August 1916)
- Mary Louise Bearpark (44), 35 Selby Street (Died from shock)
- John Charles Broadley (3), 4 Roland Avenue, Arthur Street
- Emma Louise Evers (46), 25 Brunswick Avenue, Walliker Street
- Elizabeth Hall (9), 61 Selby Street
- Mary Hall (7), 61 Selby Street (Died 9 August 1916)
- Rose Alma Hall (31), 61 Selby Street
- Charles Lingard (64), 61 Walliker Street (Died from shock 14 August 1916)
- Esther Stobbart (31), 13 Henry's Terrace, Wassand Street
- Rev. Arthur Wilcockson (86), 32 Granville Street (Died from shock)

The Zeppelin was fired at by the mobile 3in AA gun at Harpings, north of the city, without result. Only the Harpings gun was able to fire at the raider and got off eight rounds of shrapnel although hampered by ground mist. This also prevented the Sutton gun from seeing either the target or the signals of its own flank observation post, which had sighted it. The searchlights were also hampered, and only that at Harpings picked up the target, losing it as soon as it rose after the burst of the first shell.

At about 1.30 a.m. L-24 went off eastwards, being heard from Sutton at 1.30 a.m., at Hedon, Skirlaugh and Leven between 1.30 and 1.40 a.m. She came in again almost immediately further north, was heard from Bridlington and Burton Agnes, and moved out at 2.10 a.m. over Filey Bay.

L-16, under the command of Kapitänleutnant Erich Sommerfeldt, was spotted at 11.45 p.m. by the master of the *Inner Dowsing* lightship coming from the south-east, flying so high that it could only be seen with binoculars, although the night was clear and starlit.

The Zeppelin came in over the Wash, making landfall at Brancaster at about 12.30 a.m. Flying south-south-west to Fring, L-16 arrived over Dersingham at 12.40 a.m. and dropped ten HE and ten incendiary bombs which caused blast damage to the windows and ceilings of a number of private houses to the value of approximately £40. Five minutes later, eight HE and seven incendiary bombs were dropped at Wolferton, doing no damage. Having thus disposed of her bombs, the Zeppelin travelled along the coast and out to sea at Hunstanton at 1.09 a.m. No casualties were incurred.

Aftermath

The destruction caused at Hull, and the comparative impunity with which the Zeppelins flew, once again led to many protests.

At an Air Board meeting a week after the raid it was hinted that the seaplane carriers were not doing all they might to intercept Zeppelins, and that the plea of too much risk from submarines would not satisfy public opinion. The report of the meeting called forth some incisive comments from the Admiralty; these emphasised that the army was responsible for the defence against aircraft attack, and that the seaplane carriers were intended to act only as eyes of the fleet. Fleet requirements, they said, could not give way to helping the military do their work. An Admiralty report showed that, on 8 August 1916, no fewer than 114 vessels, armed with high-angle guns were situated along the east coast, from the Forth to Dover, ready and able to fire at Zeppelins.

2/3 September 1916

This raid included both army and naval Zeppelins, and was intended by Strasser to be his last 'big push' to bomb Britain into surrender. Sixteen Zeppelins set out to attack London, but their attack was badly hampered by adverse winds with belts of heavy rain and ensuing ice, which forced two of them to turn back before they made landfall over Britain.

It was also during this raid that the airship SL-11 (to be precise, it was actually a wooden-framed Schütte-Lanz, rather than a duralumin-framed Zeppelin) became the first to be shot down on British soil; landing at Cuffley, Hertfordshire, where she burned for nearly two hours after hitting the ground. The 'kill' was achieved by Lieutenant William Leefe Robinson RFC, using the new explosive and incendiary ammunition, for which he was awarded the Victoria Cross and became a national hero. The largest Zeppelin air raid of all time had been foiled and, having shown that a biplane armed with the right sort of ammunition could shoot down an airship, it was to prove the beginning of the end for the 'Zeppelin menace.'

Going, going, gone! One of numerous souvenir postcards produced to celebrate the shooting down of SL-11 by Lieutenant William Leefe Robinson RFC on 3 September 1916.

LZ–98, under Oberleutnant zur See der Reserve Ernst Lehmann, passed through the Straits of Dover, off Deal, at 11.35 p.m. and crossed the coast at Littlestone one minute after midnight. She went north-west and passed Ashford at 12.15 a.m. There, she hovered for a few minutes and then went off along the main line of the South Eastern Railway as far as Staplehurst, where at 12.30 a.m. she turned off north-west.

At 12.35 a.m. she was at Linton, and then circled south of Maidstone. She was at Otham at 12.45 a.m. and at Yalding at 12.50 a.m., after which, turning north-west she passed Hadlow at 1 a.m. and Offham at 1.05 a.m. By 1.09 a.m. she was south-west of the Hartley searchlight, which lit on the raider; the Southfleet light followed this example at 1.11 a.m. One minute later, the Southfleet gun opened fire, followed by the Dartford guns. The airship reached the railway north of Hartley at 1.13 a.m., crossed it west of Fawkham Station when under fire and, at 1.15 a.m., dropped six incendiary bombs at Longfield, immediately north of the junction of the Gravesend Branch. Twelve more incendiary bombs were dropped east of the branch line, further on. The airship then went directly over, and attacked, the Southfleet light, dropping three HE bombs just beyond it, firing a wheat stack and breaking some glass. At 1.18 a.m. the Southfleet gun ceased fire, followed at 1.19 a.m. by the Dartford guns, the target disappearing north-east behind a cloud.

She went on towards Gravesend, dropping two HE bombs at 1.20 a.m. at Northfleet Green Farm, destroying a coal shed, 'pollard house' and stable, but causing no casualties. These were followed by one HE and one incendiary

Lieutenant William Leefe Robinson VC, RFC, the first man to bring down a Zeppelin on British soil.

bomb on the Gravesend golf course. The raider then crossed the Thames east of Gravesend, dropping one incendiary bomb in the river.

The Essex searchlights were now exposed and, at 1.20 a.m., the Tilbury gun opened fire. The shooting appeared good, and the airship climbed steeply as she was crossing the river. The Fobbing gun opened fire just as she disappeared behind a cloud in a north-easterly direction, dropping two incendiary bombs at Corringham and three HE bombs on Fobbing as she passed; all to no effect.

LZ-98 next dropped eight incendiary bombs at Vange at 1.30 a.m. and went northward at high speed, passing Billericay and Chelmsford and on to Great Waltham where, at 1.45 a.m., she dropped one HE bomb, which did no damage. She then turned north-east, passing Colchester at 2 a.m. and Ipswich at 2.10 a.m. Three minutes later, she dropped a couple of HE bombs on the boundary of Rushmere and Playford parishes, destroying some crops, then passed south of Woodbridge at 2.18 a.m., south of Saxmundham at 2.21 a.m. and out to sea north of Aldeburgh at 2.35 a.m.

LZ-90 crossed the Essex coast shortly after 11 p.m., north of Clacton. At 11.20 p.m. she reached Mistley, south-east of Manningtree, and there, stopping her engines, lowered an observation car. For some reason, the wire cable from which the car hung broke. It was suggested that the pawls of the winch, by which the car was lowered, had become disengaged and had allowed the cable to run out suddenly, causing the car to fall a considerable distance below. Some marks found on the teeth of the gear wheels (which were dropped later on and found near Poslingford) might be explained as having been caused by a bar being thrust between the teeth in order to prevent the car dropping any further, as considerable damage might be caused to the ship if the cable became entangled in something on the ground. It was supposed that the rope parted suddenly under the sudden strain imposed on it when over Mistley.

The observation car itself was made of light sheet aluminium, streamlined in form, about 14ft in length, with horizontal and vertical fins at the tail in order to keep it head-on to the direction of the ship. It contained a mattress, on which the observer lay at full length and carried out his observation through celluloid windows in front of the car. Communication with the ship from the car was by means of a telephone, the wire of which was contained in the centre of the suspension cable.

The airship had stopped her engines to deal with the accident after the car fell to the ground, where it was found the following morning with about 5,000ft of wire attached. At 11.45 p.m., the airship restarted her engines and went off in a north-west direction, passing Dedham at 11.50 p.m. and Hadleigh at 12.05 a.m., and on to Foxearth, west of Sudbury, where two incendiary bombs were dropped to no effect.

Travelling westward to Wixoe at 12.30 a.m., the airship dropped twenty-one HE and sixteen incendiary bombs, surprisingly causing no casualties and breaking glass in just two houses and a school.

On turning north-east to Poslingford, the damaged winch, by which the observation car was raised or lowered, was thrown out to reduce weight at 12.40 a.m. The raider then passed north-eastwards between Thurston and Elmswell at 1 a.m., over New Buckenham at 1.15 a.m., Tasburgh at 1.20 a.m., Shotesham at 1.25 a.m., Loddon at 1.30 a.m., Cantley at 1.35 a.m. and finally out to the sea at 1.45 a.m., between Caister and Yarmouth.

SL-11, under Hauptmann Wilhelm Schramm, reached the British coast from Belgium without being detected, approached Foulness Island at 10.40 p.m., and was over Southminster at 10.50 p.m. She crossed the Blackwater at 10.55 p.m., navigating around Suffolk, Essex and Hertfordshire, until 1.20 a.m. when she dropped three HE and three incendiary bombs at London Colney, south of the Barnet–St Albans road, with no results. Five minutes later, two HE and two incendiary bombs were dropped in a wood at North Mimms, as the airship travelled due east. At around 1.28 a.m. one HE and two incendiary bombs were dropped at Littleheath, and a minute later one HE bomb fell at Northaw. At Littleheath, a water main was cut and the roofs of two houses damaged.

The airship then passed south and then north-east, approaching the gun at Temple House from the south-west. An incendiary bomb was dropped close to the Ridgeway Road, followed by two others, one on either side of the railway just north of Gordon Hill Station. At about 1.35 a.m., two HE and seven incendiary bombs were dropped on the Stud Farm at Clayhill. The HE bombs did no damage, bursting in fields, but some of the incendiaries set a row of stables on fire and killed three valuable yearlings.

The airship now turned and went off to the west, dropping three incendiary bombs at Cockfosters, near the Enfield Isolation Hospital as she went. At about 1.45 a.m. she crossed the Great Northern Railway main line south of Hadley Wood tunnel, dropped two incendiary bombs, then turned and re-crossed the line going east.

Around 1.50 a.m. three incendiary bombs were dropped in a field at Southgate, and the airship turned south, passing over Wood Green, and then east, south of Alexandra Palace. At 1.58 a.m. she was picked up by the Finsbury Park and Victoria Park searchlights, and passed over the Finsbury Park gun, which opened fire at 2 a.m..

Turning north-east over Tottenham, SL-11 came under heavy fire from other guns of the London defences, being well-lit by searchlights. At Edmonton, six HE bombs, two of which did not explode, were dropped at 2.12 a.m.; one of these dropped in the grounds of Messrs Eley's Explosive Works. Further north, at

2.14 a.m., two HE bombs fell at Ponder's End, which did slight damage to a large number of houses, broke some tramway and telephone wires and badly damaged a roadway and water main. Then, six HE bombs fell along the Enfield highway, one of which broke but did not explode. Fifteen houses and some greenhouses were slightly damaged. At 2.17 a.m. twelve HE bombs, two of which failed to explode, were dropped at Forty Hill and Turkey Street, where the backs of three houses were damaged. In spite of the populous nature of the district no casualties were caused.

From Tottenham, while the bombs were being dropped, the airship was under heavy fire from the greater part of the AA defences of north and central London, including those of Regent's Park, Paddington and the Green Park, from which she was too far distant for their fire to reach her. This no doubt contributed to the great volume of fire from London which compelled her to change course at Finsbury Park.

On her return northward she had to run the gauntlet of the north-eastern guns, and was heavily bombarded all the way from Finsbury Park to Temple House. West Ham, Clapton and Temple House guns very nearly reached their target, the latter gun firing especially well. It is, however, to the Royal Flying Corps that the credit to her destruction belongs.

Six aeroplanes had ascended in the London district, and three of these, piloted by Lieutenants Robinson, Mackay and Hunt, chased the airship. Lieutenant Robinson caught up with her when she was between Enfield highway and Turkey Street. He attacked her 'most gallantly', in spite of the fact that she was under such heavy fire and shells were bursting all around her. In consequence of his attack, as he came up from the south-east, the airship sheared off to the north-west. He fired three drums of machine gun ammunition into her and, at 2.25 a.m., a few seconds after the third drum had been fired, she was ablaze and fell headlong to the earth at the village of Cuffley. The guns were firing up to the last moment, and one or two shells are even said to have been put through the blazing mass as it fell. Being a wooden ship, she burned for nearly two hours after she reached the ground. All the bodies of the crew were completely charred and all objects on board more or less destroyed by fire.

An excerpt from the combat report by Lieutenant Leefe Robinson, for his night patrol 2/3 September 1916:

At 2.05 a.m. a Zeppelin was picked up by searchlights over north-east London (as far as I could judge). Remembering my last failure I sacrificed height (I was still at 12,900ft) for speed and made nose down in the direction of the Zeppelin. I saw shells bursting and night tracer shells flying about it. When I drew closer I noticed that the anti-aircraft aim was too high or too low; also a good many some 800ft behind – a few tracers went right over. I could hear the bursts when about 3,000ft from the Zeppelin. I flew about 800ft below it from bow to stern and distributed one drum along it (alternate New Brock and Pomeroy). It seemed to have no effect; I therefore moved to one side and gave it another drum distributed along its side – without apparent effect. I then got behind it (by this time I was very close – 500ft or less below) and concentrated one drum on one part (underneath rear). I was then at a height of 11,500ft when attacking the Zeppelin. I hardly finished the drum before I saw the part fired at glow. In a few seconds the whole rear part was blazing. When the third drum was fired there were no searchlights on the Zeppelin and no anti-aircraft was firing. I quickly got out of the way of the falling blazing Zeppelin and being very excited fired off a few Very's lights and dropped a parachute flare. Having very little oil and petrol left, I returned to Sutton's Farm, landing at 2.45 a.m.

A contemporary representation of Leefe Robinson's victory over SL-11 on a souvenir postcard produced by Walker Harrison & Garthwaite's, London biscuit manufacturers.

L-16, under the command of Kapitänleutnant Erich Sommerfeldt, crossed the Norfolk coast at 10.40 p.m. at Salthouse, passed Hindolveston at 11 p.m., Billingford at 11.08 p.m., and Mattishall at 11.15 p.m. She dropped her first bomb (an incendiary) at Kimberley at 11.28 p.m., doing no damage. At 11.35 p.m. she was over Attleborough in company with L-21, and pursued a parallel course with that of the latter for some time.

L-16 crossed into Suffolk, where she dropped three HE bombs at Little Livermere, with no results. Passing north of Bury St Edmunds at 11.50 p.m. and south of Newmarket at 12.05 a.m., she attacked London from the north-west, passing near Luton at 1 a.m.

Around 1.30 a.m. she went over Harpenden, dropping one HE bomb, followed by a further five at Redbourn which fell on fields near the Midland Railway line north-east of Redbourn. No casualties were caused, and the only damage was some broken windows in two cottages.

Her course was now changed south-eastwards, and the Zeppelin passed over Shenley at 1.45 a.m. and South Mimms at 1.50 a.m. Here, she turned northwards towards Hatfield but, on seeing the bombardment of SL-11, the commander of L-16 gave up the intention of pressing further in. Noticing that the Essendon searchlight, which had now opened, was apparently unaccompanied by any gun and offered a target involving no risk to himself, went off to bomb it. At 2.20 a.m. he circled the light, dropping nine incendiary and sixteen HE bombs on it, with the result that serious damage was done to the village. The church was badly wrecked, the choir being almost demolished. The rectory was very seriously damaged. Three cottages in the village affected and several others badly damaged. One woman (a telephone operator) and a child were killed, and a man and child injured. The light itself was not touched, though the bombs fell within 100 yards of it.

Having no desire to share the fate of SL-11, which was now falling in flames within 5 miles of him, the commander of L-16 fled as fast as possible from the neighbourhood of London, dropping an incendiary bomb on the village of Aston as he went, to no effect.

At 3.30 a.m. L-16 dropped an incendiary bomb at West Stow, north-west of Bury St Edmunds, which did no damage, and when passing over Raveningham at 4.10 a.m., for some reason a blue naval cap was dropped. At 4.15 a.m. she passed over Reedham and went out to sea near Yarmouth at about 4.20 a.m.

L-32, under Oberleutnant zur See Werner Peterson, was first heard north-west of Cromer at 9.30 p.m., going south-west. She did not make landfall, instead cruising about north of Cromer for half an hour, before arriving over Sheringham at 10.03 p.m. She passed Edgefield (10.15 p.m.), and then turned east to Erpingham at 10.35 p.m. Here, she turned south to Aylsham and at 11 p.m. the Zeppelin passed over Honingham, heading west-south-west and dropping a

Members of the RFC and RNAS sifting through the wreckage of SL-11 in a field at Castle Farm, Cuffley, Hertfordshire.

flare at Whinburgh and her first three bombs at Ovington. These were followed by a further three on Saham Toney at about 11.10 p.m.

After turning south when east of Mundford, at about 11.25 p.m., she then seemed to slow down or stop to verify her position, which she did at 11.27 p.m. and again at 11.35 p.m. At 11.45 p.m. she dropped two incendiary bombs at Two Mile Bottom, near Thetford, and headed south-west. At 11.55 p.m. she was near Mildenhall where, at Hake Bottom, she dropped a petrol tank.

Having traversed Suffolk and Bedfordshire, L-32 was heading across Hertfordshire, following the railway line past Cheddington to Tring at 2.25 a.m.; this was at the same time as the catastrophic attack on SL-11. The commander of L-32, whose aim was evidently to approach London from the north-west along the line of the LNWR, which the Zeppelin commander had found at Leighton Buzzard, now turned abruptly off to the east. He went slowly for a while, probably considering the situation, and then apparently determined to make for home after having dropped the rest of his bombs at some location which he could have plausibly represented as London.

On his return from this mission, Peterson claimed that, after leaving London on his port bow, he saw the destruction of SL-11 and then went into London and dropped bombs on 'Kensington and the City'. It is evident, from the movements of his ship, that he knew this was not the case. On a clear night, London, with its searchlights and guns in full action, would be quite unmistakable from a Zeppelin which was less than 30 miles distant from the outer ring of defences.

Instead of approaching the metropolis, he went off to the east and, in the neighbourhood of Hertford, dropped the bombs that he claimed he had thrown at Kensington. At 2.45 a.m., L-32 had passed near Redbourn and Harpenden from east to west and at 2.54 a.m. he dropped five HE and eleven incendiary bombs on Hertford Heath, killing two horses. A minute later, sixteen HE and eight incendiary bombs were dropped at Great Amwell, killing a pony and breaking the windows of three dwelling houses. About two minutes after that, two more HE bombs were dropped near Ware, to no effect.

L-32 then made off east-north-east over Suffolk. At 3.35 a.m. she verified her position and turned north-east to Stowmarket, which she passed at 3.45 a.m., followed by Bungay at 4 a.m., Somerleyton at 4.15 a.m. She went out to sea at Corton at about 4.15 a.m. The noise of her engines were heard from the *Cross Sand* light vessel at 4.20 p.m.

L-21, under Oberleutnant zur See Kurt Frankenberg, crossed the coast at Mundesley at 10.20 p.m. She was invisible in the clouds, so she was not fired at although she sounded to be just 5,000–6,000ft up. She steered a south-westerly course, passing Knapton at 10.25 p.m., between North Walsham and Aylsham at 10.40 p.m., Buxton Lamas at 11 p.m. and Drayton, north-west of Norwich, at 11.15 p.m.

At 11.35 p.m. she was at Attleborough, in company with L-16. The two, however, soon parted company, but for some time kept on the same parallel course at a distance of about 6 miles from each other.

L-21 then proceeded across Suffolk, dropping a petrol tank at Stanton, and went on to traverse Bedfordshire and Hertfordshire where, when passing Hitchin at 2.25 a.m. she saw the glare of the burning SL-11. She at once turned homewards at high speed, beginning to drop her bombs as she went. Two incendiary bombs were dropped at Dunton, east of Biggleswade, at about 2.40 a.m. and a third at Hatley Park, near Gamblingay, a few minutes later. At 3 a.m. L-21 dropped an HE bomb on North Fen, Sutton, doing no damage, and then turned north-west, dropping one HE and one incendiary bomb on Horselode Fen, Chatteris, which fell in fields and only damaged a few wheat sheaves and some mangold wurzels. She then headed north-north-east, past Doddington at 3.15 a.m. and March at 3.20 a.m.

Crossing back into Norfolk, the Zeppelin dropped one HE and one incendiary bomb at Tilney St Lawrence at about 3.35 a.m. She then dropped two HE bombs to the west and two incendiaries to the north of King's Lynn at 3.40 a.m., also with no result.

As L-21 passed over Wolferton (3.40 a.m.), she dropped a single incendiary followed by seven HE bombs. As the Zeppelin travelled away from Wolferton in the direction of Dersingham at 3.45 a.m., she was estimated to be some 6,000ft up and was engaged by a 75mm gun of the Royal Naval Mobile Brigade at Sandringham. It fired twelve rounds at her, the second of which was claimed to have been a hit.

As she passed over Dersingham, L-21 dropped seven HE and two more incendiary bombs, causing serious damage to six houses and minor damage to eight others at Dodd's Hill. Three people were injured, one of whom, Violet Ellen Dungar (36) subsequently died of her injuries. At Snettisham, four HE and three incendiaries were dropped (3.50 a.m.), followed by six HE and four incendiaries at Sedgeford (4 a.m.) and one incendiary at Thornham, before the Zeppelin finally passed out to sea.

The death of Violet Dungar, like so many details about the air raids, could not be reported in any newspapers. But the Dersingham burial register bears mute testimony to the local tributes given to her as the Reverend Arthur Rowland Grant, the rector of Sandringham and domestic chaplain to the king, conducted the service. In addition to Grant's entry in the register, the local rector, Reverend Robert Lewis, annotated that she 'Died of wounds inflicted by a Zeppelin bomb.'

L-14, under Hauptmann Kuno Manger, made landfall east of Wells at about 9.50 p.m., dropped an incendiary bomb and passed inland east of Burnham Market at 9.55 p.m. At 10 p.m. she was south of Thornham, and then seems to have either stopped her engines or circled for half an hour. She was next heard of at Ringstead at 10.30 p.m., where she dropped a single HE bomb, causing no damage.

She moved out west over the Wash to Terrington Marsh where, at about 10.55 p.m., she dropped an incendiary bomb. She then circled south of King's Lynn, near Wiggenhall at 11.10 p.m. and then travelled northwards, east of King's Lynn to Gayton, where she dropped two HE bombs at about 11.30 p.m. She turned south by Wormegay Fen, dropping an HE bomb at about 11.35 p.m., followed by three HE bombs on the Warren at Shouldham. No damage was caused by any of these bombs.

Manoeuvring south-west, L-14 passed over Downham Market at 11.45 p.m., Upwell at 11.55 p.m. and at 12.15 a.m. she was near Yaxley. She dropped an incendiary bomb which did no damage at Upwood.

At 2.25 a.m., when the destruction of SL-11 occurred, L-14 must have been between Thaxted and Dunmow, about 25 miles from the scene. Her commander must have seen what happened and immediately turned for home, dropping his bombs at random.

One HE bomb landed at Little Bardfield and one at Finchingfield at 2.30 a.m., with no results. Lavenham received two HE bombs at about 2.45 a.m., followed a few minutes later by one HE at Thorpe Morieux and one at Brettenham, none of which did any damage. At about 2.50 a.m. two incendiary bombs were thrown near Drinkstone and, at 2.55 a.m., four HE and one incendiary were dropped upon Buxhall, damaging crops but nothing further. Five minutes later five HE bombs followed near Haughley, with the same result. No more bombs were thrown on land.

L-14 passed over Diss at 3.10 a.m., Long Stratton at 3.15 a.m., Shotesham at 3.20 a.m., Blofield at 3.30 a.m., and Wroxham at 3.45 a.m. She headed back over the sea at Bacton at 4.05 a.m. pursued by an aeroplane, but she could not be picked up by the searchlights owing to the thick and low clouds.

SL-8, under Kapitänleutnant Guido Wolff, crossed the Norfolk coast and made landfall north of Holkham at 11.05 p.m. She dropped two incendiary bombs at Burnham Thorpe at 11.15 p.m., but caused no damage. She then went west, and afterwards turned south to the neighbourhood of Wendling at 11.40 p.m. and Swaffham at 12.20 a.m. At 12.30 a.m. she passed Downham Market and turned south, crossing into Cambridgeshire where she dropped an incendiary bomb at Littleport, followed by six incendiary bombs on Oxlode Fen, near Downham-in-the-Isle, with no effect.

Traversing the Fens, but dropping no bombs, it was assumed that the commander of SL-8, although some 50 miles away, must have seen the glare of the destruction of SL-11 at 2.25 a.m. Like the other airship commanders, he immediately made off at high speed and was thought to have followed the long, straight line of the Bedford Levels, the twin canals of which extend in a direct line for a distance of 22 miles from Earith Bridge, in Cambridgeshire, to Downham Market.

Once back over Norfolk, the airship started dropping bombs again, with three HE and three incendiary bombs thrown on Congham at 3 a.m., breaking windows and tiles in two cottages. One incendiary was dropped at Harpley Dams, Flitcham, two or three minutes later. SL-8 then turned north-east, to East Rudham and, at 3.05 a.m., dropped three HE bombs followed by an incendiary bomb at Helhoughton and five more HE bombs at Syderstone and South Creake, where four cottages suffered damage to windows and tiles. At 3.15 a.m. one HE bomb was dropped on Great Walsingham and two shortly afterwards at Wighton, to no effect. At 3.20 a.m. the airship went out to sea at Cley, dropping one HE bomb on land and eight into the sea as she went.

L-24, under Kapitänleutnant Robert Koch, came inland near Trimingham at about 12.30 a.m. and dropped a flare at Gunton soon after. At 12.40 a.m. she was at Blickling and turned west, passing Saxthorpe at 12.45 a.m. She dropped two HE bombs at Briston at 12.50 a.m., which caused no damage.

The Zeppelin then turned north, dropping an incendiary bomb at Plumstead, near Holt, and appearing shortly after 1 a.m. over Bacton, where she was fired upon without result. The airship moved north near the coast, and was fired at by the Royal Naval Mobile AA Brigade 3-pdr gun as the Zeppelin passed Mundesley at 1.12 a.m. Her height there was estimated at only 4,000ft. On being fired at, she circled north and dropped five HE bombs on the beach just below high tide, at the base of the cliff on which the gun stands, and within 60–100 yards of it. No damage was done.

The airship turned inland and dropped two incendiary bombs at Trunch at about 1.25 a.m., followed by thirteen HE and twenty-seven incendiary bombs at Ridlington at 1.30 a.m. Remarkably, not one of them caused any damage and nobody was injured.

It was thought that the Zeppelin had been drawn to drop its bombs at the last two locations by the flares that had been lit for the landing strip at Bacton Aerodrome. The airship then approached Bacton from the south-west, and was fired on again by the two 75mm guns and the 3-pdr of the RNAS. Visibility was, however, so bad owing to low clouds and mist, that she was only momentarily picked up and was not hit. Her height was estimated at 6,000–8,000ft, rising as she went out to sea at Bacton.

L-30, commanded by Kapitänleutnant Horst von Buttlar, arrived at Southwold, Suffolk, at 10.40 p.m. and flew over Blythburgh at 10.49 p.m. She was west of Wrentham at 10.58 p.m. when she was picked up by the Pulham searchlight and immediately turned north-east, dropping nine HE and twelve incendiary bombs as she did so. These caused damage to two farmhouses and injured one man.

Two minutes later, she passed Mettingham and dropped eight HE bombs and one incendiary bomb on Bungay Common, killing two cows and injuring three others. At 11.15 p.m. L-30 dropped six HE Bombs at Ditchingham, smashing glass and displacing tiles, followed three minutes later by four HE bombs at Broome, which smashed some windows.

The Zeppelin went off north-east to the coast, was fired at by anti-aircraft guns at Fritton, and then passed out to sea south of Great Yarmouth at about 11.25 p.m. Two direct hits were claimed by the Fritton guns, but L-30 returned to her base safely without any appreciable loss of speed.

L-11, under Korvettenkapitän Viktor Schütze, approached Great Yarmouth from the north-east at 10 p.m. and passed over the *St Nicholas* lightship at 10.05 p.m. Five minutes later, she dropped several bombs in the sea and, at 10.15 p.m., came overland, dropping one HE and one incendiary bomb about ½ mile north-west of Southtown Station. Both landed on marshy ground and did no damage. She was engaged by anti-aircraft guns which opened up on her and then went out to sea again, dropping more bombs as she went.

She moved south along the coast and, at 11.30 p.m., came in again north of Lowestoft, circling round to the northward and coming under fire from the AA guns of No. 5 Mobile AA Brigade at Fritton and Lowestoft. She at once went out to sea without having dropped any bombs, and proceeded further down the coast.

At 1.10 a.m. she dropped one white and one red rocket east of Golf House gun, Felixstowe, circled back towards Shingle Street at 1.50 a.m., was off Felixstowe again at 2.05 a.m. and off Landguard at 2.20 a.m. Turning north, she passed up Harwich harbour, dropping three HE and one incendiary bombs, all of which fell in the water. When illuminated by the searchlights and fired on at 2.30 a.m., she quickly disappeared into the clouds, turned north-east over Walton and went off up the coast, finally going out to sea north of Aldeburgh at 2.50 a.m. The only damage done at Harwich consisted of a few windows broken by concussion, and a roof penetrated by a fragment of an AA shell.

L-23, with Kapitänleutnant Wilhelm Ganzel in command, passed over the *Lynn Well* light vessel at 10 p.m., and threw a calcium flare, afterwards altering course to the south-east. She approached the Norfolk coast at Snettisham and, at about 10.15 p.m., threw several incendiary bombs which fell into the sea. Of these, three were later recovered and brought ashore at Snettisham.

L-23 then made across the Wash in a westerly direction and crossed the coast at Kirton Fen, where she dropped an HE bomb at 10.40 p.m., which slightly damaged a farmhouse and killed some poultry. This was followed by an incendiary bomb at Kirton Holme at 10.45 p.m., two at Swineshead and another at Gosberton at 10.48 p.m.

At this point the Zeppelin suddenly turned about and made northward, dropping four HE bombs and one incendiary on Boston at 10.54 p.m. There was little doubt that lights had attracted the bombs. The signal box of the Great Northern Railway station was damaged, and a workshop and office were also affected at the gasworks. One house was partially wrecked, and glass broken or doors blown in on seventy-five houses. The pier and the sluice doors of the River Witham were damaged, but not so severely as to interfere with the working of the doors. One man was killed, and two men and one woman injured. Three of the casualties were incurred out in the open.

The Zeppelin went off to the southward, past Holbeach at 11.10 p.m. and Tydd at 11.25 p.m. She then turned west, and was west of Wisbech at 11.30 p.m. and east of Spalding 10 minutes later, going round north-eastwards towards the Wash. She dropped an incendiary bomb at Weston, north-east of Spalding, which did no damage. At 11.45 p.m. she passed Holbeach again, going north-east, and at 11.55 p.m. was over the Wash on her way homeward. She dropped the rest of her HE bombs, twenty-two in number, into the sea.

L-13 crossed the Lincolnshire coast south of Cleethorpes at 10.56 p.m. After hovering a few minutes just off the town, she harmlessly dropped six HE bombs at Humberston at 11 p.m. She then turned south-west past Utterby at 11.10 p.m. and Ludford at 11.16 p.m. Here, she turned north-west to Tealby, east of Market Rasen at 11.22 p.m. After going over Walesby at 11.29 p.m., she headed south-west again to Market Rasen by 11.36 p.m.

She now resumed her westerly course and, at 11.40 p.m., dropped an incendiary bomb at Caenby which did no damage. At 11.50 p.m. she was heard at Harpswell and Hemswell, east of Gainsborough, and at 12.10 a.m., now going north, she dropped five HE bombs and five incendiary bombs at East Stockwith. Two cottages were demolished, and a woman died of shock. In the opposite village of West Stockwith a large amount of glass was broken.

The Zeppelin now turned south along the valley of the Trent, to Morton where, at 12.15 a.m. four incendiary bombs were dropped, to no effect. She then circled round to the east, and north to the neighbourhood of Epworth, where she was recorded at 12.30 a.m.

Making south-west to Bawtry at 12.40 a.m., she passed south of Tickhill at 12.47 a.m. and, bearing round to the south-east, reached and bombed Retford at 12.56 a.m. All lights had been out in the town for some time, and local opinion ascribes its discovery to the railways. However this may be, the town was found and suffered considerably. Fifteen HE bombs and four incendiaries were dropped. Two bombs fell on the Corporation Gas Works and, their fragments piercing three gasometers, ignited and destroyed them. Glass was broken in the manager's house and works office. Eight houses were seriously damaged, a small fruit warehouse demolished, and glass broken in the Weslyan schoolroom and sixty-seven other houses in various parts of the town. Three women were injured.

The Zeppelin proceeded off towards Gainsborough and, at 1.05 a.m., dropped an HE bomb at Lea, south of the town, which did no damage. Passing on eastwards, she was near Claxby and Normanby-le-Wold, north of Market Rasen, at 1.20 a.m., and then went north-east and dropped an incendiary bomb at Aylesby, west of Grimsby, at 1.25 a.m. She went out to sea at Donna Nook at 1.30 a.m.

L-22's movements are more difficult to trace, as she seems to have been flying at very high altitude. At 10.10 p.m. she was in the neighbourhood of Skegness and Wainfleet, and at 10.30 p.m. she was reported at sea off Sutton, where she dropped bombs into the sea. Traversing Lincolnshire, L-22 crossed the Humber near Killingholme at 12.35 a.m. and at 12.42 a.m. the raider was sighted east of Hull and the guns at Marfleet and Sutton opened fire on her. At 12.46 a.m. she dropped three HE bombs on open fields in the parish of Flinton, 7 miles north-east of Hull, which caused no harm. At 12.50 a.m. she passed Burton Constable, and at 12.55 a.m. left the coast at Aldbrough. At about 1.35 a.m. she was heard to drop her remaining bombs out at sea.

Aftermath

The statement from Berlin published in the German and European newspapers regarding this raid fell, as ever, rather short of the truth. After proclaiming 'another successful attack' on London it went on to claim:

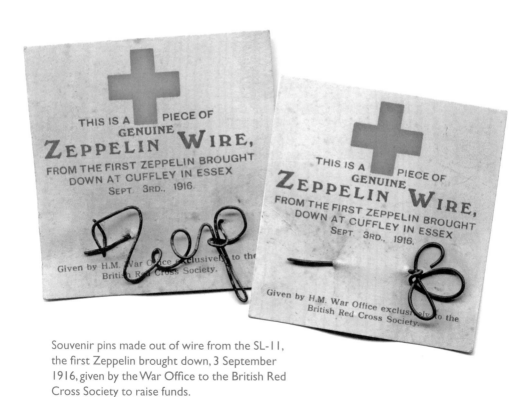

Souvenir pins made out of wire from the SL-11, the first Zeppelin brought down, 3 September 1916, given by the War Office to the British Red Cross Society to raise funds.

Other Zeppelins attacked factories and fortifications at Norwich, where strong explosions and fires were caused. Searchlight batteries and industrial works in Oxford, Harwich, Boston and on the Humber were bombed and numerous fires caused there. In Yarmouth the gas works and aerodrome were attacked and a battery silenced.

There was no mention of the loss of the SL-11.

23/24 September 1916

Eleven Zeppelins headed off on 23 September, but L-16 and L-24 had to turn back when they were halfway across. Nine Zeppelins were left to raid England between the Humber, London and the south coast of Kent. The main feature of the raid was an attack on London by the three newest Zeppelins of the German air fleet, two of which fell to the London AA defences.

L-22, the northernmost ship, was off Kilnsea at 10.25 p.m., and dropped three HE bombs in the sea. She passed over the Spurn lighthouse at 10.35 p.m. and dropped two HE bombs close to the searchlight and three near the Spurn AA gun, which had come into action. No damage was done to personnel or equipment, although one HE Bomb fell within 100 yards of the gun.

The Zeppelin then went across the estuary to the Lincolnshire coast, dropping three HE bombs in the sea as she went, and came in at Donna Nook, dropping three more HE bombs in the sea.

She passed over North Somercotes and dropped two incendiary bombs there at 10.40 p.m. Then steering west, she dropped five incendiary bombs at Grainthorpe, and one incendiary bomb at Fulstow at 10.45 p.m. She turned south-west to Utterby, where she dropped two incendiary bombs at 10.50 p.m. and went about in a northerly direction to Waltham, where she was heard at 11 p.m. She wandered around in this neighbourhood for some minutes, being heard at Holton-le-Clay at 11.05 p.m. and Laceby at 11.10 p.m. She dropped an incendiary bomb at Caistor at about 11.20 p.m. and then turned east, passed south of Donna Nook at about 11.35 p.m. and she was heard out to sea, still steering east, at 11.40 p.m.

L-24 was heard dropping some of her bombs in the sea at about 10.54 p.m. and came in over the Lincolnshire coast near Mablethorpe at about 11 p.m. She was moving in the district east of Gainsborough from 11.30 p.m. onwards, but her movements could not be definitely traced until 11.45 p.m., when she dropped an incendiary bomb at Glentham, 10 miles east of Gainsborough. She wandered between Gainsborough and Grimsby for some time, evidently looking for the latter place and for Waltham wireless station. After leaving Glentham she seems to have been near Lincoln shortly after midnight and, at 12.30 a.m., was again east of Gainsborough.

She dropped two incendiary bombs at Kingerby, north-west of Market Rasen at about 12.25 a.m. and then went off north-east towards Waltham. At 12.35 a.m., nineteen HE and eleven incendiary bombs were thrown by her at Scarthoe, south-east of Grimsby, doing no damage beyond light injury to the church roof and a few windows broken in the village. The Zeppelin was engaged by the guns at Scarthoe Top and Cleefields, after which she moved off out to sea at about 12.40 a.m.

L-14 arrived with L-17 between Mablethorpe and Skegness. One of the two ships, which is taken to be L-14, was heard 4 miles north of Skegness at 9.55 p.m., at Ingoldmells, where she came in overland. The other raider, L-17, was near Alford at 10.20 p.m. Both ships went together to Lincoln, an HE bomb being dropped at Waddingworth (probably by L-17 on the way).

They approached Lincoln at 10.45 p.m., and the Washingborough light and Canwick gun opened on L-14, which did not attempt to pass on to Lincoln but emptied her whole cargo of bombs over the light; the Zeppelin commander perhaps thought it was actually Lincoln.

The village of Greetwell received fifteen HE bombs, which broke telephone wires and killed a sheep. Washingborough got twelve HE bombs which demolished a fowl house, tore up an orchard and damaged some outbuildings, while nearby Heighington had seventeen incendiary bombs dropped. There were no human casualties. After wandering above Lincolnshire for a while, the Zeppelin passed Alford going east at 11.23 p.m., Willoughby at 11.30 p.m. and went out to sea at Sutton shortly after.

L-17, under Kapitänleutnant Herman Kraushaar, passed Lincoln at about 10.50 p.m., while L-14 was engaged with the Canwick guns; then, having verified her position, she altered course south-west in order to reach Nottingham, which was evidently her objective.

Going slowly to make sure of her course, she was close to Newark at about 11.50 p.m. She apparently circled in its neighbourhood and, at midnight, dropped an incendiary bomb at North Muskham, 3 miles to the north, which fell into the River Trent. She then went on towards Nottingham at 12.15 a.m., passing Southwell and then Gunthorpe at 12.23 a.m. At 12.34 a.m. two HE and four incendiary bombs were aimed at the Great Northern Railway goods sidings at Colwick. They fortunately fell in fields and did no damage. Immediately afterwards, six more HE bombs were thrown on fields between Colwick and Sneinton.

The AA defences of Nottingham now prepared to come into action. At 12.37 a.m. the Sneinton light was exposed but, at 12.39 a.m. a bomb fell (one of the six HE mentioned above) which severed telephone communication between the searchlight and the guns, which now had to act independently. Neither gun found the target owing to a thick ground mist. The Aspley and Wilford lights were also outranged.

Meanwhile, L-17 had dropped eight HE and eleven incendiary bombs on Nottingham, doing considerable damage, killing three people and injuring eleven more. The Midland Railway goods station was seriously damaged, while the Great Central Station and permanent way suffered slight damage. A chapel was partially wrecked and set on fire; three houses were totally wrecked, four houses and three shops and a public house seriously damaged and three others slightly affected. A large number of windows were broken.

No damage was caused to any munition or other factories doing government work. It was stated by the local authorities that the air raid on the city was directly caused by the failure to extinguish lights on the railways when Zeppelins were heard in the immediate vicinity. The point was communicated to the Railway Executive Committee and was investigated.

Leaving Nottingham at about 12.49 a.m., L-17 passed over Mapperly, dropping an HE bomb there which seriously damaged a house, and then went off on her return journey. She passed Lincoln again at about 1.20 a.m., passed Grimsby, dropped bombs in the sea and was off the Spurn, where she was fired at without result at about 2 a.m. She retired out to sea, probably at Kilnsea.

L-13, the fourth ship to visit Lincolnshire, came in north of Skegness at about 10.30 p.m., following in the wake of L-14 and L-17. She passed near Wainfleet at 10.40 p.m., and 10 minutes later was between Boston and Sibsey. She seems to have hovered in the neighbourhood of Boston for some time before making off towards Sleaford, passing Langrick at 11.20 p.m. and Swineshead at 11.25 p.m.

At 11.50 p.m. she was south of Sleaford, being engaged by an aeroplane sent up in pursuit from Cranwell Aerodrome and by the Rauceby gun. Her height was estimated at 6,000ft. On being fired at she dropped one incendiary bomb at Silk Willoughby and four at Holdingham, then thirteen HE bombs on Rauceby itself at midnight. L-13 then went about and, after dropping seven incendiary bombs at Leasingham, which did no damage, made off at high speed to the coast.

She was seen at Langrick at about 12.05 a.m. and went out to sea south of Wainfleet at 12.15 a.m., but did not head back to her base, instead heading up the coast to the Spurn where, at 12.43 a.m., she was engaged by the gun. A hit was claimed by the gunners, but this was not substantiated as the Zeppelin returned safely and at regular speed back to Germany. It is not improbable that she dropped the rest of her bombs in the sea when fired at – between 12.45 a.m. and 12.55 a.m. the explosions of eight bombs were heard from Cromer at a very great distance northwards.

L-23, under Kapitänleutnant Wilhelm Ganzel, passed near *Haisborough* light vessel at 8.35 p.m. She was heard dropping bombs in the sea north of Cromer at 8.55 p.m. then she seems to have approached the land and dropped bombs at Overstrand at 9.20 p.m., but none of these bombs fell on the shore, let alone on land. L-23 was heard to drop ten bombs in all, presumed to be HE bombs, in the sea between 8.55 p.m. and 10.05 p.m. and then returned towards her base.

L-21 was heard at sea east of Aldeburgh steering south-west, and was seen and heard coming in north of Orfordness at 9.40 p.m. After coming overland, she passed over Sudbourne Camp at a height estimated to be 8,000ft. She circled over Orford and then went west-south-west over Capel St Andrew at about 9.50 p.m.

She turned north-west to Wickham Market, where she was at 10 p.m., and a minute or two later dropped a flare at Letheringham. She then turned south towards Woodbridge and north-west again to Crowfield where, at 10.23 p.m., she threw out a petrol tank. Two minutes later, a second petrol tank was thrown at Stonham Parva. She then turned south to Coddenham where, at 10.35 p.m., an HE bomb was dropped, doing no damage.

She travelled on to the River Gipping and followed it north-west to Needham Market, where an incendiary bomb fell at 10.37 p.m. Two minutes later, she dropped two HE bombs at Badley, crossed the river northwards and was engaged by the Stowmarket guns, which fired twenty-five HE rounds at her without apparent result. In response, she dropped sixteen HE and ten incendiary bombs at Creeting St Peter and five HE and one incendiary bomb at Stowupland between 10.40 and 10.43 p.m.

At Badley, damage was done to the roof, windows and doors of a farmhouse, and the telegraph wires of the railway were broken. At Creeting St Peter, five sows were killed and farm buildings damaged. There were no casualties. Some of the bombs fell within 500–600 yards of the guns.

L-21 went off immediately east to the coast, flying high. Reported as passing over Saxmundham at 10.58 p.m., she passed along the coast between Dunwich and Walberswick at 11.05 p.m. She was seen over Henham Park, between Wangford and Wrentham, at about 11.15 p.m., having circled inland for some distance. She then went south-east out to sea at Easton Broad at 11.20 p.m., was off Great Yarmouth shortly after midnight, and then went home by way of Texel.

L-33, under Kapitänleutnant Alois Böcker, passed into the Thames estuary north of *Black Deep* lightship at 10 p.m. She was fired on by a destroyer in the Edinburgh Channel at 10.12 p.m. and came overland to Foulness at about 10.40 p.m. She went directly inland, going between Southminster and Burnham at 10.45 p.m., dropped an incendiary bomb, which did no damage, at South Fambridge at 11.05 p.m. and flew on slowly straight towards London.

She was north of Rayleigh at 11.05 p.m., south of Wickford at 11.20 p.m., west of Ingrave at 11.25 p.m. and near Billericay at 11.27 p.m. She threw a flare south of Brentwood at 11.35 p.m. and then dropped four incendiary bombs on

Upminster Common at about 11.40 p.m. Ten minutes later, she dropped six HE bombs at South Hornchurch.

At 11.55 p.m. L-33 was south of Chadwell Heath, where she dropped another flare and was picked up by a searchlight going west. The other searchlights in the neighbourhood did not pick her up and appear to have been surprised. No guns found her at this point and she went on to Wanstead where, at 11.59 p.m., she suddenly turned south-east, and at 12.06 a.m. turned south-west between the Beckton and North Woolwich guns. She afterwards headed north-west towards West Ham, where she passed over the gun at 12.10 a.m.

The guns and lights of this portion of the London defences were now well onto her, and L-33 was subjected to a continuous bombardment as she passed to the northward, dropping bombs on east London as she went. The shooting was good, West Ham getting off eighteen rounds in two minutes, and Victoria Park seven rounds in one minute. Unfortunately, both the Victoria Park guns jammed. One 4.7in shell from Beckton burst very close to the Zeppelin, and hits were claimed on her nose and tail. Wanstead fired eleven rounds in three minutes; the last three rounds observed as 'range'. It is probable that she was hit here, either by Beckton or Wanstead, but no immediate effect was caused, although shortly afterwards she began to lose gas.

L-33 began dropping bombs on Bromley, west of the Gas Light & Coke Company's works, at about 12.11 a.m. One 100kg HE and five incendiary bombs fell first in St Leonard's and Empson Streets, wrecking four two-storey houses and breaking windows in a large number of others. Six people were killed and eleven injured here. A 50kg HE bomb grazed the wall of Spratt's Dog Biscuit factory and exploded, exposing the foundations without damaging the wall. One 300kg HE bomb, one 100kg and two 50kg bombs were next dropped on the North London Railway Carriage Depot and Overtime Tractor Company's premises at Bow, causing considerable damage to sheds and a boiler house, also damaging rolling stock and permanent way.

About 12.12 a.m. a 50kg HE bomb seriously damaged a Particular Baptist chapel in Botolph Street, and a 100kg bomb wrecked the interior of the Black Swan public house at the corner of Devon's Road and Bow Road. Three women and two children were killed, and one man and three children injured in the public house. The neighbouring premises were also practically destroyed, and the windows of the County Council School and other adjoining windows were smashed. Two men, four woman and a child were also injured in the vicinity.

Crossing Bow Road, the Zeppelin dropped a 100kg HE bomb on the junction of Old Ford Road and Wrexham Road, doing considerably less property damage but injuring three women. An incendiary bomb landing on Brickfield Road caused limited damage to a house but caused no fire.

L-33 then turned north-east towards Stratford and dropped a 100kg HE bomb on Cook's Soap Works. The bomb did not explode.

The two HE bombs which fell further east on Marshgate Lane caused severe damage. The British Petroleum Company's works were severely damaged; a concrete wall was blown down and underground oil pipes were broken, an 18in water main was broken and part of a sewer bank washed away. The windows of several factories close by were broken, but there were no casualties.

Seven more 50kg HE bombs and one incendiary fell on Messrs Judd's match factory, which was set on fire and the greater part of the building, including the stock, destroyed.

The Zeppelin, now finding the fire to which she was subjected dangerous to her safety, and possibly having been seriously hit, was reported to have disappeared 'in a cloud' (probably released ballast) at 12.19 a.m. The engines were heard to be running badly and making a pounding noise, no doubt owing to the propeller having been hit.

At 12.23 a.m. she was picked up by the Kelvedon Common searchlights again and, at 12.25 a.m., she was fired upon at an estimated height of 9,000ft by the 13-pdr guns on the common.

There can be little doubt that, during the bombardment over east London, one propeller had been damaged and a shell had passed through her body. She also began to lose gas, though not rapidly, through punctures caused by shell splinters.

At 12.30 a.m. L-33 was attacked by Second Lieutenant Alfred de Bath Brandon RFC (while he was out on patrol from Hainault to Sutton's Farm), but his bullets failed to bring her down. The majority of the prisoners of L-33 were later vague as to the aeroplane attack, which apparently passed more or less unnoticed by them; but they are unanimous that the Zeppelin was pierced by a shell shortly after leaving the river, i.e. over east London.

Brandon continued his close pursuit of the Zeppelin which, at 12.35 a.m., passed south of Ongar and at 12.45 a.m. was north of Chelmsford, Essex. It was here that L-33 started to jettison cargo. At Broomfield, spare parts, two aluminium cartridge boxes and a leather machine gun case were thrown out, and at Boreham, a mile or two further, a machine gun followed.

At 12.55 a.m. the Zeppelin was between Witham and Maldon. She now began to labour considerably as her speed and height diminished. Around 1 a.m. she dropped a second machine gun in the grounds of 'Moncktons' at Wickham Bishops. A third machine gun followed at Gate House Farm, Tiptree. At 1.05 a.m. she was near Tolleshunt Major and gradually drifting nearer the ground owing to her loss of gas. Here, a fourth machine gun was thrown out at Gate House Farm.

At 1.15 a.m. she went out to sea near West Mersea. However, Böcker (her commander), in view of her condition, preferred to land and be taken prisoner rather than run the risk of drowning with his crew, like the men of L-15. So he almost immediately returned to the coast and, at 1.20 a.m., brought his ship down to earth in a field between Little Wigborough and Peldon, some 3 miles inland, north-east of Mersea.

The GHQ Intelligence Report stated:

> Slight explosions took place and the ship took fire but owing to her great loss of gas comparatively little damage was inflicted upon her, only the outer casing being destroyed and the front gondola suffering severe damage. The framework partly collapsed amidships when the casing took fire.

In reality, L-33 had had a forced landing and the crew fired it. Kapitänleutnant Alois Böcker and his crew were all taken prisoner shortly after the landing.

Early in 1918, Böcker was returned to Germany in an exchange of prisoners, with the usual stipulation that he should not fly in combat again. He did end the war in the air though, in his old Zeppelin L-14 as director of airship training at Nordholz.

L-32 passed through the Straits of Dover between 9 p.m. and 10.30 p.m. She then felt her way along the coast to Dungeness where she came in at 10.50 p.m. and dropped six HE bombs, wrecking one house and seriously damaging another. Two bombs fell in the sea.

She then hovered for a time, and went inland towards Appledore, worked round south-west to Peasmarsh and then off towards Hythe, appearing again at Lydd at 11.45 p.m. Finally, at about 11.45 p.m., she went off by Appledore and Cranbrook to Tunbridge Wells, where she arrived at 12.10 a.m.

She now turned north and at 12.30 a.m. dropped an incendiary bomb at Ide Hill, near Sevenoaks, doing no damage. At 12.50 a.m. she reached Crockenhill and Swanley Junction, and dropped seven HE bombs near the Crockenhill searchlight, which had opened on her. No damage was done, but windows were broken at Swanley.

Passing over Dartford, L-32 crossed the Thames east of Purfleet at 1 a.m. and was at once picked up by the lights and guns north of the river. There seems to have been a mist south of the river, which rendered the Zeppelin difficult to see and prevented the guns and lights of the Woolwich subcommand from dealing satisfactorily with her.

North of the Thames, L-32 was engaged and dropped bombs in answer to the attack. She passed between Beacon Hill and Tunnel Farm, under fire from the guns at the latter place, which claimed two distinct hits. Immediately afterwards, at 1.03 a.m. she dropped nine HE bombs (two weighing 100kg) and six incendiary bombs at Aveley, doing no damage. She then came under fire from Belhus Park, Tilbury, Shonks and Fobbing.

Aerial view of the fired wreckage of Zeppelin L-33 shortly after its crash landing at Little Wigborough on 24 September 1916. (RAF Museum, Hendon)

At about 1.05 a.m. the Zeppelin got rid of the rest of her bombs, dropping twenty-three HE bombs (including two of 300kg weight and eight of 100kg) and twenty-one incendiaries at South Ockendon, which miraculously only broke a few windows and injured two horses. The bombs between Aveley and South Ockendon were thrown in a straight line from a point about 3,000 yards north-east of Aveley to within 400 yards south-west of North Ockendon – a distance of approximately 3 miles.

At 1.10 a.m. the Zeppelin was found and attacked by Second Lieutenant Frederick Sowrey RFC, who had been out on patrol from Joyce Green to Sutton's Farm. His third drum of ammunition set her on fire in several places and she began to fall slowly, finally crashing to the ground in flames at Snail's

Views of the fired wreckage of L-33 at Little Wigborough, Essex.

Hall Farm, Great Burstead, south of Billericay. The wreckage burned for forty-five minutes. Second Lieutenant Brandon, who saw the combat from his aeroplane some distance off, said of Sowrey's firing that the Zeppelin looked as if it were 'being hosed with a stream of fire.' The fireball of the burning L-32 was seen from a great distance away, even from a British submarine in the Straits of Dover.

The combat report of Second Lieutenant Frederick Sowrey for the night of 23/24 September 1916:

At 11.25 p.m. I received orders to patrol between Sutton's Farm and Joyce Green and at 11.30 p.m. I left the aerodrome. The weather was clear with a few thin clouds at 3,000ft. At 4,000ft I passed another machine proceeding in a northerly direction. I was then flying due south. I continued climbing as hard as possible and at 12.10 a.m. I noticed an enemy airship [L-33] in a southerly direction. It appeared to be over Woolwich. I made for the airship at once but before I could reach it the searchlights lost it. I was at this time at 8,000ft. There was a certain amount of gun fire but it was not intense. I continued climbing and reached a height of 13,000ft. I was still patrolling between Sutton's Farm and Joyce Green. At 12.45 a.m. I noticed an enemy airship [L-32] in an easterly direction. I at once made in this direction and manoeuvred into a position underneath. The airship was well lighted by searchlights but there was not a sign of any gun fire. I could distinctly see the propellers revolving and the airship was manoeuvring to avoid the searchlight beams. I fired at. The first two drums of ammunition had apparently no effect but the third one caused the envelope to catch on fire in several places in the centre and the front. All firing was traversing fire along the envelope. The drums were loaded with a mixture of Brock, Pomeroy and Tracer ammunition. I watched the burning airship strike the ground and then proceeded to find my flares. I landed at Suttons Farm at 1.40 a.m. 24th instant. My machine was BE-2c 4112. After seeing the Zeppelin had caught on fire, I fired a red Very's [sic] light.

L-31, under the command of Kapitänleutnant Heinrich Mathy, came overland at Rye at about 11 p.m. and pursued a west-north-west course over Hawkhurst, Horsmorden, and Tunbridge Wells to Caterham at 12.15 a.m. Here, the Zeppelin first came into contact with the south-eastern defences which, however, gave her little trouble. This was due either to inexperience on the part of those in charge of the lights, or to a new manoeuvre on the part of the Zeppelin commander who – intentionally or not – blanketed or blinded the searchlights by dropping illuminating flares. Only one gun, the 4in QF gun at Croydon, opened fire and that only at the rate of one round per minute.

Second Lieutenant Frederick
Sowrey RFC.

The Zeppelin was flying unusually high (estimated at 12,000ft) and fast. She was ably manoeuvred by her commander, and succeeded in bombing London, crossing the city from south to north, and in getting clear away without injury.

Coming up from Caterham, L-31 dropped four HE bombs at Kenley at 12.25 a.m., severely damaging three houses and injuring two people. These were evidently her trail bombs aimed at the first light seen. She was next directly over the Purley light at 12.30 a.m. which had failed to pick up the target before, owing to a defective mirror. The Croydon light now picked up the ship for a moment, when two flares were dropped, which caused it to lose her. She was picked up again directly afterwards at 12.36 a.m. and the gun opened fire for two minutes when she dropped another flare which blanketed the searchlight. The Dulwich and Streatham searchlights did not find the Zeppelin at all. The Dulwich gun saw her for a second by means of other lights, but did not open fire.

Meanwhile, after dropping two HE and two incendiary bombs on two farms at Mitcham, the concussion of which damaged a few houses slightly, the Zeppelin had begun her bombing of London. Going straight on northwards, at about 12.35 a.m., Streatham was severely bombed, with ten HE and twenty-two incendiary bombs falling in quick succession. Streatham Common railway station was damaged, and the permanent way and some rolling stock and a signal box broken. Four houses were wrecked, and ten shops and ten other houses were severely damaged. There was the usual large amount of broken glass. In all seven people were killed and twenty-seven injured; of these six were killed in a tramcar which was hit by the fragments of a 300kg HE bomb on Streatham Hill. The railway station at Streatham Hill was also damaged.

The Zeppelin was now on the line of the main road through Streatham and Brixton Hill to Kennington, and followed it accurately as far as the latter place. On Brixton, six HE (one of them 100kg) and seventeen incendiary bombs were dropped, wrecking a house, and seriously damaging a garage, twenty-one houses and twenty-one shops, and slightly affecting forty more houses. Seven people were killed and seventeen injured.

Finally, at Kennington, a single HE bomb was dropped which caused extensive damage to windows.

L-31 then flew over the Thames east of London Bridge, being clearly seen from the Embankment at about 12.40 a.m., and six minutes later threw ten more HE bombs on Lea Bridge Road and Leyton. These caused serious damage to about a dozen houses, and lesser damage to a large number of others, besides killing eight and injuring thirty-one people. Mist had risen, and she was not seen well enough for any of the north-eastern gun stations to open fire on her.

Her subsequent course lay east of Chingford at 12.50 a.m., over Buckhurst Hill and near Waltham Abbey at 12.59 a.m. Chingford searchlight picked her up from 1.03 a.m. to 1.07 a.m. Then the Zeppelin made a course over Harlow to Takeley, east of Bishops Stortford, at 1.15 a.m., Haverhill about 1.30 a.m., Bury St Edmunds at 1.45 a.m., Diss at 1.55 a.m., Bungay at 2.05 a.m., north of Beccles at 2.10 a.m., and finally moved out to sea south of Great Yarmouth at about 2.15 a.m., being fired at both from the shore (Fritton and naval gun) and by a warship. At 2.20 a.m. she passed over the *Cross Sand* lightship.

Amid the victories in the air there was a tragedy …

Reports of a dozen Zeppelins having crossed the North Sea intent on attacking London and the east coast were rife, and numerous aircraft went up on patrol to try to intercept them. Two Royal Flying Corps aircraft from 51 Squadron were dispatched to go on anti-Zeppelin patrol from Thetford (Mattishall). Tragically, one of the pilots of these aircraft, Lieutenant Michael Thunder, failed to gain enough height on take-off and crashed as a result. Badly burned, he was removed to the Norwich Hospital, but died of his injuries later on 24 September.

Lieutenant Michael Hubert Francis Thunder RFC (1874–1916), the first pilot to be killed in 51 Squadron, was given a funeral with full RFC honours and six of his fellow officers acting as pall-bearers, at Ramsgate (St Augustine) Roman Catholic churchyard on 29 September 1916.

25/26 September 1916

L-16 and L-14 came in almost simultaneously at about 10.05 a.m. over Bridlington Bay, with L-16 to the northward at Barmston. She passed near Driffield between 10.10 p.m. and 10.20 p.m., was at Huggate at 10.27 p.m. and south of Malton at 10.45 p.m. and then returned to the coast.

At 10.50 p.m. an incendiary bomb was thrown on Velmire Farm, Whitwell-on-the-Hill, near Malton. By 10.55 p.m. she was at North Grimston, and at 11 p.m. dropped another incendiary bomb at Langtoft, which did no damage. L-16 then turned northward in the direction of Scarborough, dropped flares and then returned southward, dropping a third incendiary bomb at North Burton, south of Hunmanby about 11.30 p.m.

L-16 then went in the direction of Driffield, crossing the railway at Nafferton at 11.35 p.m. and turning sharply north again. At 11.40 p.m. she was engaged by the Bessingby gun, which fired a single round at her, without effect. One shell was afterwards picked up at Kilham, 6 miles west. Five minutes later, the Zeppelin was heard west of Bridlington, at 11.50 p.m. she went directly over Hunmanby and at 11.55 p.m. went out to sea over Speeton, in Filey Bay, making her way back to base. She probably dropped the rest of her bombs at sea, but there was only one doubtful report of a single bomb being heard.

L-14 came in at Atwick, passed north of Beverley at 10.20 p.m., near Market Weighton at 10.30 p.m., south of Pocklington about 10.35 p.m. and Stamford Bridge at 10.40 p.m., following the railway line to York.

At 10.45 p.m. she approached the city from the north-east and dropped an HE bomb at Heworth. She then passed south over the eastern outskirts of the city dropping seven HE and two incendiary bombs en route, which wrecked one house and did extensive window damage, but caused no casualties. Still going south, she dropped two HE and five incendiary bombs at Fulford which merely damaged some telephone wires.

L-14 was then engaged by the 3in gun at Acomb, which fired five rounds at her without effect, but with the result of bringing about her retreat. She passed again east of the city, going northwards and, shortly after 11.15 p.m., dropped an incendiary bomb on Pilmoor. The Zeppelin then turned south-west and dropped an HE bomb at Newby with Mulwith, south-east of Ripon, and four HE bombs at Ripon Rifle Ranges, Wormald Green, doing no damage.

Turning south, she passed over Killinghall Camp near Harrogate, shortly after midnight, and dropped four HE bombs at Dunkeswick at 12.15 a.m. on the RFC landing ground to no effect. At Harewood, at 12.20 a.m., thirteen incendiary bombs fell, which did slight damage to an outhouse.

The Zeppelin, now going east, was picked up by the mobile searchlight at Collingham. On getting to the beam she steered directly for the light, dropping three HE Bombs, one of which severed telegraphic communication between the gun and the light. Nine rounds were fired by the gun, and the Zeppelin went off north-east. She was next reported passing over the North Eastern Railway main line at Tollerton at 12.50 a.m. At 12.55 a.m. she was between Easingwold and Strensall, and at 1.15 a.m. over Rillington, east of Malton. At 1.30 a.m. she went out to sea at Scarborough.

L-21, under Oberleutnant zur See Kurt Frankenberg, arrived over Sutton-on-Sea at 9.45 p.m. At 9.50 p.m. she was at Alford, and south of Wragby at 10.15 p.m. From there she travelled onwards to the north, passing round the northern outskirts of Sheffield, crossed the Peak District and on to Todmorden at 11.55 p.m. At she was at Bacup and went north-west to Lumb, between which place and Newchurch she dropped two incendiary bombs.

She then turned south-west and made for the railway at Rawtenstall, where she dropped two HE bombs. She moved onwards west to Haslingden, following the railway south, and dropping five HE and two incendiary bombs at Ewood Bridge, which caused slight damage to the sewage farm, railway line and smashed some windows. Still following the railway, L-21 dropped seven HE bombs at Holcombe, which damaged the post office, injured the postmistress Mrs Elizabeth Hoyle, damaged telegraph wires, and broke windows in the church and stopped its clock.

At Ramsbottom, further down the valley, two HE bombs were dropped, followed by two incendiary bombs at Holcombe Brook. L-21 was reported as having approached Bolton from Astley Bridge and Sharples at 12.20 a.m., dropping a bomb that narrowly missed the Eden Orphanage. Travelling in a south-westerly direction the raider went across Halliwell where the blast from an HE bomb smashed widows on Darley Street. A bomb then destroyed a terraced house in Lodge Vale. Fortunately the three women inside escaped with shock and minor injuries. Incendiary bombs also fell on Waldeck Street and Chorley Old Road. L-21 then made for the centre of town, travelling over Queen's Park. Another incendiary fell on Wellington Street, which set a house on fire. Fortunately the fire brigade were on hand in time to rescue a woman and two children, who had been trapped inside, from an upstairs window.

The worst of the damage was then suffered from the bombs dropped on Kirk Street and John Street; six houses were demolished, six more were badly damaged and many of the others suffered doors blown in, windows shattered and their brickwork gouged by fragments of bomb casing hurled around in the blasts.

Sixteen people were pulled out of the wreckage alive, but two died on the way to hospital, bringing the death toll to thirteen.

Those killed were:

- Mr James Allison, 64 Kirk Street.
- Mrs James Allison, 64 Kirk Street.
- David Davis, Lodger, 64 Kirk Street.
- Mrs Elizabeth Gregory (42), 66 Kirk Street.
- Miss Ellen Gregory (17), 66 Kirk Street.
- Frederick James Guildford, Lodger, 64 Kirk Street.
- Mrs Bridget Irwin (44), 58 Kirk Street.
- Ellen Margaret Irwin (2½), 58 Kirk Street.
- Mrs Anne McDermott (36), 62 Kirk Street.
- Mary Ellen McDermott (5), 62 Kirk Street.
- Mr William McDermott (42), 62 Kirk Street.
- Mrs Martha O'Hara (41), 60 Kirk Street.
- Mr Michael O'Hara (42), 60 Kirk Street.

Passing over Great Moor Street, Spa Road, Moor Lane and Marsden Road, the raider then turned south, passed over Gilnow Mill, crossed the railway lines and dropped bombs on Rope Walk in Washington Street and the Co-op Laundry on Back Deane Road. Properties were damaged but no one was injured.

Crossing Deane Road, the raider flew over Quebec Street and Cannon Street before arriving at Ormerod and Hardcastle Mill in Daubhill. Here an incendiary fell and started a fire, but it was swiftly extinguished. Another bomb fell on Parrot Street and Apple Street, damaging privies and smashing windows. Turning North, an HE bomb was dropped on Trinity Church, causing some damage, but failed to explode.

The final three bombs fell around the Town Hall, on Mawdsley Street, Ashburner Street and Mealhouse Lane. At 12.45 a.m. the raider left, travelling in the direction of Darwen. During her time over Bolton L-21 dropped a total of nine HE and eleven incendiary bombs. No damage was done to the Bessemer Steel Forge or public buildings.

From Bolton she turned due north, was south of Blackburn at 1.05 a.m., and at 1.30 a.m. she was near Skipton. At about 1.35 a.m. she dropped an HE bomb at Bolton Abbey, which failed to explode. Passing over the North Yorkshire Moors, she apparently had some difficulty locating her position, and it was not until 3.05 a.m. that she passed out to sea at Whitby.

L-22 came in over the Lincolnshire coast at 10.30 p.m., and dropped an incendiary bomb at Maltby-le-Marsh shortly afterwards. Going north-west, she was seen from Louth at 11 p.m. and at 11.13 p.m. dropped another incendiary bomb south of Market Rasen. Heading in the direction of Rotherham, at 12.15 a.m., she dropped seven incendiary bombs near Tinsley Park Colliery, doing no damage.

L-22 then followed the railway to Sheffield, approached from the south-east and passed diagonally across the Attercliffe and Brightside districts at the east end of the city, over the quarter in which all the great armaments works were situated. She dropped fifteen HE and fifteen incendiary bombs in total but, by good fortune, no damage was done to any of the munitions works except in the case of Messrs John Brown & Co.'s Atlas Works, where an incendiary bomb dropped through the roof of a machine shop causing a slight fire. With this exception, all the damage was done to a small cottage property nearby, and private houses in the district, with resultant loss of life. In all, twenty-eight people were killed and nineteen injured. Nine houses and a chapel were demolished and damage of varying amounts occurred to sixty-two other dwellings.

The Zeppelin was at a great height, and there was a considerable amount of mist that prevented her from being seen. The first five HE and three incendiary bombs dropped near the gun position at Manor Oak, one HE falling within 50ft of the gun, which did not fire owing to the fog and the height of the Zeppelin. The 3in gun at Shire Green fired two rounds in the direction of the Zeppelin, although unable to see it. On being fired at the raider fled north-east and looked for a route seaward. She was south of Barton-on-Humber at 1.30 a.m. and at 1.40 a.m. she was engaged by the Chase Hill gun with one round.

Two minutes later she was spotted over the Humber, where a round was fired at her by HMS *Patrol*. The Zeppelin hurried inland to avoid the warship and crossed over the river at Immingham at 1.48 a.m., dropping bombs which fell in the water.

At 1.50 a.m. she passed south-west of the Sutton AA gun, travelling north, and was engaged by the Sutton, Marfleet and Harpings guns. Sutton fired four rounds, Marfleet the same number and Harpings eight. Only Harpings saw the target, which was flying at a height estimated at 10,000–11,500ft. She passed west of Grimston Hall out to sea at about 2.05 a.m.

L-23 was about 30 miles out to sea off the Norfolk coast at 8.00–8.30 p.m. when five bombs were heard from Cromer at a great distance. The Zeppelin does not appear to have come overland but went back to base after discharging her bombs in the sea.

L-30 was also believed to have been off Cromer at 8.15 p.m., and approached Yarmouth at about 8.50 p.m. She wandered up and down the coast for some time, dropping a large number of heavy bombs in the sea at about 10.25 p.m.

L-30 was reported in the southern area of the North Sea at midnight, and an hour later on the Dutch coast off Helder on her way back to base. It was not thought that either Zeppelin crossed the coastline, but reports of 'strange airships' came from various places in Norfolk during the night and were dismissed as improbable. Another 'phantom airship', which seems to have been taken more seriously, was also reported from the Spalding and Boston district between 8.30 and 11 p.m.

L–31 followed the new line of attack inaugurated by her in concert with L–32 two days before. She came through the Straits of Dover, but without attacking the defences, and was off Dungeness at 9.35 p.m. At 10.05 p.m. she was close to Hastings and at 10.15 p.m. was heard from Bexhill going west out at sea. She passed over Selsey Bill at 11.06 p.m. and then headed across to the Isle of Wight after dropping a flare to ascertain her whereabouts. At 11.30 p.m. she was over Sandown Bay, practically stationary, and then after passing south–west of the Culver battery (which did not open fire on her) went inland over Sandown, was seen over Yaverland at 11.40 p.m. and over Ryde five minutes later, going north and making straight for the entrance of Portsmouth harbour, flying very fast at a great height. She was well seen by the searchlights and all the AA guns opened on her. Curiously, the guns firing on the Zeppelin suffered a great number of misfires, owing to faulty ammunition, and accidents to guns were reported. However, not a single bomb could be traced as being dropped in the town of Portsmouth.

L–31 passed on directly over the entrance to the harbour at 11.50 p.m. and moved up the middle of it to Porchester where, at 11.55 p.m. she turned, probably to avoid fire, and drifted over Fort Southwick at midnight. Veering round to the east from Fort Southwick, she reached the neighbourhood of Burgess Hill. L–31 remained stationary for a time while her commander verified his position, of which he was evidently uncertain. Having obtained his bearings he went off along the line of the railway (where a train was said to be passing at the time) and the South Downs, arriving at Bexhill at 1.45 a.m. She dropped a flare and went on northwards in the direction of Petersfield and eventually went out to sea in the direction of Hastings. She touched the land again at Rye at 2 a.m. and then went up the Straits towards Dover, where she passed at 2.25 a.m. At 2.30 a.m. she seems to have dropped bombs, probably at some warships, and then returned to base, passing through Belgium.

Aftermath

The German communiqués after this raid declared that Leeds had been bombed, and it was not improbable that the Zeppelin commanders claimed to have achieved that target in their reports, but the nearest any of them came to Leeds was L–21, which was, at one point, 20 miles west of Leeds, having missed Sheffield on the way.

'... POOLS OF BURNING RED FLAME'

Decorated Zeppelin Commander Horst Freiherr Treusch von Buttlar evokes the atmosphere of missions at the height of the campaign in *Zeppelins over England* (1931):

> Our raids over England were much alike as the airships which carried them out, and which, as they hung in the sky along the German coast at sunset, could hardly be distinguished from one another.
>
> Even the course our raids pursued was always the same. The scene at our departure, the scene when twilight began to fall, when the darkness of night spread over the sea and the first lights of the English coast began to gleam in the distance – all this was always the same, even to the great island suddenly plunging itself into the deepest gloom the moment the news of our raid had been reported to the authorities.
>
> Then came the same gleaming white searchlights, exploring every corner of the heavens for airships, until one here and there suddenly shone out all milky-white; the same white clouds of shrapnel smoke, thinning out into diaphanous veils; the same red lights from the gun flashes below on the ground, the same fires kindled by our bombs, which poured out pools of burning red flame.
>
> Sometimes five, sometimes nine, sometimes as many as seventeen ships received orders to carry out a simultaneous attack on England and drop bombs on some particular area of the country. Whether they went to the North, to the Midlands or to the South, with London as their main objective, depended entirely upon the weather conditions to be expected in the west.
>
> The first ship went up at twelve noon. As soon as she was up, the ground staff hastened to deal with the next, and thus one ship after another left her shed and steered a westerly course.
>
> It was always the same routine. We flew at a height of a few hundred feet over Heligoland Bight and exchanged signals with outpost aeroplanes. We could see our friends come from Tondern in the north, observed the airships ascending and setting out as we passed Wittmundhafen and Hage, and saw the Ahlhorn airships arriving from the south and steering north towards us.
>
> We never flew in squadron formation or in line ahead or anything of that sort. Each of us flew independently, though there were always some ships which kept close to others. This was the case chiefly with the beginners, who always clung to the airship commanded by an experienced man. For the great problem was, in the first place, to determine the position of one's ship and next to discover a spot along the English coast which would enable one, as far as possible, to pass unobserved.

1/2 October 1916

Eleven Zeppelins left their shed to raid England on 1 October. Of these, L-13, L-22, L-23 and L-30 failed to make the crossing and fly over Britain. Those that did make landfall mostly did so between Cromer, Norfolk and Theddlethorpe, Lincolnshire, between the hours of 9.20 p.m. and 1.45 a.m.; the exception being L-31, under the command of Kapitänleutnant Heinrich Mathy, which made landfall off Lowestoft and was brought down over Potter's Bar, Hertfordshire, by Second Lieutenant Wulstan Joseph Tempest RFC. Some of the Zeppelins had come in close enough to attract the attention of the coastal guns, such as L-22, which was spotted off the coast of Haisborough about 9.30 p.m. It was fired at from Bacton, dropped four HE bombs in the water in reply then sheered off from the coast and back out to sea.

L-31 passed south of the *Cross Sand* light vessel at 7.45 p.m. and the *St Nicholas* at 7.55 p.m. She crossed the coast at Corton, passed over Lowestoft and was at Wrentham at 8.05 p.m., Framlingham at 8.30 p.m. and Needham Market at 8.50 p.m. She travelled on into Essex, where she was picked up by searchlights at Kelvedon Hatch and went off north-east and on into Hertfordshire. At Hertford she sounded as if she had shut off her engines and drifted slowly with the slight north-north-west wind in the direction of Ware. At 11.30 p.m. she started again, and was spotted next at 11.40 p.m., coming under heavy fire from the guns at Newmans and Temple House. The problem was the large number of blind shells fired from the 3in 20cwt guns at Temple House and Newmans. Of fifty rounds fired by the Temple House gun, only twenty bursts were observed, and of forty-four fired by Newmans, 75 per cent failed to burst.

Despite this, the Zeppelin commander abandoned the idea of reaching London and dropped most of his load of bombs, some thirty HE bombs and twenty-six incendiaries, on nearby Cheshunt. They seriously damaged four houses and slightly affected over 300, as well as breaking a great many glasshouses covering an area of 6½ acres, but luckily only injuring one woman.

After throwing the first dozen HE bombs, the Zeppelin turned to starboard with extraordinary suddenness. This was deduced from the position of the remaining HE and the incendiary bombs dropped by her, which fell as she swung round on an even keel to the west. Of these, five, which fell on the Recreation Ground, were of the largest size weighing 300kg. They did no damage. Turning westward the Zeppelin was seen twisting and turning, rising and falling, to avoid the lights and pursuing aeroplanes. Her speed was now much diminished. On the way she dropped an HE bomb near Potter's Bar which did little damage.

Second Lieutenant Wulstan Tempest left the RFC ground at North Weald in BE-2c 4577, at 10 p.m., to patrol between Joyce Green and Hainault. Sighting L-31 at 11.40 p.m. he immediately pursued her; when at a height of 12,700ft over Potter's Bar he fired one drum, which was effective, whereby the airship fell in flames at 11.54 p.m. Tempest's aircraft was wrecked on landing at North Weald at 12.10 a.m., but fortunately Tempest walked away unharmed.

Two other aeroplanes were in the locality and one reported that the Zeppelin broke into two parts as she fell, probably owing to the explosion of two or three HE bombs which she had not dropped. She was said to have fallen slowly, leaving a trail of burning fragments above her, while heavier flaming objects fell faster to the earth below her.

The combat report of Second Lieutenant Wulstan Joseph Tempest RFC, 1/2 October 1916 was as follows:

About 11.45 p.m. I found myself over south-west London at an altitude of 14,500ft. There was a heavy ground fog and it was bitterly cold, otherwise the night was beautiful and starlit at the altitude at which I was flying. I was gazing over towards the north-east of London where the fog was not quite so heavy when I noticed all the searchlights in that quarter concentrated in an enormous

Second Lieutenant Wulstan Joseph Tempest RFC. (RAF Museum, Hendon)

pyramid. Following them up to the apex I saw a small cigar-shaped object, which I at once recognised as a Zeppelin, about 15 miles away and heading straight for London. Previously to this I had chased many imaginary Zepps only to find they were clouds on nearing them. At first I drew near to my objective very rapidly (as I was on one side of London and it was the other and both heading for the centre of the town) all the time I was having an extremely unpleasant time as to get to the Zep I had to pass through an inferno of bursting shells from the AA guns below. All at once, it appeared to me that the Zeppelin must have sighted me for she dropped all her bombs in one volley and swung round, tilted up her nose and proceeded to race away northwards climbing rapidly as she went. At the time of dropping her bombs I judged her to be at an altitude of about 11,500ft. I made after her at all speed at about 15,000ft altitude, gradually overhauling her. At this period the AA fire was intense and I being about five miles behind the Zeppelin had an extremely uncomfortable time. At this point misfortune overtook me, for my mechanical pressure pump went wrong and I had to use my hand-pump to keep up the pressure in my petrol tank. This exercise at so high an altitude was very exhausting, besides occupying an arm, thus giving me 'one hand less' to operate when I commenced to fire. As I drew up with the Zeppelin, to my relief I found I was free from AA fire for the nearest shells were bursting quite three miles away. The Zeppelin was now nearly 15,000ft high and mounting rapidly. I therefore decided to dive at her for though I held a slight advantage in speed she was climbing like a rocket and leaving me standing. I accordingly gave a tremendous pump at my petrol tank and dived straight at her, firing a burst straight into her as I came. I let her have another burst as I passed under her and then banking my machine over, sat under her tail and flying along underneath her, pumped lead into her for all I was worth. I could see tracer bullets flying from her in all directions but I was too close under her for her to concentrate on me. As I was firing, I noticed her begin to go red inside like an enormous Chinese lantern and then a flame shot out of the front part of her and I realised she was on fire. She then shot up about 200ft, paused and came roaring down straight on to me before I had time to get out of the way. I nose-dived for all I was worth with the Zep tearing after me and expected every minute to be engulfed in the flames. I put my machine into a spin and just managed to corkscrew out of the way as she shot past me, roaring like a furnace. I righted my machine and watched her hit the ground with a shower of sparks. I then proceeded to fire off dozens of green Very's lights in the exuberance of my feelings. I glanced at my watch and saw it was about ten minutes past twelve. I then commenced to feel very sick and giddy and exhausted and I had considerable difficulty in finding my way to ground through the fog and landing, in doing which I crashed and cut my head on my machine gun.

Above left: Some of the wreckage of L-31 shortly after it had been brought down in a field near Potters Bar, on 1 October 1916. **Right**: The impact mark left by Zeppelin Commander Heinrich Mathy after he jumped from the burning L-31 as it headed for the ground. Astonished villagers ran over and found him to still be breathing, but he died almost immediately afterwards.

L-24 was off the Norfolk coast at Weybourne at 10.05 p.m., and came in at about 10.15 p.m. on a south-west course. She passed Salthouse at 10.20 p.m., near Field Dalling at 10.30 p.m. and Thursford at 10.45 p.m.; Stoke Ferry at 11.30 p.m. and between Ely and Mildenhall at 11.45 p.m. Five minutes later, she passed over Soham and was over Wicken Fen at 11.55 p.m.

At midnight, she travelled over Waterbeach going south. Here, she suddenly altered her course to the west and slackened speed. It was probable that this was in direct consequence of the catastrophe to L-31, which must have been seen and correctly assessed by the commander of L-24. There can be no doubt he saw the glare, since it was visible from Bartlow, only 12 miles south of him, and was reported there as 'lighting up the whole country.'

L-24 went slowly west, passing Cottenham and St Ives and on into Hertfordshire, where she was attracted south-east by the flares of the night landing ground at Willians, east of Hitchin. At 1.14 a.m. she dropped her first bomb there, letting fall twenty-eight HE and twenty-six incendiary bombs between Willians and a point 2½ miles to the eastward. Miraculously, no damage was done, but a soldier of the Royal Defence Corps, a member of the night landing guard at Willians, was killed on the landing field.

At Weston, 2 miles east of Willians, one of the Zeppelin's crew dropped his cap overboard, while the bombs were being thrown. Having got rid of his bombs, the commander of L-24 made headway to return to base, crossing Suffolk and going out to sea at Kessingland at 2.35 a.m.

L-34, under the command of Kapitänleutnant Max Dietrich, was first heard travelling south-east of the *Haisborough* light vessel at 9.36 p.m. and made landfall at Overstrand at 9.42 p.m. She travelled west-south-west, passing Felbrigg at 9.45 p.m. and Melton Constable at 10 p.m. L-34 turned south-west, passing west of Foulsham at 10.05 p.m. and dropped a flare over Kempston at 10.15 p.m. Maintaining the same course, she passed over Swaffham at 10.25 p.m. and Stoke Ferry at 10.35 p.m.

Altering course westward, she passed West Dereham at 10.45 p.m. and Downham Market at 10.50 p.m., crossing over into Cambridgeshire via Outwell at 10.55 p.m. On the way into Northamptonshire, the Zeppelin was picked up by the searchlight and gun detachment at Corby, which immediately fired upon her. Steering directly south-west for the guns, she dropped seventeen HE bombs on a curving line between Kirby Hall and the southern entrance to Corby Tunnel, then went off sharply north-east, throwing thirteen incendiary bombs in a line east of the road from Rockingham to Gretton.

The Zeppelin commander was, somehow, completely deceived and, having thrown his bombs on what he considered was some important defended place, went off towards his base at great speed. The bombs fell in woods and fields, no damage being done except to a single railway telegraph wire.

After leaving Gretton, the Zeppelin passed Easton-on-the-Hill, near Stamford, at 12.15 a.m., Wisbech at 12.50 a.m., crossed into Norfolk, going out to sea between Palling and Horsey at 1.40 a.m., dropping three or more HE bombs in the sea between 1.55 a.m. and 2.30 a.m.

L- 21, under the command of Oberleutnant zur See Kurt Frankenberg, came overland at Weybourne at 9.20 p.m. and skirted the coast going west, passing Warham at 9.30 p.m. and on to Burnham Overy Staithe at 9.40 p.m. Here, she turned south-west, passing Docking at 9.50 p.m. and, when over Heacham at 10 p.m., dropped two incendiary bombs, which did no damage.

Turning south and following the coastline to the Wash, past Wolferton at 10.10 p.m. to King's Lynn at 10.15 p.m., she again turned following the coast north of Sutton Bridge at 10.25 p.m. and on into Lincolnshire, dropping a flare at Gosberton at about 11.05 p.m.

While verifying his position over Oakham at midnight, the commander probably saw the catastrophe of L-31 and went off in a north-east direction at high speed through Lincolnshire, giving up the search for his objective. Passing near Donnington at 12.10 a.m., at about 12.30 a.m. an HE bomb was dropped at South Kyme, killing a sheep.

L-21 went out to sea near Donna Nook at 1.10 a.m. and was heard dropping bombs in the sea, from the Spurn lighthouse shortly afterwards.

L-16 made the Lincolnshire coast at Theddlethorpe, where she dropped one HE bomb shortly after midnight and then turned southward, following the coast at some distance inland. At 12.45 a.m. she dropped an incendiary bomb at Huttoft and at 12.50 a.m. an HE bomb at Willoughby, neither of which did any damage.

She pursued her course parallel with the coast until she reached the neighbourhood of Wainfleet and there turned sharply inland in a north-westerly direction at 1.10 a.m. She almost immediately dropped three HE bombs between Westville and Stickford and, going on in the same direction, released two incendiary bombs at East Kirkby at 1.15 a.m. followed by four HE and two incendiary at Fulletby at 1.30 a.m. No harm was caused by any of these bombs, except at Hameringham where a cow was killed and two horses injured.

The Zeppelin then turned off to the eastward, being heard south-west of Alford about 1.40 a.m., and then south-eastwards going out to sea near Wainfleet at 2 a.m. She made straight across to the Dutch coast and, shortly before 5 a.m., violated Dutch territory at the Helder before reaching the German Islands and the Bight of Heligoland.

L-14, commanded by Hauptmann Manger, made her landfall on the Lincolnshire coast near Friskney at 12.45 a.m. Passing slowly down the coast, she was near Old Leake at 1.10 a.m. and in the vicinity of Boston for nearly an hour, but finally went inland in a north-easterly direction, passed Coningsby about 2.20 a.m. and at 2.30 a.m. was north of Billinghay. At 2.40 a.m. she dropped five HE and seven incendiary bombs at Blankney Dales, doing no damage, and then turned sharply east to Kirkstead, where at 2.45 a.m. an HE bomb fell, breaking a window.

Turning northward at 2.50 a.m., L-14 dropped eleven HE bombs at Stixwould, five of which failed to explode. One horse and three sheep were killed, but no further damage done. A HE bomb was dropped at Bucknall, where her course was altered north-east, and an incendiary bomb was dropped in the neighbouring village of Horsington at 2.55 a.m., to no effect.

Continuing on her north-east course, the Zeppelin dropped four HE at Goulceby and one HE at Stenigot at 3.05 a.m., doing no damage. Her course was then altered eastward and another HE bomb was thrown in Burwell Wood, south of Louth at 3.10 a.m., doing no harm.

Having now got rid of all her bombs, the Zeppelin made for the sea and went out at Mablethorpe at 3.20 a.m.

L-22 passed east of *Haisborough* light vessel at 9.10 p.m., going south-west, and dropped four HE bombs in the sea off Bacton on being fired at from that place. She sheered off at once from the coast in a north-west direction, and appears to have returned back to her base, going not by the direct route but straight across to the Dutch coast first, near the island of Texel, and following the Frisian Isles. It was thought that she had been hit by the Bacton gun and received sufficient injury to return by the safest route.

L-17, under the command of Kapitänleutnant Hermann Kraushaar, dropped a flare at 1.35 a.m. to enable sight of the coastline, followed by three HE bombs off the Norfolk coast between 1.40 a.m. and 1.43 a.m. She then came in over Weybourne at 1.45 a.m., passed over Baconsthorpe at 1.50 a.m. and Hindolveston at 2.10 a.m., where she altered course to the eastward to Guestwick. Turning south-west near Reepham at 2.20 a.m., she was east of Dereham at 2.30 a.m. and Shipdham at 2.35 a.m., and remained in that area for some time.

L-17 was next spotted over Binham, where she turned north-east and dropped a single HE bomb when passing over Marlingford at 3.10 a.m., quickly followed by another that landed at Easton, neither of which did any damage. Now flying eastward fast, she was north of Norwich at 3.12 a.m. passing near Blofield at 3.16 a.m., Salhouse at 3.20 a.m. and, while over Martham at 3.30 a.m., she was seen to change course south-east and passed out to sea at Caister five minutes later. Passing over *Cross Sand* light vessel at 3.45 a.m., she was heard to drop seven HE bombs into the sea five minutes later.

FIGHTING ZEPPELINS: A PILOT'S PERSPECTIVE

This fascinating account was written by a pilot who had flown in action during the great Zeppelin offensive of 1916. It is also worthy of note that he placed more store in bombs than machine guns (or did not wish to give away the new secret weapon for Zeppelin killing – the Brock-Pomeroy exploding phosphorous bullets). Sadly he remains anonymous, for the account was published under the name of 'An Air Pilot' in 'BP', the works magazine of Boulton & Paul Ltd, Norwich, in December 1916:

FE-2b night fighter ready for a Zeppelin patrol. (RAF Museum, Hendon)

'Fourteen or fifteen airships participated in the attack on Great Britain last night; two of the raiders were brought down.' Hard official words these, that read in cold black and white of print, fail entirely to bring to the reader's mind a true idea of the romance, the danger and nerve wracking conditions under which this novel form of warfare is fought out.

Let us imagine, if we can, the difficulties the aeroplane pilot has to face. It is dark, pitch dark, sky and earth are alike indistinguishable. Flying at the best of times contains more than a comfortable element of danger and in the darkness is accentuated. The darkness deprives the air pilot of all senses of direction and of locality, greatly hampers him in the manoeuvring of his craft and renders unpleasantly possible a collision with another aeroplane on similar errand bent.

Starting out there are a hundred and one small details to be attended to, as the testing of the engine, the trying of elevators and ailerons and the examination of petrol and the oil tanks in order to ascertain if there is a sufficiency of both to last a two or three hour trip. All this to be performed in the pitch dark, with the engine screeching loud so that a man may not hear a word and the attendant mechanics indistinguishable in the gloom. Fortunately for the pilot a small dry-cell electric lighting set is installed in the body of every machine and by this means the pilot is able to distinguish his instruments – a most necessary adjunct to safe flying – as the altimeter which records the height, the revimeter which indicates the speed of the engine and the compass, more necessary than any other instrument for night flying.

Getting off from the ground is by no means a pleasant sensation. There are hangars, high roofs and chimney stacks all waiting to be collided with, patches of thin and rarefied air, which will bump the machine down as much as 30ft at a time; the ever present danger of engine failure, necessitating a descent to the darkened earth beneath, always so full of death traps for the airman and his craft.

Clear of the earth at about 1,000ft there are, here and there, faint patches of light and dark grey and the subdued reddish glow of the distant metropolis; the locomotive of a passenger train, bright as a searchlight for a brief moment, then passing away into the outer darkness. Higher and yet higher and the sensation! The mind of a Jules Verne or a H.G. Wells could not imagine a feeling more eerie, more strange than this. Noise and darkness, the incessant deafening purr of the engine, the pitch blackness on all sides, relieved by the one tiny light inside the fuselage as welcome and cheery to the airman as a distant lighthouse to a sailor in a storm.

Then the searchlights begin to blaze, creeping up across the sky in ribbons of shining brightness. One plays for a moment on the machine, the pilot is almost blinded, before it passes on, in its strange search across the heavens. For an encounter with the raiding airship is not at all probable at an altitude of below 6,000ft and from that height up to 15,000; the only likely encounter is with the observation car of a Zepp. This car is usually suspended hundreds of feet beneath the mother craft by means of a stout aluminium cable or cables; is about 7ft by 5ft, composed entirely of aluminium and contains sufficient space for one observer, who is in telephone communication with the commander.

At last the pilot of the aeroplane has a feeling; he cannot hear, because of the noise of his own engine and he cannot see because of the intensity of the darkness all around him that the Zeppelin is near at hand.

The combat between the aeroplane and the Zeppelin might be compared to that between a British destroyer and the German Dreadnought in the recent Jutland battle; for once above the Zeppelin, the aeroplane pilot can use his bombs, which are considerably more effective than a machine gun, and the broad back of the gas-bag offers a target which can hardly be missed.

With regard to the matter of manoeuvring, the aeroplane has the great advantage of being remarkably quick both in turning, climbing and coming down, whereas the zeppelin is a slow and clumsy beast at the best of times. The zeppelin is very susceptible to flame and explosion of any kind; the gas in the envelope, a mixture of hydrogen and air, forms an extremely explosive mixture. The aeroplane, owing to the fabric of which is composed and the petrol needed for propulsion is, to a certain degree, inflammable, but not nearly to the extent of the airship. Per contra the airship possesses a distinct advantage in that it is able to shut off its engines, and to hover, which is impossible for an aeroplane to do. Again in the matter of speed in a forward direction, and for that matter backwards also, for the Zepp engines are reversible – the aeroplane holds palm.

The combat finished the aeroplane pilot has yet to make a landing, surely the most dangerous and tricky manoeuvre of the whole flight. The difficulties and dangers thus encountered are too obvious to need explanation, further than to say that the landing has to be effected in the dark, with only an electric ground light for guidance.

27/28 November 1916

Ten airships left north Germany in the early afternoon of 27 November, under orders to attack the Midlands and Tyneside. L-24 and L-30 failed to make landfall over England.

L-34 attacked the Durham coast in company with L-35 and L-36. She came in over Black Halls Rocks at 11.30 p.m. and went inland over Castle Eden, where she was picked up by the Hutton Henry light at 11.34 p.m. When just beyond the light she steered south-south-east, and arrived over Elwick, turned east-south-east and dropped thirteen HE bombs on the searchlight. This demolished a cow shed, injuring two cows, and cut the telegraph wires, but resulted in no damage to the light, although three bombs fell in the same field about 150 yards from it.

At this point, the Zeppelin was attacked by Second Lieutenant Ian V. Pyott RFC, who had taken off from Seaton Carew in BE-2c 2738 and saw her between Sunderland and Hartlepool, in the beam of the Hutton Henry light, in which she was held throughout her course. L-34 was going south towards him. At this moment, the aviator was at 9,800ft and the Zeppelin seemed to be a few hundred feet below him. She then rose rapidly.

Pyott flew towards her at right angles and then passed underneath her amidships, firing as he went. He swung quickly round again to follow the Zeppelin, which was by that time turning east and bombing the Elwick light, probably in an attempt to elude the aeroplane. Aeroplane and Zeppelin then flew on parallel courses for about 5 miles towards the sea, 200 yards apart – L-34 flying at a speed of approximately 70mph. Pyott then concentrated his fire on a single spot on her port quarter, firing in all seventy-one rounds, until at 11.46 p.m., when directly over West Hartlepool, a small patch of the Zeppelin's envelope became incandescent. The fire spread rapidly, until the whole Zeppelin was in flames. The aviator dived instantly to avoid the fall of the blazing L-34 but, although at a distance of 300 yards from her, his face was somewhat scorched by the heat.

The Zeppelin fell almost vertically. Some large object, probably a gondola, was seen to fall from her centre and then the envelope separated from the framework and both fell flaming into the sea well clear of the town, about 1,800 yards east of the Heugh lighthouse. The precise time of the fall of L-34 may be fixed at 11.50 p.m. A portion of her was still burning on the water after midnight but, in half an hour, no trace of her was left but a scum of oil on the surface of the water.

Just before catching fire, L-34 had begun dropping bombs on West Hartlepool. Sixteen HE bombs fell, which did considerable damage but caused few casualties. A large number of windows in houses and shops were broken, and a grandstand in a football field was demolished. Two women died of shock, and three men, six women and four children were injured – one of these men and one of the women subsequently died.

As the raider passed over the town, she came under heavy fire from the AA guns at Hartlepool and Seaton Carew. Those at Port Clarence and Billingham also fired a few rounds, but these guns, which were 6 miles south of Hartlepool and Elwich respectively, were too far off to be able to reach their target. The Hartlepool Cemetery gun, on the other hand, no doubt got very near her, and it was thought possible that at least one shell hit L-34 at about the same time she was set on fire from the machine gun fire of Second Lieutenant Pyott.

3787 D   ZEPPELIN RAID ON THE N.E. COAST. NOV. 27th. 1916.   ROTARY PHOTO. E.C.

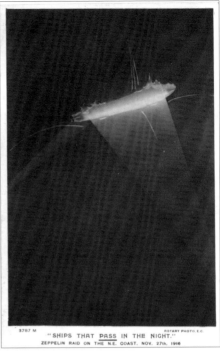

3787 M   "SHIPS THAT PASS IN THE NIGHT."   ROTARY PHOTO. E.C.
ZEPPELIN RAID ON THE N.E. COAST. NOV. 27th. 1916

3787 K   "THE STRAFER STRAFED"   ROTARY PHOTO. E.C.
ON THE N.E. COAST. NOV. 27th. 1916.

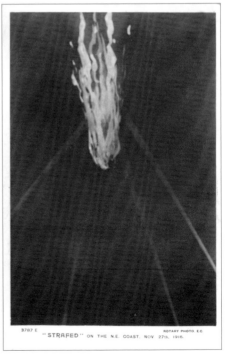

3787 E   "STRAFED" ON THE N.E. COAST. NOV. 27th. 1916.   ROTARY PHOTO. E.C.

Four postcards depicting the stages in the destruction of L-34, 27 November 1916.

The monitor *Prince Rupert*, lying off the Tees, prepared for action but was unable to fire owing to the position of the Zeppelin in relation to the town. A naval aircraft, piloted by Flight Lieutenant de Roeper, went up from Redcar in pursuit of the raider but arrived on the scene after the destruction. Had he pursued his course out to sea he might have caught up with L-35.

The flare of the burning Zeppelin was seen as far north as Matfen, near Morpeth, and as far south as Poppleton, north of York, from the ground. From the air it was seen by a pilot flying near Melton Mowbray, 136 miles away.

L-35 came in at Hawthorn at 11.36 p.m. and went inland for 10 minutes, passing over Hesledon. She then turned round and headed out to sea again north of Easington at about 11.45 p.m. It is thought that she fled upon observing the aeroplane attack on L-34 about 10 miles to the south of her. She returned home at once, at high speed, without having dropped any bombs. She had been picked up and fired upon by the Seaham gun, which confused L-35 with L-34 and claimed the fall of the latter for their gunnery. In fact L-34 was 13 miles away – only L-35 was within their range.

L-36 never came overland. She was about 15 miles off the mouth of the Tyne at 11.20 p.m. and no doubt intended to raid the Sunderland/Newcastle district, but the catastrophe of L-34 occurred before she reached the coast. On seeing the burning ship she immediately went away northwards, dropped an uncertain number of HE bombs in the sea off Alnmouth at 12.30 a.m. and then made her way back to base.

L-13 came in overland in company with L-22 at Flamborough Head at 10.05 p.m. The two ships passed Bridlington at 10.07 p.m. and Burton Agnes at 10.10 p.m. They then separated, with L-13 passing Langtoft at 10.30 p.m. and going south-west, being next heard of at Pocklington about 10.50 p.m.

Shortly before 11 p.m. she dropped an incendiary bomb at Yapham, 10 miles east of York, followed by twenty-one HE and three incendiary bombs in open fields in the parish of Barmby Moor, 1 mile nearer York, doing no damage and causing no casualties. At 11.05 p.m. L-13 passed Stamford Bridge, and at 11.15 p.m. approached York from the south-east. She was engaged by the Acomb gun, which drove the Zeppelin off to the east, but not before she had dropped two HE and twenty-one incendiary bombs on the outskirts of the city, wrecking one house and damaging several others, and slightly injuring a man and a woman.

At 11.35 p.m. the Zeppelin passed Pocklington going east and, between 11.55 p.m. and midnight, was near Driffield. At 12.10 a.m. she passed Langtoft. She dropped five incendiary bombs at Wold Newton, fifteen minutes later and then went north-west to Snainton. She was seen north of East Ayton at 12.45 a.m., and finally went out to sea north of Scarborough under fire from the AA guns, at 12.50 a.m., dropping bombs in the sea as she went.

L-22, after separating from L-13 near Burton Agnes, passed Driffield around 10.25 p.m., going south, and passed between Market Weighton and South Cave at about 10.40 p.m. At 10.45 p.m. she moved north of Holme-on-Spalding-Moor, and at 10.52 p.m. was picked up by the Willingtoft light and gun. After two rounds had been fired, she suddenly turned north-west, circled round to the north-east and then went off in a south-easterly direction.

Her commander seems to have been uncertain of his whereabouts, and proceeded slowly, probably drifting with engines shut off. Crossing and re-crossing over the Humber at 12.15 a.m. she passed near Skirlaugh, going east, and appears to have gone out to sea at about 12.20 a.m. at Hornsea, under fire from the guns at Cowden.

No bombs were dropped on land by L-22. Her dip southwards over the Humber with engines cut off was thought to have been caused by the appearance of aeroplanes, which she was anxious to avoid.

L-14 remained just a short while overland. She came in at Tunstall and then turned north-east towards the coast at 9.20 p.m. She was engaged by the mobile AA guns at Cowden. She at once dropped a large number of bombs in answer to the guns, while the latter were firing. In all, eighteen HE and twenty-six incendiary bombs were thrown, falling just north of Cowden, between the village of Mappleton and Rowlston Hall, and all to no effect.

The Zeppelin immediately went off westward at high speed and made a wide circle northward towards Barmston where, at 9.28 p.m., the mobile AA guns came into action against her. She turned west again and then south, dropping on her way two petrol tanks, one at Moortown about 9.50 p.m. and the other at Brandesburton, north of Leven.

The Zeppelin apparently shut off her engines and drifted south-east towards Hull, before being picked up and fired upon by the Sutton gun at 10.15 p.m. She passed on to the southward, and then turned east and out to sea north of the Spurn at approximately 10.25 p.m.

Around 11.45 p.m. she was thought to have been responsible for dropping bombs in the sea off Cromer that were heard at that time. Her return to the coast with engines shut off, as silently as possible and without attempting to pursue her journey inland, very soon after she had bombed Mappleton, was ascribed to the appearance of the aeroplanes that had gone up from Elsham. These aircraft also considerably influenced the movements of L-22 and L-21.

L-16 crossed the coast at Filey Bay at 9.20 p.m. and proceeded directly south-west, passing Langtoft at 9.32 p.m., Huggate at about 9.45 p.m. and south of Pocklington at 9.55 p.m. At 10 p.m. she was heard north of Holme-on-Spalding-Moor, and at 10.10 p.m. reached the line of guns between Howden and Selby, being fired upon by the Hemingborough Grange gun.

She turned north-west, and was immediately picked up by the Cliffe gun, which opened fire. She changed course again, south-west, and at about 10.15 p.m. came under fire from the Woodhouse gun. The Zeppelin then passed over Barlow Aircraft Works, turned south-east and increased her speed considerably in order to get out of this dangerous neighbourhood as soon as possible. She did not attempt to drop bombs, as her commander evidently wished to avoid attracting aeroplanes in his vicinity. When out of range, she turned west again passing over Temple Hirst at 10.20 p.m. and then south-west, passing Wormersley around 10.25 p.m.

At 10.30 p.m. she approached the city of Wakefield, coming down to a low level. She passed over Heath Common and off south-east to Sharlston where, at 10.32 p.m. three HE and six incendiary bombs were dropped, doing no damage and causing no casualties. At 10.35 p.m. she passed Hemsworth, turned south and south-west between Grimethorpe and Great Houghton, then round to Cudworth where, at 10.40 p.m., an incendiary bomb was dropped.

She headed westward to Monk Bretton, where two HE bombs were dropped with no effect. Then, turning off abruptly to the north-north-east at 10.45 p.m., she dropped two HE and two incendiary bombs, neither of which ignited, at South Hiendley, again doing no damage whatever. Continuing on her course, she reached the neighbourhood of Pontefract shortly before 11 p.m. and dropped four HE and four incendiary bombs near the town, causing no casualties and only breaking a few windows. The incendiary bombs failed to ignite.

L-16 then turned south-west towards Featherstone where she threw six HE bombs, which did no damage. Then, after circling in the direction of Knottingley, she resumed her northerly course. At 11.15 p.m. she dropped an incendiary bomb at Lumby and another at Monk Fryston, west of Selby, doing no damage. She passed Sherburn-in-Elmet between 11.15 p.m. and 11.20 p.m. and Tadcaster at 11.27 p.m. Here she veered off north toward York, and at 11.35 p.m. approached the city from the south-west. The Acomb guns came into action, and the Zeppelin made no attempt to reach York, but at once sheared off in a north-west direction towards Beningbrough.

No further bombs were dropped until about 12.15 a.m. when she dropped one HE bomb at Birdsall's West Farm, Helperthorpe, and one at Jepson's farm, Boythorpe. Neither of these exploded as they had no fuses. Five HE bombs were then thrown at Foxholes, and these last damaged a house roof and broke a few windows.

The Zeppelin then proceeded north-east to Seamer and went out to sea south of Scarborough, under fire from AA guns at 12.35 a.m.

L-21, under the command of Oberleutnant zur See Kurt Frankenberg, made its first landfall at Atwick, East Riding of Yorkshire, at 9.20 p.m. and immediately came under fire from the Barmston guns. After steering north-east and out to sea she soon returned and evaded more anti-aircraft fire as she moved up to the Potteries. At 11.34 p.m. she approached the Leeds defences, and the Brierlands gun opened fire on her at that time. Her height was estimated at 12,000ft. Owing to a ground mist, the searchlight had difficulty picking her up but the target was easily visible over open sights and ten rounds were fired. The Zeppelin made no attempt to reply but went on in her south-west course, and was engaged a few minutes later by the gun at Rothwell Haigh, which fired four rounds at her under difficult circumstances, the target being only faintly visible for a few seconds.

L-21 then turned away from the gun and Leeds, going south at great speed. At 11.48 p.m. she dropped her first bombs, one HE and two incendiaries at Sharlston, east of Wakefield, where an hour before L-16 had also dropped bombs. Lights of some kind were probably visible as a result of the previous attack. The HE bomb fell on a siding, doing no damage.

At 11.53 p.m. the Zeppelin passed Royston going south-east, and was hovering over Barnsley at 12.10 a.m. At 12.15 a.m. she dropped one HE and two incendiary bombs at Dodworth, 2 miles to the south-west. She then went in the direction of Penistone, and was next reported at Derwent at 12.30 a.m., passing south-west over the peak.

Around 12.50 a.m. she threw a single incendiary bomb at Pott Shrigley, 4 miles south of Disley, which fell near a brickworks, doing no damage. The Zeppelin was probably attracted by the brick kilns which may have shown some light. The Zeppelin continued to travel in a southerly direction and was heard passing over Bollington at 12.55 a.m. to Macclesfield, where she was seen at 1 a.m. She hovered over the town, which was in total darkness, for two minutes and then went on over Congleton at 1.05 a.m. to Birchenwood near Kidsgrove, where a single HE bomb was dropped at 1.10 a.m., which did no damage.

Next she dropped three HE bombs at Goldenhill and three at Tunstall. These last wrecked two houses, demolished three outhouses and damaged three others. Glass was broken in sixty cottages and a man was injured.

The Zeppelin, attracted by the glare of the slagheaps, then went off westward over Chatterley to Chesterton where, at 1.20 a.m., sixteen HE and seven incendiary bombs were dropped, doing no damage beyond the breakage of a little glass, and causing no casualties. Having passed by Newcastle-under-Lyme and Hanley, the raider dropped four incendiary bombs aimed at the collieries between Fenton and Trentham, three of which failed to ignite.

Of the bombs dropped, one 50kg HE bomb at Goldenhill did not explode. The bombs at Chesterton were more than likely attracted by a number of ironstone-burning hearths which were showing a good deal of light.

L-21 had, by this time, reached the limit of her westerly course and turned for home. Passing over Castle Donnington towards Nottingham, she shut off her engines and drifted in a south-easterly direction over Beeston at 2.11 a.m. and over Ruddington at 2.18 a.m., where a searchlight picked her up, with the result that she immediately started her engines again and went off to the east.

At about 2.50 a.m. she approached the searchlights north of Peterborough, heading over Buckminster at 2.58 a.m., and was found and chased by two aeroplanes, which first drove her in the direction of Bottesford, then down towards Bytham, then north again, towards Grantham and south to Essendine where, at about 3.25 a.m., she finally escaped to the eastward. Both aeroplanes were very near the Zeppelin, but did not succeed in getting quite close enough to engage her with any likelihood of success, though one actually fired on her. The Zeppelin also used her machine guns.

L-21 regained her direction after she had shaken off her assailants and made off eastwards, passing Spalding at 3.35 a.m., Wisbech at 3.45 a.m. and King's Lynn at 3.50 a.m. Here, her commander slowed down and verified his position, probably recognising the proximity of the Wash, and intending to go out to sea by it. Proceeding slowly to Narborough at 4.14 a.m., he went north to Hillington at 4.20 a.m.

Perhaps wondering if his petrol supply would carry him across the North Sea, Frankenberg changed direction and appeared to be making for Belgium. The north-west wind carried the airship along at a slow rate in an east-south-east direction, past Swaffham at 4.40 a.m. and East Dereham at 4.55 a.m.

About this time, L-21 was almost caught by an aeroplane piloted by Lieutenant W.R. Gayner of RFC Marham, who had been attracted to the Zeppelin by a light which she was showing. She was at a height of 7,500ft but, when almost within striking distance of her, the aviator's engine revolutions dropped and he was compelled to land at Tibenham. L-21 had had another lucky escape and made off north-eastward in the direction of Reepham, where she arrived at 5.05 a.m.

The Zeppelin then drifted slowly over Taverham at 5.15 a.m., and was then spotted north of Norwich at 5.30 a.m., Wroxham at 5.40 a.m., and Acle at 5.50 a.m. She headed to the coast, where she appeared north of Great Yarmouth at 6 a.m., drifting at a great height over the post at 6.05 a.m., when she was fired upon by the anti-aircraft guns at Bradwell and by a monitor in the Roads.

At 6.18 a.m. she passed east of Lowestoft, still under fire, and hovered near the coastline for some minutes. A certain amount of firing took place from a 6-pdr naval gun mounted at Lowestoft, and also from the 12-pdr naval field guns mounted for high-angle firing at the north end of the town. Fire ceased as soon as it was realised that the Zeppelin was not within effective range. Owing to the failure of the searchlight on land, the height of the Zeppelin could not be gauged properly. Twenty-three rounds were fired and a hit was claimed with the second, when the raider was over Lowestoft. The Zeppelin quickened her pace and went out to sea at 6.30 a.m. at a speed which had increased to 35mph.

As day was breaking, the Zeppelin was visible by ordinary observation without the aid of the searchlights as she made out to sea at 6.30 a.m. Meanwhile, both military and naval aeroplanes were on her track. Owing to the approach of daylight they could clearly follow the Zeppelin, which was steering east-south-east from Lowestoft. At 6.35 a.m., two naval aeroplanes, a pair of BE-2cs – machine no 8625, piloted by Flight Lieutenant Egbert Cadbury, and machine no 8420, with Flight Sub Lieutenant G.W.R. Fane – had risen from RNAS Burgh Castle on the approach of the airship from the west, and overtook her a few miles east-south-east of Lowestoft.

The destruction of L-21 made the cover of *The Sphere* on 9 December 1916.

The following description of the ensuing combat is taken from the reports of the aviators: Lieutenant Cadbury got under her (L-21) at about 700ft distance and fired his Lewis gun into her after part, being at the same time under heavy machine gun fire from the airship. His first magazine not having the desired effect, he changed magazines and repeated his action, still without effect. He put four magazines through his Lewis until his ammunition was exhausted.

The airship commander increased his speed to 55mph on being attacked. On perceiving that Cadbury had exhausted his ammunition, Flight Sub Lieutenant Fane then approached to within 100ft of the starboard side of the airship and tried to open fire, but his Lewis gun jammed owing to the cold and the oil having frozen.

Flight Sub Lieutenant Pulling, who had been on patrol in BE-2c 8626 from RNAS Bacton, witnessed the two previous attacks and then approached to with 60ft of the airship. He opened fire under a continuous fusillade from those of the airship's machine guns that could bear on him. After he had fired ten rounds the airship caught fire, and in a few seconds was 'nothing but a fiery furnace.' Pulling immediately dived to starboard to avoid falling debris and, even though they were ablaze, the crew of the airship continued to fire at him for some appreciable time.

The Zeppelin then fell stern foremost into the sea, where she sank at once, shortly after 6.40 a.m., leaving no trace but a large area of oil-covered water about 8 miles east of Lowestoft. There were no survivors. L-21 had been attacked at 8,200ft and she took a little over a minute to fall from that height into the water.

The Bradwell guns had begun firing at her again at 6.35 a.m. although she was nearly 16,000 yards away, and claimed to have hit her with their twentieth round which was fired at 6.40 a.m., just before she fell. However, if the position given by the pilots was correct, some 15 miles off, there is no doubt that this round did not hit her, and that these guns had no part in her destruction.

With the crash of L-21 into the sea came the final conclusion to a dogfight that had been watched by thousands of spectators who gathered along the sea fronts of Great Yarmouth and Lowestoft. The *Eastern Daily Press* captured the atmosphere on the ground:

There were no signs of the invaders in this district until about six o'clock in the morning when the people were disturbed by the booming of guns. Shortly afterwards the Zeppelin, a huge monster, could be seen with the naked eye blotting out the dazzling stars as she was speeding her way seawards … The hundreds of people who were gazing, particularly those who had the use of glasses, got a fine view of the airship as the stars disappeared and dawn approached. Now and again the guns would cease and there were mingled feelings among the onlookers. Some tried to shout because they thought she was hit, others groaned, for they feared she would escape.

Flight Lieutenant Cadbury (left) and Flight Sub Lieutenant Pulling.

On land, the spectators were not sure if the gunners had scored a hit or the aircraft had achieved a similar effect, but there came a moment when it appeared that the Zeppelin had become disabled, nor was she at any great altitude or elevation; another correspondent who witnessed the scene noted:

There she hung, going a little one way and then another but never making any headway. Suddenly a bright flash came from her and the assembled crowds

The end of a Zeppelin in the North Sea Nov. 28. 1916.
(passed for publication by the press Bureau Nov. 29:16)

E & H Scoones

A photographic artist recreated the destruction of L-21 for this postcard.

shouted 'She's hit.' They were right and in the second or so succeeding the flash flames burst all along the Zeppelin, which dropped into the sea in a cascade of red and gold, leaving a trail of smoke behind.

There had been a spell of the gravest anxiety, and then the whole town rang out with one great spontaneous cheer and the ships sounded their sirens freely, for, as the *Eastern Daily Press* reporter concluded:

> The air monster, which had given the gunners the hardest twenty minutes of their lives, burst into flames and fell into the sea. How far she was away did not matter; our seaplanes had fixed her and all that could be seen a few minutes later was a large cloud of black smoke which gradually gave way to the grey dawn.

28 November 1916

The First Aeroplane Raid on London

The first raid by an aeroplane on London was made in a LVG, with a 225hp Mercedes engine, piloted by Deck Offizier Paul Brandt and with Leutnant Walther Ilges as his observer.

The aeroplane left Mariakerke on the morning of 28 November with the object of attacking the Admiralty offices in Whitehall. The weather was fine, but hazy and the enemy aeroplane, flying high, was reported by no more than two people and these did not identify it as German.

The first that was known of its offensive nature was when bombs began to fall, just before midday. Six 10kg bombs exploded between Brompton Road and Victoria Station. They did little damage, but injured ten people.

The observer took twenty photographs of aerodromes, military camps and ammunition works on his way to London. These photographs were destined not to be developed. On the homeward journey the engine gave trouble, the camera and plates were thrown overboard and the aeroplane was eventually landed near Boulogne, where it was captured.

THREE

14 February 1917

Just after 8 a.m. on 14 February an LVG or Albatross-type aeroplane was observed from Deal AA gun station, who fired eighteen rounds of HE at the raider from their 12-pdr AA gun. The aircraft had been flying very high and came out from behind a cloud. Fire was opened at 8.08 a.m. and ceased at 8.12 a.m., the target was receding, at an estimated height of 10,000ft. No hit was observed, but the shooting was reported to be very good. Shipping in the Downs also opened fire. No naval or military aeroplane ascended in pursuit.

16/17 February 1917

LZ–107 was first located over France at 12.05 a.m., when her engines were heard at sea from Dunkirk. She dropped the majority of her bombs near the aviation ground of Les Baraques, to the south of the Calais–Sangatte road, and a second group in the sand dunes near the remount depot of the Belgian Army. A further eight incendiary bombs were also dropped on the shore just below high–water mark.

She was last seen on the French side of the channel at 1.40 a.m. moving towards Dover, and was next sighted by a trawler off the Goodwins and reported to be 5 miles off Dover. She was seen over Deal and Kingsdown at about 2 a.m., and last heard from Ramsgate and Foreness a quarter of an hour later. No bombs were dropped on English soil and the Zeppelin was not fired upon.

16/17 March 1917

Five of the latest naval Zeppelins left to conduct the raid on 16 March 1917, namely L-35, L-39, L-40, L-41 and L-42.

L-42, commanded by Kapitänleutnant Dietrich, did not attempt to raid but hugged the Dutch coast and then crossed Belgium, returning to base after cruising for nearly two hours in the vicinity of Zeebrugge.

The four remaining Zeppelins attacked Kent and Sussex. The effect of high winds during the raid seems to have completely upset the German plan of attack; the raiding Zeppelins lost their true sense of direction and blundered badly in making their way home.

L-39, under the command of Kapitänleutnant Robert Koch, was first to reach the coast, approaching from the direction of *Kentish Knock* and coming in over Margate at 10.20 p.m., going west. Her rate of progress was first slow. At about 10.30 p.m. she passed Birchington, going south-west and then, quickening her

Kapitänleutnant Robert Koch and the crew of L-39, who perished after their raid on England on 17 March 1917.

pace, passed over Upstreet to Bekesbourne where, at 10.50 p.m. she dropped a single HE bomb at Hode Farm, slightly damaging two cottages.

Going on south-west she dropped five HE bombs and one incendiary at 11 p.m., between Waltham and Sole Street, doing no damage. The raider then gathered speed. She passed Ashford at 11.03 p.m. and was over St Leonards at 11.40 p.m. recorded as 'making a terrible noise.' At 11.52 p.m. she was over Bexhill and at midnight over Pevensey Bat where she went out to sea and is believed to have been heard ditching some of her bombs while out at sea, possibly aiming them at some vessel from Cuckmere Haven.

Travelling over France, it was at 5.25 a.m. that L-39 was observed north of Estrées, where she dropped three HE bombs. Five minutes later she was stationary over Compiégne where she came to a standstill, either by a failure in her petrol supply or an engine breakdown. Here she was bombarded for fifteen minutes by three batteries of French AA guns which, owing to the difficulty of observing their fire in the half-light, fired as many as 106 rounds of incendiary shells at her.

L-39 came down in flames from a height stated to be over 10,000ft at 5.55 a.m., when only 8 miles from the German lines. The entire crew perished. It was a serious blow to the German air fleet, Kapitänleutnant Koch and the crew of L-39 possessed over twenty months experience of Zeppelin work, having previously manned L-24 and SL-3.

L-35, commanded by Kapitänleutnant Herbert Ehrlich, crossed the coast at Broadstairs at 10.40 p.m. and pursued much the same south-westerly course that L-39 had taken. At 10.55 p.m. she dropped her first bomb, an HE at Ickham, which did no damage. She followed it up at 11 p.m. with one HE and one incendiary at Nackington, also to no effect. L-35 passed Ashford at 11.15 p.m., but her commander then either lost his way or considered that a speedy return was advisable on account of the high winds.

She circled north-west in the direction of Charing and then round to the east, passing over Wye Aerodrome and, at 11.35 p.m., dropped five HE bombs and one incendiary at Crundale. These brought down a ceiling in a cottage ¼ mile away.

Some of the crew dropped facetious postcards near Elham, which eventually proved of interest in confirming the identification of the Zeppelin. L-35 then made off eastwards, and at midnight dropped five HE bombs, two of them 300kg in weight, in the parish of Swingfield. Slight damage was done to a few windows and ceilings in two farms.

Going on in the same direction, four incendiary bombs were dropped at Houghton, one 300kg HE bomb at Whinless Down, ½ mile from Dover, and one incendiary, also near Dover, none doing any damage.

The raider then went out to sea west of Dover at 12.15 a.m. She crossed the French coast between Calais and Gravelines at about 12.40 a.m. and went home over Belgium, passing Brussels at 3 a.m. and making for the neighbourhood of Koblenz. She finally reached Dresden in a crippled condition, one of her engines having met with a serious mishap.

L-40, commanded by Kapitänleutnant Erich Sommerfeldt, seems to have followed in the wake of L-39, at a gradually increasing distance, as her speed was slower. At 12.50 a.m. she was reported off the coast at Herne Bay and at 1 a.m. she was distinctly seen there coming in over the coast. Shortly after 1.40 a.m. she dropped a single HE bomb at Nackholt, south-east of Rye. The concussion knocked a few tiles off the roof of a cottage.

L-40 then went on southwards to Newchurch, in Romney Marsh, where at about 2 a.m. five HE (one of 300kg and four of 100kg) and three incendiary bombs fell, damaging one pane of glass in a farmhouse. The raider travelled directly south, dropped fourteen incendiary bombs near Little Appledore Farm, Newchurch, doing no damage, followed by one HE bomb (100kg) at Melton Farm, Ivychurch, which killed four sheep.

One HE bomb of 50kg weight then fell at Yokes Court, St Martin's, and one 100kg and one of 50kg in a field at St Martin's, followed by one 300kg, two 100kg and one 50kg in field ½ mile north-east of Old Romney. The raider then turned due east, and at about 2.15 a.m. went out to sea again near New Romney.

L-41, commanded by Hauptmann Kuno Manger, came overland at 1.20 a.m. at Pett Level, south of Winchelsea. As she came in she dropped eight HE and two incendiary bombs. Two of the HE bombs appear to have dropped in the sea.

Kapitänleutnant Herbert Ehrlich and his air crew and ground crew for L-35, 1917.

The others did some damage by concussion to two empty bungalows and two farmhouses in the neighbourhood. The Zeppelin then went slowly north-east and when near Rye, turned south-east along the course of the River Rother, dropping seven HE and six incendiary bombs on Camber Marsh between 1.40 and 1.45 a.m. Of these, two HE bombs and one incendiary fell on the right bank of the river between Rye and the Chemical Works, and the rest on the left bank near Rye harbour. One of the HE bombs broke the windows of a couple of bungalows, otherwise no damage was done.

The raider passed sharply north-east past Guldeford and seems to have followed the Royal Military Canal for some distance, afterward turning east across Romney Marsh to the coast and going out to sea near Dungeness at 2.05 a.m.

Numerous aircraft from the RNAS and RFC took to the air that night, none of them managed to spot the raiders. Second Lieutenant David Dennys Fowler took off in his BE-2e 7180 from Telscombe Cliffs at 12.01 a.m. About ten minutes later Fowler's plane crashed 1½ miles from the aerodrome, killing him. He was buried with full RFC honours at Rottingdean (St Margaret) churchyard. He was just 20 years old.

1 March 1917

At 9.40 a.m. on 1 March 1917, a hostile aeroplane appeared over Broadstairs. It dropped three bombs on the sea, about 50 yards from Clockhouse Point, followed by six others on land. A cottage in Gladstone Road was wrecked, and four houses and the county school in the same street were slightly damaged. Several other houses, chiefly in Grosvenor Road and King Edward Avenue, had their windows broken. One woman and five children were slightly injured next to the County School. The aeroplane went west for about 1 mile and returned out to sea.

The raider was not fired at from Thanet. However, owing to information not having been communicated to the London AA defences about British aircraft being on patrol in the area, one was fired at by the Crayford Gun, which was twice in action between 11 a.m. and 1.30 a.m.

16 March 1917

At 5.20 a.m. on 16 March a hostile aeroplane, which must have crossed the coast unobserved near Ramsgate, suddenly appeared out of low clouds at Garlinge, near Margate, coming from the direction of Hengrove. It came down comparatively low as the clouds were not too much above 1,000ft, and dropped twenty-one bombs, one of which failed to explode. They did no damage beyond breaking glass at various farms and cottages, and caused no casualties. The aeroplane, which was clearly seen, was a captured British machine of Handley Page type and was described as still bearing its British markings.

The first two bombs fell close to Dent-du-Lion Farm; the third close to Mutrix Farm; the fourth 150 yards east of Westgate Aerodrome; the fifth between the roadway and the railway west of Mutrix Farm; the sixth close to the railway embankment and then ten landed within 100 yards of one another in a field, 700 yards south-east of Westgate Station.

The aircraft then turned north-west, and dropped its seventeenth bomb on a lawn in front of Street Court School; the eighteenth on a greenhouse, 250 yards north of Westgate Church; the nineteenth 20 yards south of the bandstand on the front at Westgate and the twentieth and twenty-first in the sea within 300 yards of the shore. Glass was broken in nineteen houses, the greenhouse and the bandstand. The damage was estimated at £35–£45. All the bombs were small, of 5kg weight.

5 April 1917

On 5 April at 10.03 p.m. a hostile seaplane was heard north-north-east of Deal, going north-west. At 10.12 p.m. it was heard south-west of Ramsgate going north-east and about 10.30 p.m. it came over the coast between Broadstairs and Ramsgate, dropping four HE bombs, which merely broke some panes of glass in a house, doing some damage to the value of £4.

The seaplane then went over the eastern end of the town of Ramsgate, dropping one HE bomb, which fell in a garden, doing no damage. It then appeared near Sandwich and dropped three HE bombs between Stonar Camp (Inland Water Transport) and Shingle End coastguard station, which also did no damage. The raider then made off seawards. No casualties were caused, and no action was taken owing to the suddenness of the attack and low-lying clouds.

19 April 1917

Six hostile aeroplanes or seaplanes approached the North Goodwins at 6.38 a.m. Three of them carried a torpedo each, the other three were apparently intended as escorts. The first torpedo was dropped on the North Goodwin Drifter Division. The six machines then headed off, although one almost immediately returned to renew the attack but was driven off. The seaplane that had dropped its torpedo then went home, accompanied by its escort.

The remaining four aircraft were seen by the armed trawler *Verinie* off the North Brake Buoy at 6.40 a.m., flying very low. One attacked the SS *Nyanza* at 6.45 a.m., dropping one torpedo. Two then appeared off Ramsgate, circled round HMS *Marshal Ney* and discharged one torpedo, which entered Ramsgate harbour and became embedded in mud. No damage was done by the three torpedoes, and no bombs were dropped by the hostile machines. Owing to thick fog, low clouds and bad visibility at all stations, it was impossible to send up machines in pursuit.

7 May 1917

A single aeroplane appeared over the metropolis after midnight on the morning of 7 May, and dropped five 12½kg bombs, which fell between 12.40 and 12.50 a.m. in a straight line between Hackney and Holloway.

The first bomb fell at 12.40 a.m. on Hackney Marshes, to no effect. The second landed at 12.45 a.m. on Newington Green Mansions, exploding in a top flat, and killing a man and seriously injuring a woman. Two rooms were badly damaged. The nose-cap of the bomb passed through three floors and was picked up in the ground floor flat.

The third fell at 12.47 a.m. in the gravel path adjoining 19 Aberdeen Park, Stoke Newington. No damage was done, except to a fowl house and one or two windows. At 12.48 a.m. the fourth bomb dropped in Highbury Fields, halfway between Highbury Terrace and Highbury Place. The bomb burst a water main. The last bomb fell at 12.50 a.m. in Eden Grove, Lower Holloway, but failed to explode after damaging the roof of a bathroom.

The aeroplane was not seen over London, but was heard, and it is thought to have made off in a north-westerly direction, afterwards doubling back to the coast. It was claimed to have been heard in various places in Essex along the probable line of its inward and outward route, but none of the evidence was very definite or satisfactory, except from Romford and Hornchurch, where it seems to have been heard going south-east towards the river. It was supposed that the pilot followed the north shore of the Thames. The bombs were small, 12½kg types.

23/24 May 1917

Six Zeppelins were sent to 'attack south London' on 23 May:

L-44, under Kapitänleutnant Stabbert and carrying Zeppelin chief Strasser, came in at Lowestoft at 2.23 a.m. but rapidly developed engine failure, had to abort the mission and was fortunate to get back safely.

L-42, commanded by Kapitänleutnant Max Dietrich, hovered about between Harwich and the *Sunk* lightship for some time, and finally came overland near

Walton-on-the-Naze about 12.20 a.m. She went westwards at slow speed and at 12.45 a.m. was near Colchester, then reported towards Sudbury and then Halstead where, at 1.30 a.m. she harmlessly dropped an incendiary bomb in a field about 1 mile north-west of the town.

She now turned south-west, and was near Braintree at 1.45 a.m., afterwards going off north-west. At 2.05 a.m. the raider dropped another incendiary and one HE bomb at Radwinter, 500 yards from the church, between it and the vicarage, also doing no damage.

L-42 went off northwards, passing near Haverhill at about 2.15 a.m., near Stetchworth ten minutes later and Newmarket shortly afterwards and dropped one HE bomb at West Row Fen, 1 mile east of Mildenhall at about 2.35 a.m.. It fell ¼ mile away from any inhabitation and did no damage.

At 2.38 a.m. another HE bomb was dropped at Pitts Drover, near Lakenheath, on a grass track about 1½ miles from the latter place, doing no damage.

Crossing into Norfolk at 2.40 a.m. she dropped five HE bombs at Hockwold, causing slight damage to two farm buildings and two fields. At 2.42 a.m. two HE bombs were dropped on Weeting Heath, doing no damage, followed by two more (one of which did not explode) at Cranwich. Soon after, another incendiary fell at Ickburgh, with two more that landed in field at Hilborough. No further damage was done. The Zeppelin then bore off north-east, passing Swaffham at 2.50 a.m., East Dereham at 2.55 a.m., Reepham at 3 a.m. and was seen heading south-east near Hindolveston at 3.13 a.m.

L-42 then turned north, and was spotted at Briningham, Stody and Brinton. It passed between Holt and Bodham at 3.20 a.m. and went out to sea between Weybourne and Sheringham five minutes later.

L-45, under the command of Kapitänleutnant Waldemar Kolle, made landfall at Hollesley Bay, Suffolk, at about 1 a.m. and traversed the county, passing over Wickham Market and Stradbroke. She then came over the Norfolk border, dropping an incendiary bomb at Banham on the way.

L-45 was next reported over Swaffham and Litcham. The weather had taken a turn for the worse, and there was heavy rain and thunder as the Zeppelin passed over Docking at 2.20 a.m. where she dropped one HE and one incendiary bomb on the village, to no effect. The Zeppelin was last heard passing out to sea at Thornham.

On her way back across the sea, L-45 was pursued and attacked by a seaplane (Large America No 8666) from RNAS Great Yarmouth, piloted by Flight Lieutenant Galpin and Flight Sub Lieutenant Leckie. This was at about 6.38 a.m., when she was 10 miles north-east of Terschelling (where they had destroyed L-22 on 14 May). They fired one magazine of ammunition at a range of 300 yards, before the Zeppelin disappeared into the clouds and could not be found again. Galpin and Leckie were also unfortunate enough for their seaplane to run out of petrol on its homeward flight. Safely landed on the sea at Cromer Knoll, both men and machine were returned in tow.

L-40, under Kapitänleutnant Erich Sommerfeld, was first heard off Lowestoft and Southwold at 12.10 a.m. and came inland over Kessingland at 12.18 a.m. Dense, low lying cloud prevented her being spotted or fired upon as she flew overhead. Pursuing a course in the direction of Norwich, the Zeppelin passed Loddon at 12.35 a.m. and Rockland St Mary at 12.40 a.m. Travelling eastward of Norwich at 12.45 a.m., and probably believing he should be over the city, he dropped a 300kg bomb on Little Plumstead. The only damage consisted of glass broken in two cottages and a greenhouse, value estimated at £3 2s 6d.

Heading northwards, the Zeppelin dropped two petrol tanks, one at Horstead at 12.50 a.m. and the other at Worstead at 12.55 a.m. Passing North Walsham at about 1 a.m., L-40 dropped an HE bomb at Knapton, which demolished some telegraph wires for a distance of 50 yards but did no other damage.

Passing out to sea at Mundesley at 1.05 a.m. she dropped the remainder of her bombs in the sea as she went. Fourteen explosions were heard from the *Cockle* and *Newarp* light vessels, growing gradually fainter as the airship proceeded north-east.

L-41, commanded by Hauptmann Kuno Manger, came overland at 2.23 a.m. at Lowestoft. Travelling north inland, at 2.37 a.m. she was heard at Lound and Blundeston, and at about 2.45 a.m. went out to sea again south of Yarmouth, being last heard going east at 2.50 a.m.. She had dropped no bombs on land. Owing to the thick low clouds, it was probably as impossible for the commander to know where he was, as it was for British AA guns to see and open fire on him.

L-43, under the command of Kapitänleutnant Hermann Kraushaar, made landfall at Hollesley Bay, turned north when off Bawdsey and at about 2.20 a.m. came overland near Hollesley. She went inland north-west, past Woodbridge and, at 2.32 a.m. was at Grundisburgh. Pursuing the same course, at 2.40 a.m. she was at Mickfield and Little Stonham, east of Stowmarket, and ten minutes later, between Eye and Mellis. She crossed into Norfolk via Redgrave at 2.55 a.m., dropping a couple of petrol tanks at South and North Lopham soon after. Shortly after 3 a.m. L-43 passed East Harling, going north, and dropped five incendiary bombs at Wrentham five minutes later. No damage was caused.

At 3.10 a.m. another incendiary bomb fell at Tottington, followed by a flare thrown out at Saham Toney. The Zeppelin now circled to the east and south, dropping another petrol tank at Carbrooke at about 3.20 a.m. Steering west towards Little Cressingham, two incendiary bombs were dropped there at 3.25 a.m.

The Zeppelin then resumed her northerly route, dropping an HE bomb at Houghton-on-the-Hill at 3.30 a.m. and two more at North Pickenham, causing damage to crops and to the windows and tiles of a farm house. At 3.35 a.m. six HE bombs fell at Little Dunham, breaking seven panes of glass in a cottage window. At 3.40 a.m. two incendiary bombs were dropped at West Lexham, both of which failed to ignite, and another incendiary was dropped at Weasenham St Peter at 3.45 a.m. to no effect.

Three HE bombs were then dropped at Wellingham, damaging a farmhouse, five cottages and a chapel. Unfortunately, farm labourer Frederick Pile (45) had just called his employer to warn him of the approach of aircraft, but a moment after he left the house a bomb fell on the public road demolishing a stone wall. Mr Pile's body was discovered on the road a few yards from the crater.

About two minutes later, three HE and two incendiary bombs were dropped while the Zeppelin was over East Raynham, breaking the glass in the windows of three houses and seven cottages; blowing tiles off roofs and smashing 250 panes of glass in greenhouses at Raynham Hall; damaging trees in the park and uprooting a large white hawthorn and depositing the main portion of the tree 50 yards away. One bomb, which landed in a small wood, felled a tall oak tree at a considerable distance from the spot where it exploded, while another bomb killed two fine carthorses in a meadow.

A further five HE bombs and two incendiaries were dropped at South Raynham, where one cottage was wrecked, several windows and doors of others were blown out and the windows of the church, the vicarage, a private house and fourteen cottages were broken.

In all these villages, the cottages affected by the blast from the bombs not only lost windows, but it was also 'no uncommon thing' for their ceilings to fall in too. It was reported that one unfortunate girl of 15 was in bed when the ceiling fell and buried her under the debris. She was lucky to escape unhurt. A man in another cottage was blown out of bed, and fell heavily on the floorboards. He was injured too, with a slight cut to his upper lip from a small piece of shell.

It was now daylight, and having thus deliberately bombed two unprotected villages and caused around £500 worth of damage, L-43 made for the sea. Despite being warned of the danger, many people came out onto the street upon hearing its approach or in the hope of spotting it. The Zeppelin passed between East Rudham and Fakenham, South Creake and Walsingham and finally, out to sea between Wells and Burrow Gap at about 4.05 a.m., under a heavy stream of fire from the Mobile Anti-Aircraft 3-pdr Vickers QF gun at Holkham, where it eventually disappeared into the clouds out at sea.

Aftermath

Frederick Pile was laid to rest at Wellingham on 26 May 1917, in a service conducted by the local rector, the Reverend L.K. Digby, who annotated the entry he made in the Dersingham burial register 'Killed by a bomb dropped from German Zeppelin.'

L-43 returned to its base safely, but a few days later on 14 June 1917 fell victim to a RNAS seaplane in the North Sea. There were no survivors.

16/17 June 1917

On the afternoon of 16 June 1917, four Zeppelins (L-41, L-42, L-44 and L-48) left the north German sheds to raid England. The shortness of the night, there being only four hours of darkness, rendered it improbable that the raiders would attempt to penetrate far inland. Indeed, it was curious that the enemy should have considered a raid worthwhile at this season of the year, taking into consideration the defenceless condition of Zeppelins in face of attack by aeroplanes in the beam of a searchlight or in daylight.

L-42, under the command of Kapitänleutnant Dietrich, was 35 miles north-east of Margate at 11.30 p.m. At 1.20 a.m., she was sighted by the *Tongue* light vessel coming from the north-east and steering south-west The Zeppelin cruised in the neighbourhood of the light vessel for about three quarters of an hour. Her height was reported as 5,000–7,000ft and she was said to be cruising very slowly. It was suggested that L-42 could have been waiting for the night to get darker. For some time she was over the Elbow buoy. She remained off Margate, being repeatedly reported from both the North Foreland lighthouse and the *Tongue* light vessel, until 2.05 a.m., when she came in over the North Foreland. At 2.08 a.m., the NCO (non-commissioned officer) in charge of the Ramsgate searchlight reported that the sound of a Zeppelin could be heard to the north-east. Two minutes later, the raider was sighted and the searchlight exposed, being quickly followed by the Hengrove, St Peter's and Cliffsend lights. The Zeppelin was picked up for a few seconds and then lost again.

L-42 appears to have steered south-west towards Ramsgate, parallel to the coast, passing the Marina Pier and then circling inland over the harbour. The first bomb was dropped at about 2.15 a.m. in the sea about 400 yards south-west of Marina Pier, and the second about 150 yards further on off the Pavilion. The third bomb dropped on the Naval Ammunition Store, 20 yards south of the clock tower in Ramsgate harbour. A fourth and a fifth fell in Albert Street, and a sixth on the north-east side of Crescent Road. In addition, one bomb was dropped in the grounds of Southwood House, one on the Manstone Road just beyond the railway crossing and two in a field north-west of Nethercourt Lodge, about 500 yards from the railway bridge.

The raider then proceeded inland to Manstone, where three HE and two incendiary bombs were dropped, without causing damage beyond a few panes of broken glass in nearby houses; there were no casualties. L-42 then turned north-east and dropped two incendiary bombs at Garlinge, doing no damage. She was sighted east of Margate at 2.20 a.m. and passed out to sea over North Foreland at 2.24 a.m., having been overland for only 14 minutes.

Considerable military damage was achieved. One bomb made a direct hit on the ammunition store and blew it up, and also demolished the buildings of the naval base. Thousands of windows were broken in the town, principally owing to the explosions caused. The two bombs which fell in the town demolished three cottages, and more or less seriously damaged sixteen houses and shops. Two men and a woman were killed, and seven men, seven women and two children were injured, although these were the only casualties incurred in the whole raid. Of the seven men injured, one was a naval rating, one a RFC Lieutenant and another a policeman. The rest were civilians. The Ramsgate AA guns fired twenty-seven rounds.

A certain amount of confusion seems to have been caused by the attack. All reports agree that the sound of the Zeppelin's engines seemed much less than has hitherto been the case, so that when she was heard she was believed to be much further off than she actually was. The searchlight reported that she was not in range of the Ramsgate guns, when in all probability she could have been. Owing also to the belly of the Zeppelin being painted black, she was almost invisible in the beams of the searchlight when it was exposed. The target was only held by the light for about 15–30 seconds and then lost again and, while they were still searching for it, the Zeppelin successfully dropped a bomb on the ammunition store extinguishing the light. The din of the series of explosions, and breaking of glass that followed, further rendered it impossible for the light director to make himself heard.

Four aircraft of the RNAS went up at 12.40 a.m. from Manstone, and returned at 3.35 a.m. with nothing to report. The Zeppelin seems to have eluded their observation successfully, though it was reported by the North Foreland lighthouse that from 1.20 a.m. to 2.32 a.m., when she was lost in cloud to the north-north-east, she was within sight of the RNAS watchers to Westgate, giving her position with every movement. From first to last, the North Foreland lighthouse seemed to be used by the Zeppelin as a centre from which she took her bearings for Ramsgate, and to which she returned to take her bearings for her departure.

Going northwards at considerable speed, she was on the latitude of Saxmundham when the catastrophe to L-48 took place, and about the same time was herself attacked by RNAS machines from Yarmouth. One seaplane, flown by Flight Sub Lieutenant Bittles, went up to engage her at 11,000ft, about 30 miles east of Lowestoft. On being attacked, she at once rose to 15,000ft and the seaplane was unable to follow her to this height. Flight Lieutenant Egbert Cadbury in his Sopwith Scout, then engaged her at 15,000ft, at which point she rose further to 16,000ft and he was compelled to break off the action, owing to a fracture in his petrol pipe.

The Zeppelin was further pursued for an hour and a half by a Curtis seaplane, flown by Flight Commander V. Nicholl and Flight Sub Lieutenant Leckie, but the speed of L-42 was too great for the seaplane, which did not overtake her and she was able to return safely to Germany.

L-41, commanded by Hauptmann Kuno Manger, was observed apparently drifting with engines cut off at about 2.20 a.m., off the coast near Martham. She appears to have dropped some bombs in the sea between 2.30 and 2.50 a.m., and was finally driven away shortly after 3 a.m. by a RNAS aircraft from Burgh Castle, piloted by Flight Sub Lieutenant Walker, who pursued her 30 miles out to sea but could not overtake her.

L-48, under Kapitänleutnant der Reserve Franz Georg Eichler, was the flagship of the raiding squadron, having Korvettenkapitän Viktor Schütze, Commodore of the North Sea Airship Division, on board. She was located 40 miles north-east of Harwich at 11.34 p.m. After this, she approached the coast slowly, arriving nearly an hour and three quarters later, some 20 miles north-east of Harwich. At 1.24 a.m. she was located 6 miles east of Orfordness, and ten minutes later was again located slightly north of this position.

It was not until 2 a.m. that L-48 crossed the coast just south of Orfordness. Steering west, she was reported near Bromeswell at 2.15 a.m. Later, at 2.25 a.m. she was sighted from Woodbridge, apparently hovering over Little Sutton, and at 2.30 a.m. she was sighted again in more or less the same position, after circling in an attempt to pick up her bearings. Shortly afterwards, the raider appears to have picked up the bearings for Harwich for, at 2.42 a.m., she was sighted approaching

Zeppelin L-48.

the Trimley Heath gun from the direction of Martlesham, where she dropped three HE bombs, which broke a few panes of glass.

Fire was immediately opened by all seven guns of the AA defences. Guns from the ships in Harwich harbour and 6-pdrs mounted on minesweepers lying off Bawdsey also engaged the Zeppelin. As it seems improbable that L-48 approached closer to Harwich defences than Kirton, it is improbable she ever came within range of more than four of the AA guns, and that many rounds would have fallen short of the target. On the other hand, the fire of so many guns created a formidable barrage through which the Zeppelin would have had to pass before she could reach Harwich. This barrage undoubtedly deterred her from approaching the defences more closely. In all, 569 rounds were fired, of which 498 were from 3in 20cwt guns. The night was exceptionally clear, and the raider could be seen clearly with the naked eye.

Harwich Garrison reported: 'soon after the opening of fire, shells from the AA guns were observed to burst close to the Zeppelin and fragments of shell hit her. The Zeppelin altered her course and was observed to be in difficulties; she circled round and was evidently out of control.'

On the other hand, all reports sent in by units of the 72nd Division at Ipswich, who were observing the action from a flank, state that the Zeppelin was not observed to be hit by AA shells, all of which seemed to fall short of her. And this is confirmed by the pursuing aviators, who observed that all the AA fire was too low.

Leutnant zur See Mieth, who was saved from the wreck of L-48, stated under cross-examination that the ship was out of control between 12.30 a.m. and 2 a.m. (BST) owing to engine trouble. The engines were repaired, after which the attack on Harwich was undertaken. Mieth stated emphatically that L-48 had not been hit by the fire of the AA guns, being well out of range the whole time.

When fire first opened on the raider from the Harwich defences at 2.42 a.m., L-48 was about a mile north of Kirton. Thence she seems to have steered south-east right across Parker's Lane and west of Kirton Lodge, dropping twelve bombs, one of which did not explode. Tiles were blown off a barn, but otherwise no damage was done except to fields. A mile south of Kirton Lodge, L-48 turned due east, passing south of the village of Falkenham, and dropping nine bombs at regular intervals of 300–400 yards. Of these, two were of the largest size (300kg) and two of intermediate size (200kg), but no damage was done. Just beyond Falkenham she turned north-east and crossed the River Deben.

She seems to have been travelling very slowly, for she was sighted just north of Kirton at 2.42 a.m., and the Harwich guns continued to fire on her until about 3.17 a.m. when she was lost to sight north of the River Deben. At this, she proceeded north at considerable speed but was evidently still in trouble, as her nose was observed to be constantly slewing round to the left and she was steadily losing height.

At 3.20 a.m. she was reported in the neighbourhood of Orford, and at 3.28 a.m. she was brought down in flames by Second lieutenant L.P. Watkins of the Canadian Army and the 37th Squadron, Home Defence Group, and fell in a field at Holly Tree Farm, near Theberton, north-east of Saxmundham. She had then traversed a distance of about 12 miles in as many minutes. The Zeppelin was attacked at the same time by another aeroplane of the RFC, piloted by Captain R.H.M.S. Saundby of the Experimental Station at Orfordness, but the coup de grace appears to have been given by Lieutenant Watkins.

The light of the burning Zeppelin was seen as far north as the *Haisborough* light vessel. Lieutenant Watkins states that L-48 was flying at a height of about 13,700ft when he attacked. She had apparently descended considerably from the height she was originally at when engaged by the Harwich guns, and her gradually decreasing height was noted by the pilot of one of the pursuing aeroplanes, Lieutenant F.W. Holder, who stated that when he was at 14,200ft near Harwich L-48 was 2,000ft above him, after which she 'was observed to lose height while proceeding in a northerly direction at about 50mph.'

L-48 came down more slowly than had been the case in previous catastrophes of the same kind, the fall taking three to five minutes. The envelope was stripped off by flames and blew away as it had on previous occasions, appearing as a blazing mass above the heavier body of the Zeppelin and thus giving rise to the statement that she broke in two parts.

The three survivors owe their lives to the comparatively slow fall. The Zeppelin came down stern first, at an angle of about 60°, smashing the whole after part, including the rear gondola, all the occupants of which were killed. The front gondola was badly damaged, but one of its occupants survived, though terribly hurt – Leutnant zur See Otto Mieth was rescued from the burning Zeppelin by the local constable. Kapitänleutnant Eichler and four of the crew jumped from the Zeppelin, but were killed outright by the fall. Korvettenkapitän Schütze was burnt alive.

Of the two men in the two side gondolas, both survived. Maschinistenmaat Ücker, who was in the port gondola, was severely injured, but the other, Maschinistenmaat Ellerkamm, who was in the starboard gondola, came to earth absolutely unhurt, and was standing by the wreck when arrested by a petty officer of the Royal Navy and the local constable.

L-48 hit the earth obliquely, so that the starboard gondola, in which Ellerkamm was, remained high in the air and practically undamaged. Both side gondolas seem to have been slung higher than was the case in other Imperial German airships. There was no doubt that L-48 had some difficulty with her engines that compelled her to drift for a period variously estimated from fifteen minutes, to the hour and a half given by Leutnant zur See Mieth. It was not likely that this trouble was caused by the AA gunfire from Harwich.

Aerial reconnaissance photograph of crashed L-48 at Theberton, taken by an aircraft from RFC Norwich on 17 June 1917. Note the crowd of onlookers and the ring of soldiers surrounding the site.

The original graves of the sixteen members of the crew of L-48 in St Peter's churchyard extension, Theberton, Suffolk.

SHOT DOWN BY THE BRITISH

Leutnant zur See Otto Mieth wrote an article that evocatively described the last operational flight of L-48, originally published in German in *Frankfurter Zeitung Illustriertes Blatt* on 28 February 1926 translated into *Shot Down by the British*, it was published in *The Living Age* on 17 April, 1926:

16 June 1917 was a bright beautiful summer day. Our naval airbase, Nordholz, near Cuxhaven, lay embosomed in idyllic heath country and amid clumps of pines and birches. Its gigantic sheds and grounds basked in the sunshine as if there were nothing but peace and goodwill on earth. Suddenly a wild, warlike shriek, beginning with a deep rumple and rising into a long, shrill tremolo, rent the dreamy atmosphere. Thrice did the siren call.

Thus Mars suddenly strode into the tents of peace, for this was the summons for a raid against England. Files of attendants rushed out of the barracks to the airship sheds, whose doors suddenly yawned wide open as if they had been burst out by the rising roar of the motorists within. A moment later two giant Zeppelins slowly emerged. One was L.48, the newest airship in the navy, to which I had been assigned as watch officer.

As I directed the operation of bringing her out, I studied with proud delight the slender, handsome lines of the giant, six hundred feet long and sixty feet through at its greatest girth. Four gondolas, one on either side, and one fore and one aft in the centre, were suspended below its body. They contained five motors, while the front gondola was reserved for the steersmen and their instruments. Our regular crew consisted of twenty men, including two officers, but today we carried an attack commander, Captain Sch— [Korvetten- Kapitän Viktor Schuetze].

Black is the colour of night and black was the colour of our ship. Our shield was darkness, for when she enwrapped the earth and nature and man on moonless nights she announced the hour for us to rise to lofty altitudes and to attack the enemy behind his ancient walls of water.

We did not look forward expectantly to the devastation we planned to wreak. That was in the line of duty, for which we risked our lives. But the real joy in our service was, after all, the charm of nature, the sense of isolation in infinite space for our fragile ship – alone with the heavens above and the waters beneath the earth.

As soon as I boarded the ship, our mooring lines were loosened, propellers began to whirl and the L.48 rose quickly but majestically into the air. A last wave of the hand, a shout of 'Back tomorrow!' and the North Sea rolled beneath us.

Our course lay due west. We were in the best of spirits and though our sailors were superstitious, no-one recalled the fact that this was our thirteenth raid. Our sealed orders were opened. They read briefly: 'Attack South East England – if possible London.' Willhelmshaven appeared on our port side. The vessels of our high sea fleet, lying on watch at Schillig Reede, signalled, 'A successful trip.'

The North Friesland Islands came into sight and disappeared behind us. We pushed steadily onward. Slowly the homeland sank into the misty distance and over Terschelling we found ourselves already in the enemy zone of operation. Only a few days before, the British had surprised and destroyed two of our reconnoitring airships at this point. We rose to the three thousand metre level scanning the air anxiously in all directions but discovered no sign of the enemy.

On and on. Our motors hummed rhythmically, our propellers whistled. It gradually became darker. The last rays of the sun gilded the waves and a light mist spread like a thin veil over the earth making it difficult to pick up our bearings. We had gradually risen to five thousand metres and were close to the southeatern coast of England. But it was still too light for our purpose, so we were forced to bear away from land and wait for darkness. Suddenly a heavy thunder storm swept over England. Flashes of lightning a kilometre long rent the clouds. This wonderful scene lasted but a few minutes and then passed on but then we resumed our course we discovered that there had been a violent atmospheric disturbance and that the direction of the wind had suddenly changed and we were bucking a strong southwest gale that impeded our progress. By this time it was perfectly dark and we crossed the English coast in the vicinity of Harwich. Silver-white streaks of surf were clearly visible beneath us, so that we could easily follow the contours of the coast. But everything else was absolute blackness; not a light was visible.

We knew, therefore, that an alarm had been. Millions of people were aware of our coming and were preparing to give us a warm reception. We made our last preparations. Signals rang through the ship, 'Full speed ahead,' 'Clear ship for battle.' Now for the luck of war!

By this time it was bitterly cold, the temperature having fallen seventy-two degrees since we left Germany and we shivered even in our heavy clothing. At our high altitude, moreover, we breathed with great difficulty and in spite of our oxygen flasks several members of the crew became unconscious. Nevertheless we pushed on steadily against the southwest wind, driving our machines at their full power. But June nights are short in England and our chances of reaching London grew constantly less. Suddenly a starboard propeller stopped and an engineer reported that the motor had broken down.

As our forward motor was knocking badly, we had to give up London. Thereupon one bit of bad luck followed another. Our compass froze and we had great difficulty in keeping our bearings. At length we decided to attack Harwich, which lay diagonally ahead of us wrapped in a light stratum of fog so we made for the leeward side of the town in order to cross over quickly with the wind behind us. It was 2.00am and our altitude was 5600 metres, or nearly eighteen thousand feet.

When we swung around and pointed directly for Harwich it was still as death in the gondola. All nerves were tense. The only sounds that broke the silence were low orders to the steersman from time to time. Suddenly somebody work up below us. Twenty or thirty searchlights flashed out in unison, thrusting long, white groping, luminous arms into the air. They clutched hastily and nervously, crossed each other, passed so close to us that our gondola was as bright as day. Yes, they even flickered across the ship itself without detecting us. Meanwhile we drew closer and closer to our goal, sliding between the shafts of light with humming propellers. For several minutes this game continued. Then one searcher picked us up and held us fast in his circle of light. Thirty white arms grasped greedily at us as if they would tear us out of the air with their eager clutches. Our slender black ship was flooded with their radiance, which it reflected in jetty sparkles from its glittering body. Instantly it began to thunder and lighten below as if all inferno had been let loose. Hundreds of guns fired simultaneously, their flashes twinkling like fireflies in the blackness beneath. Shells whizzed past and exploded. Sharpnel flew. The ship was enveloped in a cloud of gas, smoke and flying missiles. Hissing like poisonous serpents, whistling, howling, visible during their whole trajectory, blue-white uncanny fire-shells and rockets sang past us. *Peng! peng!* bellowed the English guns in their sharp staccato, like a great pack of hounds at the heels of a stag. But we kept steadily forward into this witches' cauldron. Every man stood at his post with bated breath. The weariness, the cold and the rarefied air had been forgotten. Our beating hearts fairly drummed against our sides. I kept my eye glued on my vertical glass, my right hand on the lever of the electric bomb-release. Gradually our target came into the field of vision until it reached the point set. I pressed the lever and at fixed intervals, one by one, the bombs fell. A new sound now punctuated the incessant roar beneath – the dull throbbing **boom! boom!** as our missiles struck the earth. The whole thing lasted only a minute or two, but in that brief interval was concentrated the experience of an ordinary lifetime. We steered straight ahead across the area of fire. To be or not to be was now the question. Were a single one of the countless shells that flew past us to strike our six hundred feet of unprotected body, our gas would be aflame in an instant and our fate would be sealed.

It seemed a miracle that we ever emerged from the tumult. The firing got weaker and at length ceased. The searchlights were extinguished.

Night embraced us again and covered also the land with its opaque blanket. Only dull-red glowing spots far behind us marked the points where our bombs had started conflagrations.

It was half past two and from our altitude the pale glow of England's midsummer dawn was already visible. So it was high time to get back over the open sea, for one there our principle danger would be over. But our frozen compass was our undoing. Instead of steering to the east, we inadvertently headed towards the north and before we discovered our error we had lost valuable time. Added to that, our forward motor also failed us, so that our speed was sensibly diminished.

I had just returned to my station after dispatching a radiogram reporting the success of our raid and was talking with Captain Sch——, when a bright light flooded our gondola, as if another searchlight had picked us up. Assuming that we were over the sea, I imagined for an instant that it must come from an enemy war-vessel but when I glanced up from my position, six or eight feet below the body of the ship, I saw that she was on fire. Almost instantly our six hundred feet of hydrogen were ablaze. Dancing, lambent flames licked ravenously at her quickly bared skeleton, which seemed to grin jeeringly at us from the sea of light. So it was all over. I could hardly credit it for an instant. I threw off my overcoat and shouted to Captain Sch—— to do the same, thinking that if we fell into the sea we might save ourselves swimming. It was a silly idea, of course, for we had no chance of surviving. Captain Sch—— realised this. Standing calm and motionless, he fixed his eyes for a moment upon the flames above, staring death steadfastly in the face. Then, as if bidding me farewell he turned and said 'It's all over.'

After that, absolute silence reigned in the gondola. Only the roar of the flames was audible. Not a man had left his post. Everyone stood waiting for the great experience – the end. This lasted several seconds. The vessel kept an even keel. We had time to think over our situation. The quickest death would be the best; to be burned alive was horrible. So I sprang to one of the side windows of the gondola to jump out. Just at that moment a frightful shudder shot through the burning skeleton and the ship gave a convulsion like the bound of a horse when shot. The gondola struts broke with a snap and the skeleton collapsed with a series of crashes like the smashing of a huge window. As our gondola swung over we fell backward and somewhat away from the flames. I found myself projected into a corner with others on top of me. The gondola was now grinding against the skeleton, which had assumed a vertical position and was falling like a projectile toward the earth. Flames and gas poured over us as we lay there in a heap. It grew fearfully hot. I felt flames against my face and heard groans. I wrapped my arms around my head to protect it from the scorching flames, hoping the end would come quickly. That was the last I remember.

Our vessel fell perpendicularly, descending like a mighty column of fire through the darkness and striking stern first. There was a tremendous concussion when we hit the earth. It must have shocked me back to consciousness for a moment for I remember a thrill of horror as I opened my eyes and saw myself surrounded by a sea of flames and red hot metal beams and braces that seemed about to crush me. Then I lost consciousness a second time and did not recover until the sun was already high in the heavens.

Gradually I collected my thoughts. How did I get here in these strange surroundings on this litter? It was like a dream. I half raised myself painfully and saw that my legs were wound in thick, bloody bandages. I could hardly move them, for they were broken. Then I made a new discovery, my head and legs were covered in burns, my hands were lacerated and when I breathed I felt as if a knife were thrust into me. I thought to myself 'Am I dreaming or awake?' Just then a human voice interrupted my groping thoughts: 'Do you want a cigarette?' And a Tommy stuck a cigarette case under my nose with a friendly grin. So it was no dream. I was a prisoner.

I now learned what had happened. An English aviator had crept up on us unobserved and had managed to fire our ship. We fell in an open field near Ipswich. All our crew were killed except myself and two subordinate officers, one of whom died later from his wounds. The other was in one of the side gondolas, which chanced to be out of reach of the flames and though he became unconscious for a moment he was not injured. The moment we struck ground he clambered out and ran away as if the Furies were after him but a person must be excused for losing his head under such circumstances. I have never been able to understand just how I personally escaped. Probably my comrades who fell on top of me when the ship settled aft shielded me from the flames, for I was not seriously, even though painfully, burned. When we struck, stern foremost, the light skeleton of the long vessel telescoped and this broke my fall and the prow stood upright above the debris, so that I was not hit by flying beams.

When I asked how my English captors found me, they said they heard me groaning and were able to pull me out of the flames before it was too late. I soon recovered from my shock and wounds, survived my long imprisonment and have even become accustomed to having everyone who meets me, who knows of my experience inquire solicitously, 'Do you feel any bad effects?'

German propaganda postcard showing a Zeppelin squadron attack, proclaims: 'Towards England!'

21/22 August 1917

Eight naval Zeppelins left their north German sheds on 21 August, with Zeppelin chief Strasser aboard L-46. The whole squadron kept together more than had usually been the case, but its action was more than usually indecisive.

L-42, under the command of Kapitänleutnant Dietrich, approached Hull and tried to bomb the city. Dietrich was a man of determination, and it had been he who had attacked Ramsgate on 16/17 June and achieved there one of the very few successes of real military importance that can be credited to the Zeppelin raids, when he blew up a naval ammunition store. On this occasion near Hull, he only succeeded in blowing up a Methodist chapel.

L-42 came in over the coast at 1.03 a.m. near Tunstall, going south-west. At 12.12 a.m. she passed 2 miles north by east of Halsham Camp, and then seems to have gone north-east. At about 12.20 a.m. L-42 dropped an incendiary bomb at Elstronwick, which ignited but did no damage. It was thrown in order to ascertain whether the Zeppelin was overland or not. This point settled, the Zeppelin went south-west towards Hull.

At 12.30 a.m. she was hovering with engines shut off between Ryhill and Paull. The searchlight at the latter place was exposed and the Zeppelin tried to dodge the beam, at the same time approaching from the south-east and beginning to drop bombs slowly. At 12.48 a.m. the searchlight picked up the target, and two minutes later the gun opened fire. The raider then made directly for the gun at 50mph, dropping bombs as she came, but at 12.53 a.m., when immediately above it, turned suddenly off north-east, disappeared behind a cloud and shut off her engines, then drifted.

The lights and guns at Marfleet and Chase Hill Farm had also both opened on her at 12.50 a.m., and her commander evidently considered discretion the better part of valour in view of the possibility of aeroplane attack while so brightly illuminated. Marfleet held the target until 12.55 a.m., when it finally disappeared to the north-east. Seven bombs had been dropped, which caused no casualties and did no damage, beyond breaking the gun telephone lines to the flank observer.

At 1 a.m. the Zeppelin appeared over Hedon and dropped five HE bombs in the Baxtergate and on the south side of the Hull Road. A Primitive Methodist chapel was wrecked, and the doors and windows of eleven cottages and a Roman Catholic chapel in Baxtergate blown in. A YMCA hut on the Burstwick Road was also seriously damaged. One bomb fell on a grass field 400 yards due south of the church.

Going in the same north-easterly direction, L-42 next dropped two HE and twelve incendiary bombs 1 mile east of Preston, followed by a third HE bomb ½ mile further on in a wheat field. No damage was done. The Zeppelin then turned to the south, and at 1.10 a.m. was again picked up by Marfleet light, going southward very unsteadily and steering as much as possible behind small clouds. At 1.15 a.m. she dropped a single HE bomb at Thorngumbald, which did no damage, and at the same time she was fired on by Marfleet and Paull guns, both of which continued firing for five minutes.

Kapitänleutnant Martin Dietrich, commander of L-42.

She was now retiring in a south-easterly direction, going out over the Humber towards Immingham where, at 1.19 a.m. she was picked up by the Immingham Halt light and gun, followed a minute later by those at Killingholme Marsh and Chase Hill Farm. Caught in the beams of these lights she stopped, and was seen from two of the guns to turn round in her own length. She travelled south-west toward Killingholme for a short distance and then, apparently daunted by the gunfire, or fearing aeroplane attack, rose 'almost vertically' to a great height (estimated at 18,000ft by the Marfleet gun), and went off abruptly north-east. At 1.22 a.m. the gun at Sutton opened on her at extreme range, and at 1.24 a.m. the New Holland gun also got off one round at her. At 1.25 a.m. she was lost by all the lights, having apparently gone behind a dark cloud.

L-42 was chased out to sea by an aeroplane, piloted by Lieutenant Hubert Solomon of No. 33 HD Squadron. When in the neighbourhood of Beverley, at a height of 15,000ft, he saw the Zeppelin but could not keep pace with it, and at the same time keep climbing. He followed her, however, out to sea for 20 miles, firing three bursts from his Vickers gun at long range, on the off-chance of hitting her. As he could get no nearer, he returned, landing at Elsham at 2.20 a.m.

Another pilot of the same squadron, Lieutenant Walbank, while west of Hessle, saw the Zeppelin momentarily in the searchlight beams.

In all, nineteen aeroplanes took off, two of which crashed on landing (both were BE-2e aircraft, one piloted by Lieutenant J.A. Dales, who was seriously injured, and the other by Lieutenant E.D. Hall, who was uninjured). The RNAS sent up a Large America seaplane from Killingholme at dawn, which returned at midday without having sighted the enemy.

L-44, commanded by Kapitänleutnant Franz Stabbert, cruised northwards along the coast and was reported off Scarborough at 2.30 a.m.

The rest appeared to have done nothing. Strasser and the rest of the fleet simply hung about off the coast and went off home about an hour before L-42.

'Mythical' Zeppelins were reported approaching the Kent and Suffolk coasts, others (by the Lancashire & Yorkshire Railway) inland in Yorkshire as far as Featherstone and Pontefract, and actually in Lancashire, at Rochdale. Mythical bombs were reported from Doncaster. The origin of those rumours in the south of England was hard to explain, it was certain though, that no Zeppelins came up anywhere near that part of the coast.

Aftermath

An elaborate communiqué from the German Admiralty was more than usually absurd. It claimed that the squadron 'under the proved leadership of Fregattenkapitän Strasser' bombed Hull, warships in the Humber and industrial establishments, which were observed collapsing.

THE PIGEON THAT SAVED SIX LIVES

On 5 September 1917, Squadron Commander Vincent Nicholl was flying a DH-4 aircraft with Flight Lieutenant Trewin from RNAS Great Yarmouth, on an anti-Zeppelin patrol, when its engine seized and the aircraft came down in a rough North Sea and soon sank.

Bob Leckie, who was piloting flying boat 866, spotted their predicament, and went to their rescue. His crew managed to haul the two nearly drowned airmen out of the water. There were now six men on the flying boat, and all Leckie's attempts to get her off the sea failed, so he was faced with the task of taxiing back in an overloaded boat, with a heavy following sea. They decided to launch four pigeons carrying messages of their location and situation. Eventually both engines ran out of petrol. Their situation was desperate, the men were very ill with sea-sickness, but they just had to keep bailing the plane out otherwise it would have sunk.

At 11.30 a.m. on 8 September, the message Nicholl had sent three days earlier was delivered to RNAS Great Yarmouth. Pigeon NURP/17/F16331 had been found dead with exhaustion on a beach a few miles away, by locally based soldiers. The bird and the message were taken to the War Signalling Station and the message was telephoned through. The search, which had been abandoned with all hope lost, was immediately resumed and all six crew were rescued in the nick of time. The pigeon was stuffed and displayed in a case in the mess; upon the case was fixed a brass plaque bearing the inscription 'A very gallant gentleman.'

The 'very gallant gentleman' pigeon. (NURP/17/F16331)

IN THE FORWARD CONTROL GONDOLA OF A ZEPPELIN

The above picture of the interior of a Zeppelin is reproduced from a recent aviation number of the "Illustrierte Zeitung," and according to the artist both the after and forward gondolas would appear to be remarkably spacious. This is due to a certain amount of artistic exaggeration. The movements of the Zeppelin are controlled and directed from the forward gondola by means of speaking tubes and telephones communi-cating with other parts of the airship. An officer is busy with a chart, above which are placed the engine telegraph dials, and is giving his instructions to a man at one of the wheels. A member of the crew, spanner in hand, is descending the ladder which leads from the gondola to the "catwalk," which is the communication way of the ship and is a V shaped passage built inside the main framing.

THE AFTER ENGINE GONDOLA OF A ZEPPELIN—From the "Illustrierte Zeitung"

The above depicts the scene in the after engine gondola of a Zeppelin, and is stated by the "Illustrierte Zeitung" to be drawn at the moment the airship is "passing through enemy air defences after a successful attack on England." As the picture depicts the airship calmly sailing in broad daylight this title is obviously absurd. On the right one of the sliding windows is wide open and the machine gunner keeps a wary eye aloft. The drum containing the cartridges is placed at right angles to the barrel. Behind his arm is the "breather" for cooling the crank case of the engine, above his head a radiator, with the water-pipes connecting it to the engine. In the foreground two mechanics are attending to one of the three large 250-h.p. motors which are to be found in the after gondola, and in the background an engineer oils the overhead valves of another of the engines, whilst his left hand is apparently placed on top of the cylinders, which should not be a very comfortable position for any length of time.

19/20 October 1917

'The Silent Raid'

The 'silent raid' on 19 October 1917 was dubbed so because the AA gun defences were muzzled lest they guide the raiders to London, and because the Zeppelin engines were almost silent.

The sky was cloudy in Lincolnshire and much of East Anglia, and free of cloud in Essex, the London area and Kent, but very misty. Acoustic conditions were peculiar; sound was not carrying very far.

The engines of the Zeppelins were almost inaudible, and this gave rise to the erroneous supposition that they were drifting with the wind with engines cut off, in order to escape notice. This, however, was impossible, as at the heights at which they were flying, their engines would have frozen instantly if stopped, and actually did so in several cases when through accident or negligence they failed.

The sound of exploding bombs was also deadened in such a way that they were often supposed to be much further off than was actually the case. This was so in the London area, where the explosions of the bombs at Hertford were only faintly heard at neighbouring gun stations, whilst at Theydon Bois they were not heard at all. At Harwich, the bombs exploding at Wix, 8 miles away, were not heard, and at Great Oakley, 3 miles away, they were supposed to be as far off as the neighbourhood of Colchester.

Eleven naval Zeppelins participated in the raid. The objective was 'middle England,' specifically the industrial regions of Sheffield, Manchester and Liverpool. It was known from prisoners captured in France on the following day that Strasser did not accompany the raiding squadron as on previous occasions, but followed the proceedings from Ahlhorn. Some of L-45's crew expressed surprise that he did not order all Zeppelins to return as soon as the strength of the wind had been reported to him. One prisoner cynically remarked that, since Strasser, the 'FdL' (Führer der Luftschiffs) had recently been decorated with the Pour le Mérite he no longer had any occasion to risk his life on a Zeppelin.

Their mission was to bomb the industrial centre of England. If the total number of raiders, their objective and their courses over England are taken into account, it may well be claimed that this was to be the biggest raid attempted by the enemy against the Midlands since 31 January 1916. In the total number of Zeppelins employed it is only surpassed by the attack of 2 September 1916, when every naval and military airship fit for the journey on that day went up to attack London.

The raiders crossed the North Sea and headed for a rendezvous to the east of Flamborough Head. They gradually rose on approaching England but, when at an altitude of about 12,000ft, began to meet an unexpected northerly wind which caused the Zeppelins to struggle against the wind without success. The cold was intense and led to trouble with the cooling system of the motors. The height at which they flew over England appears in no case to have been less than 16,000ft. Indeed a few Zeppelins attained nearly 20,000ft. Height sickness affected the crews to a considerable extent, the results being faulty navigation, bad wireless telegraphy work and carelessness in the management of the engines.

L-41, under Hauptmann Kuno Manger, had been fired on by HMS *Albion* as she approached the mouth of the Humber at 6.55 p.m. At 7.02 p.m. the flash of a bomb dropping into the sea was seen at Saltfleetby. At 7.05 p.m. she passed the Spurn and ten minutes later went inland over Cleethorpes. At 7.18 p.m. she passed Waltham, at 7.25 p.m. Caistor and, while over Holton-le-Moor at 7.30 p.m., she dropped a petrol tank.

She passed on south-westwards, and at 7.40 p.m. threw two 50kg bombs at North Carlton, north of Lincoln, killing two sheep. She passed near Lincoln, going south-west, and half an hour later was near Derby, where she circled about slowly for an hour and a half. At 8.20 p.m. she was between Derby and Burton, and then at 9.05 p.m. she was back again near Derby. At 9.25 p.m. she was over Ashby-de-la-Zouch, going north-west. At 9.40 p.m. she was north of Burton and at 9.50 p.m. again at Derby.

She was not observed again until an hour later, when she suddenly appeared at Netherton, west of Birmingham, and dropped a series of HE bombs and incendiaries between Netherton and Barnt Green. The first dropped were three 50kg HE bombs and two incendiary bombs that fell at Rough Hill in Rowley parish. These were followed by two incendiary bombs near Dudhill Farm, west of Rowley Regis, and by one 50kg and five incendiaries near the Eagle Colliery, east of Old Hill. Two incendiaries then fell on a hill known as 'the Tump', close by. Only slight damage was done to glass from concussion. One of the 50kg bombs did not explode, and two of the incendiaries did not ignite.

One 100kg, which failed to explode, and one 50kg bomb then fell at Mucklow Hill, north-east of Halesowen, followed by five incendiaries which fell between The Leasowes and Frankley, and four, which did not ignite, at Bartley Green. None of these bombs did any damage.

Seven HE bombs fell at Longbridge, five of them (two 100kg and three 50kg) near the Austin Motor Works, which were fully lit up as they had not received warning of the raid until the moment at which the bombs were dropped. Of the five bombs, one 100kg and two 50kg did not explode. The two others did slight damage (estimated at £500) to the boiler house, aeroplane shop and engine shelter. One man was slightly injured. The bombs at Longbridge were of 50kg weight and fell on a farm, doing no damage.

Finally, a 50kg bomb fell at 11 p.m. at Rednal, Cofton Hackett, to no effect. The Zeppelin then went over Redditch, which had received warning of the raid at 10.58 p.m. After leaving Redditch, the Zeppelin disappeared and was not observed again in the district south of Northampton. She reappeared near Towcester at Field Burcote, where at 11.50 p.m. she dropped two 100kg bombs, neither of which exploded.

At midnight she passed Easton Thorpe near Wolverton, and at 12.21 a.m. was south of Hitchin, passing Standon at 12.35 a.m., Rochford at 1.05 a.m. and heading out over the estuary at Shoeburyness five minutes later. At 1.25 a.m. she came over the Kentish coast at Whitstable, at 1.35 a.m. was at Betteshanger, west of Deal, and went out to sea at Kingsdown at 1.40 a.m.

L-44, under Kapitänleutnant Franz Stabbert, entered the Wash and turned north over Frampton Marshes at 7.30 p.m. At 7.45 p.m. she turned west near Old Leake, and then went south-west passing west of Boston, and at 7.53 p.m. was heard at Kirton, south of Boston. She then followed the Great Northern Railway line southward, passed Spalding at 8 p.m. and, after hesitating for some time, was immediately west of Peterborough at 8.30 p.m.

She headed directly southwards to Bedford and, at 9.08 p.m. dropped ten 50kg bombs in a line between Elstow and Kempston, passing over the crossing of the Midland main line and the Cambridge branch of the London & North Western Railway, narrowly missing a large ammunition dump placed between the two lines.

The Zeppelin was right over the ammunition dump, and bombs fell within 150 yards on either side of it. Two of them failed to explode. The glass roof of an engineering works close by was smashed by concussion, and two men were injured. The works were fully lit up, and so was a military training school close by, no warning having been given of the raid. The warning was not received until 9.15 p.m. when the Zeppelin had dropped her bombs and was already at Newport Pagnell.

Near Wolverton at 9.20 p.m., she turned south-east along the London & North-Western Railway main line and, at 9.40 p.m. dropped one 300kg, four 100kg and four 50kg HE bombs, besides one incendiary at Heath and Reach, near Leighton Buzzard. No damage was done, beyond the breaking of glass in some cottages, and there were no casualties.

At 10.40 p.m. she was north of Maidstone and then, attracted by the flares of Detling Aerodrome, she turned south to bomb it. At 10.42 p.m. she dropped one 100kg and one 50kg bomb at Milgate Park, Bearsted, breaking the windows and doors of Milgate House and two cottages, and killing a sheep. Then, at 10.45 p.m. a second 300kg bomb was dropped at Leeds, doing ten shillings worth of damage to crops. The aerodrome was entirely missed.

Ten minutes later, L-44 was north of Ashford and at 11 p.m., west of Canterbury. Passing Elham at 11.20 p.m., she went out to sea between Folkestone and Dover at 11.30 p.m. under fire from the guns at Lympne, Cheriton and Cauldham. Her height here was estimated at 12,000ft.

Twenty minutes later, she passed Boulogne under fire from AA guns and was carried inland on a south-south-east course. L-44 was brought down by French AA fire near Lunéville, when three tracers were fired at 19,000ft, one of which pierced the envelope of the Zeppelin aft. She immediately caught fire, the flames spreading in a few seconds from one end to the other, and fell to the ground close to Chenevières. One of the cars became detached and dropped to the ground 100ft away from the main body which had been reduced to a mass of half-calcined aluminium girders. There were no survivors.

L-45, commanded by Kapitänleutnant Waldemar Kölle, left Tondern at 11.25 a.m. on 19 October carrying sufficient petrol for a flight of twenty-two hours. Apart from this load, every precaution had been taken to lighten the ship. Thus, in place of her full cargo of bombs, equivalent to 2½ tons, she took less than 1¾ tons. Her load was claimed to have been two 300kg bombs, fourteen 50kg and ten incendiaries. It would appear that, at the last moment, two 100kg and two 50kg bombs were added to that load. The crew was also reduced.

Although it was usual for raiders to cross the North Sea at a fairly low altitude, L-45's commander seems to have made a considerable height shortly after leaving the German coast. It was suggested that visibility was poor even at the start, and that he was anxious not to run any risks either from British aircraft or from surface craft. As a result of these tactics, many of the crew were already suffering from strain due to the rarefied atmosphere as soon as the English coast was reached at 8 p.m.

Before reaching England, the navigation of L-45 seems already to have suffered from the storminess of the weather. A strong northerly wind was encountered, increasing in strength (60–75ft per second) as the ship rose to greater altitudes. The cold was also intense, and is given as 7°F over the North Sea and as low as 15°F when the Zeppelin was flying at a height of 19,000ft.

After she crossed the coast between 8 and 9 p.m., several searchlights were seen and crew state that the ship was fired upon – this was not the case. After hesitation for some time out to sea, the Zeppelin came in over the Yorkshire coast near Withernsea at 8.20 p.m. Her course appears to have been considerably interfered with at first, by the presence of a number of British aeroplanes, which had gone up about an hour earlier in pursuit of L-41 and the other Zeppelins. She was, however, able to avoid them by rising to a great height, probably 19,000ft, at which she stayed for most of the time she was over England.

At 8.40 p.m. she was going down the Humber, past Grimsby and then turned up river again. At 9.04 p.m. she was near Immingham, and at 9.10 p.m. back again south of Grimsby. During these movements she was apparently being chased and fired at by aeroplanes. She now went off southward, passed Louth at 9.15 p.m., then bore S.S. and at about 9.30 p.m. dropped a petrol tank at Bracebridge, south of Lincoln.

Half an hour later she was in the neighbourhood of Leicester, where she was picked up by an aeroplane piloted by Lieutenant G.H. Harrison, 38 Squadron RFC, who was flying at about 14,000ft, L-45 being 1,000ft above him. He got directly under her tail, fired three bursts and saw the shots hit the raider. After his first burst of fire, L-45 made a sharp turn and apparently fired her forward machine gun. Lieutenant Harrison's gun then jammed and while trying to clear it he followed the Zeppelin, which was going off north-east. He was unable to put his gun to rights, and at 10.25 p.m. left L-45 in the neighbourhood of Melton Mowbray and returned to Stamford aerodrome.

After his disappearance, the raider attempted to resume her south-west course, but was carried by the wind directly southward and at 10.50 p.m. reached Northampton, where she dropped her first bombs. There is no doubt, from the statements of her crew captured next morning in France, that all on board thought that the Zeppelin had penetrated much further west than was actually the case. They had been fighting the whole time to get westward against the north-west wind, and when they bombed Northampton they thought they had reached Oxford and damaged 'factories' there.

They state that they were fired upon, which was untrue. The statement was probably an invention intended to justify the dropping of bombs. The damage done at Northampton was confined to a few cottages, and no factories or military objectives were touched. Two bombs, probably 50kg, were dropped in a stream at Kingsthorpe doing no damage, then three 50kg bombs were dropped at Dallington, immediately north of the town, followed by nine incendiary bombs on Northampton itself. These fell along the line of the London & North-Western Railway, in close proximity to Castle Station. Two fell on the line itself north of the station, and two in Parkwood Street, immediately west of it, wrecking a cottage (No. 46) killing Mrs Eliza Gammons (51) and her twin daughters Gladys and Elsie (13), who died in hospital from shock caused by the severe burns they had suffered. Glass was also broken in a number of other cottages.

Six 50kg bombs fell at the northern end of the railway tunnel under Hunsbury Hill, doing no damage. One 100kg bomb landed at Wootton Hill Farm, breaking glass to the value of £5, and one 50kg bomb, which did not explode, at Brainhill Farm. A 50kg bomb followed at Preston Deanery and one at Piddington, on the border of Salcey Forest. No damage whatever was caused by these last bombs and one of the bombs at Hunsbury failed to explode.

From Northampton, the raider followed the main line of the London & North-Western Railway right into London. At 11.05 p.m. she passed Leighton Buzzard, at 11.15 p.m. Watford and by 11.25 p.m. she had reached the north-western suburbs of London. Here she dropped two 50kg bombs at Hendon, one in Colindeep Lane and the other in the grounds of the new Grahame White Aerodrome, doing slight damage to several houses and cottages.

She dropped one 100kg and one 50kg bomb on the Midland Railway sidings immediately south of Cricklewood Station. The railway track, telephone wires and some rolling stock were damaged and the glass in the Haberdasher's School, Westbere Road, and a large number of houses was broken. One man was injured.

L-45 then passed over London unheard and unseen, and when over the centre of the metropolis at 11.30 p.m. releasing a 100kg bomb that fell in Piccadilly, opposite Messrs Swan & Edgars. It made a large hole in the roadway, affecting a gas main, and broke a large amount of glass, besides damaging a few shop fronts. The glass roof of the Geological Museum in Jermyn Street was wrecked by the concussion. Five men (three of them soldiers) and two women were killed, and ten men (including two soldiers, a sailor and a police constable) and eight women were injured.

The casualties would probably not have been so large, had not a number of people who had originally taken cover come back out into the streets. The long period that had elapsed since the air raid warning, without anything subsequently happening, caused a great many people to suppose that the raid was over.

The raider went south-east and next dropped a 300kg bomb in Camberwell. It fell in the rear of a premises in Albany Road near the junction of Calmington Road. Three shops – substantial buildings – were demolished, and a large number of shops and houses were more or less seriously damaged. A sailor and a mechanic of the RNAS were killed, as were four women and four children; nine men, five women and nine children were injured.

Finally, another 300kg bomb fell at Hither Green, Lewisham, destroying three houses and damaging many others. Five women and nine children were killed, and two men, three women and two children injured.

Although some of the crew must have known that the 'huge city' (as they described it) over which they passed was London, the majority do not seem to have realised the fact until they had landed in France and exchanged notes with one another. Some of the men then boasted to their French captors that they had bombed the British metropolis, and were very proud of the fact.

L-45 had actually not been under orders to attack London, and only passed over the city as the result of circumstances which could not be controlled. The senior warrant officer, who was responsible for the navigation, seems to have had no clear idea as to his whereabouts at the time, and he pointed out afterwards that, owing to the haze, not even the Thames was visible. Incidentally, not one of the crew was aware that their Zeppelin had succeeded in dropping a bomb in the heart of the West End of London, and had flown over the Admiralty and the War Office.

L-45 went on eastward, passing Sidcup at 11.40 p.m., and shortly after midnight, when just south of the mouth of the Medway, she was attacked by an aeroplane of 39 Squadron, piloted by Second Lieutenant T.B. Pritchard RFC.

She was flying at a height of 13,000ft and estimated to be 2,000ft above him and about 150 yards in front. He opened fire, which the Zeppelin did not return. As she ascended rapidly and changed her course from east-south-east to west-south-west, going at a speed of about 55mph, Pritchard attempted to keep her in fire but could not do this on account of her height.

He followed her, however, for thirty-five minutes on her course to the coast. She passed out to sea near Hastings shortly before 1 a.m., with the British aviator still in pursuit and climbing steadily. When out to sea, however, the Zeppelin outdistanced him and became lost to view. He decided to return to base as his petrol was getting low, and steered north-north-west for the coast. At 1.15 a.m. he crashed in a field at Hooe, near Bexhill, and was severely injured.

The maximum height attained by L-45 after her encounter with Pritchard certainly exceeded 19,000ft. She touched 20,000ft sometime later. Owing to this great height the engine revolutions fell off from a normal working maximum of 1,200rpm to 1,000–1,100rpm. It was, however, emphatically denied by the prisoners that the motors were stopped and that the ship was allowed to drift with the wind, as reported in some quarters. Such tactics would inevitably have resulted in the radiators freezing, and in the ship becoming helpless. In point of fact, one engine – that in the portside car – failed after bombs had been dropped and while the ship was still over England. There was also trouble with the spark plugs, owing to their 'sooting' – probably as a result of a new kind of oil that was being used – and after stopping, the engine very quickly 'froze up'.

In spite of the attempt to steer an easterly course from London onwards, the raider was steadily being driven south. No mention was made, during interrogation of the crew, of the attack by Pritchard that caused the Zeppelin commander to give up his attempt to beat eastward and drove him directly south to the Channel. They merely state that it was very cloudy, and that the Zeppelin continued to navigate at her maximum altitude in the hope of being able to regain Germany by passing over Belgium or the occupied parts of France.

Shortly after dawn, L-45 came under fire, but the crew were satisfied that they were well out of range as they saw flashes on the ground but spotted no bursts in the air. This was probably at Vonges, in the Côte d'Or, where the anti-aircraft guns that were stationed at a munitions factory reported firing at a Zeppelin, with the result of it making her increase her height.

At around 7.30 a.m. L-45 passed Mâcon, and at 8 a.m. flew over Lyon and Meyzieux where the Zeppelin was fired at by an anti-aircraft gun, at a height of 12–13,000ft. It was learned, after the examination of the prisoners that, while over France, L-45 was navigated at her maximum altitude in order that she might regain Germany without molestation, but the petrol supply for L-45 was becoming exhausted. The commander was hoping he could at least land in Switzerland, and he brought the Zeppelin down to about 2,000ft with a view to locating his whereabouts but the attempt was not successful. No maps were carried of southern France, Switzerland or Italy, and this added to the difficulties of the situation.

Most of the crew were convinced that they were over Switzerland, because the mountains in the department Hautes-Alpes, over which L-45 was then flying, were covered with snow. In any case, the men were certain that they could not be over Germany as a large number of people came out of their houses to stare at the Zeppelin – in Germany, the passage of a Zeppelin in the early morning was taken more or less for granted.

Strenuous attempts were made to get the engines, three of which failed, to work properly. The engines in the port car had failed while L-45 was still over England. At about 10 a.m., L-45 had reached Sisteron at the confluence of the Rivers Buëche and Durance, and Kapitänleutnant Kölle decided at this juncture that he must land. He circled slowly over the town to the intense excitement of the population and of the German officer prisoners of war held in Sisteron Castle. By a curious coincidence, amongst these prisoners there were the officers of the military airship LZ-85, which had been brought down at Salonika eighteen months before.

The Zeppelin moved off north on sighting some Annamite labourers dressed in blue uniforms; the commander appears to have been alarmed. He then realised that he was still over France, but only had a few gallons of petrol left and so a landing was inevitable. He ordered the manoeuvring valves of the gas bags to be opened, and the two emergency landing flags to be displayed.

The Zeppelin was over the wide stony bed of the River Buëche, which was almost dry, and it was a good spot for landing. As soon as L-45 touched the ground she was caught in an eddy of wind which heeled her over. This resulted in the port wing car being torn off and left on the ground with its two occupants. Two other men jumped out from the after car. Relieved of this weight, the Zeppelin rose and was again caught by the wind and swung through an arc of 180°. She was dashed against the eastern side of the valley with her bows pointing north, at a spot some 600 yards from the site of her intended landing. The remainder of the crew jumped to the ground having, in some cases, dropped 10–20ft. The front car seems to have been badly smashed at the second concussion, but its occupants were not injured. The four men who had been left on the ground with the portside car could not rejoin their comrades as there was an intervening stream, so they gave themselves up to some local peasants.

As soon as the main party was clear of the ship, Kölle drew them up, collected all their papers and made a pile of them in the ship, which was then fired by the navigation warrant officer and, as she still held a good deal of gas and had 3–4kg of petrol in each tank, she burned readily and there were even some explosions. The crew of L-45 saluted their ship as she was burning and then marched off to the nearest farm. Here, they were advised to surrender by a German sergeant major of the infantry who happened to be there in charge of a working party of German prisoners, and the party were soon interned.

L-46, under Kapitänleutnant Heinrich Hollender, came in over the coast at Bacton, Norfolk, at 10.30 p.m. and immediately dropped ten HE bombs at Walcot, one being of 100kg weight and the rest of 50kg. Two horses were killed, and the roof and glass of some farm buildings damaged. Five minutes later, ten more 50kg bombs were dropped at East Ruston, merely breaking a guinea's worth of glass in a cottage.

The Zeppelin then went off south-east, passed Burgh at 10.45 p.m. and out to sea between Yarmouth and Lowestoft at 10.50 p.m. She neared the Dutch Island of Zeeland shortly before midnight, and then headed back to her base.

L-47, with Kapitänleutnant Michael von Freudenreich in command, came over the English coast at about 7.45 p.m. near Sutton-on-Sea, and went along the coast to Skegness. At 7.55 p.m. she dropped a 50kg bomb at Ingoldmells, which failed to explode. She then moved south-west, and was south of Holbeach at 8.10 p.m. Turning east, she was north of Spalding at 8.15 p.m. and near Castle Bytham at 8.30 p.m. The Zeppelin circled and hovered in this neighbourhood for some time, and then, at 8.55 p.m., went off south-west dropping two 50kg bombs near Wittering aerodrome, which did no damage.

She then took a steady south-east course, and dropped two incendiary bombs at about 9.20 p.m. at Ramsey, doing no damage. She passed north of Newmarket between 10 p.m. and 10.05 p.m. At 10.28 p.m. she dropped a 50kg HE bomb at Raydon, close to Hadleigh Aerodrome, to no effect, followed at 10.30 p.m. by ten 50kg bombs at Great Wenham, which caused only slight damage to farm buildings and cottages. At 10.32 p.m. the raider harmlessly dropped a 100kg bomb at Chattisham.

The Zeppelin then crossed the Orwell near Wrabness at 10.35 p.m. and went out to sea at Walton-on-the-Naze at 10.40 p.m. An hour later, she was at sea off Ostend, and followed the Dutch coast back to her base.

L-49, under Kapitänleutnant Hans Geyer, left Wittmundhafen at 7 p.m. for a raid over the Midlands. The objectives in her orders were the 'centre' of England – Sheffield, Manchester and Liverpool. She was to carry a full crew, but at the last moment one engineer from the starboard wing car was left behind. She carried a load of 2 tons of bombs and petrol for a twenty-two hour flight.

The journey over the North Sea was accomplished without incident and at a good speed. The height at which the ship travelled was not clear from the statements of the crew, but she probably started at about 3,000ft and increased her height gradually as she crossed the sea and approached the English coast. At about 7.30 p.m. her speed was greatly reduced when the Zeppelin was near

England, and she had risen to a height of about 14,000ft. The commander desired to wait for complete darkness before crossing the coast, perhaps passing over the North Sea rather more quickly than he had expected. As a result of this check, the forward motor was accidentally put out of action. The engineer, in restarting the propeller, appears to have stuck the throttle valve and stopped his engine and, before he could start it again, the radiator had frozen by 8 p.m.

The incident dampened the ardour of the commander and, although he subsequently changed his mind, he actually contemplated abandoning the raid at this juncture. Moreover, whilst moving at a slow speed for some three quarters of an hour he had, without knowing it, drifted very far south from the appointed rendezvous of the squadron at Flamborough Head.

The commander claimed to have thrown bombs between 8 p.m. and 9 p.m. on military establishments in the neighbourhood of Hull, Bridlington, Scarborough and Flamborough, and to have made hits, one of them on a train and a station. He had, in reality, come overland at Holkham, Norfolk, at 8 p.m. and subsequently headed south-south-east, dropping one 300kg, one 100kg and one 50kg bomb on a farm near East Dereham, breaking £10 worth of glass.

Next, he dropped nine 50kg bombs at Yaxham, doing £24 worth of damage to glass, followed by two 50kg bombs at Thuxton, damaging farm buildings, then another 50kg bomb at Coaton and eight incendiary bombs at Hardingham, doing no damage.

At 8.30 p.m., three 50kg bombs and three incendiaries fell at Kimberley, killing three horses and injuring one seriously. Cattle sheds were demolished and a farmhouse was damaged. An incendiary bomb fell at Runhall, which did not ignite, followed by eight incendiaries at Wicklewood and three at Suton, near Wymondham, none of which did any damage.

The raider had followed the railway all the way from East Dereham. The commander stated afterwards that he felt confident he had bombed two aerodromes; that he had recognised the landing flares and had distinguished the red and green lights which were fired from the aeroplanes. The navigating warrant officer, however, was more sceptical and said that these lights were part of a large railway yard. The first officer of the Zeppelin was even less pronounced in his views, but he admitted that, owing to the height and prevailing wind, none of the crew could be sure of where they stood, in fact, he would not go any further than to say the bombs were thrown 'at some lights'. He claimed, however, to have recognised railway lines and trains.

L-49 continued to follow the line past Forncett St Peter, where an incendiary bomb was dropped, to no effect, and Diss, which was passed at 8.45 p.m., then on to Haughley and Finborough where, at 8.55 p.m. she turned east. His bombs being expended, and the commander still uneasy at having lost the use of one of his motors, he decided to return to his base via Belgium. At 9.02 p.m. the raider was north of Ipswich and then went south, passing out to sea near Walton-on-the-Naze at 9.20 p.m. At 9.48 p.m. she flew across the eastern end of Thanet, going south-east and then hovered over the downs and off Deal for nearly an hour.

By this time, owing to the height at which he was flying he had, unknown to himself, run into a very strong northerly current of wind which was driving the ship very much out of her course. Meanwhile, there was a haze over the south-east of England and the dead reckoning made by the navigator was becoming very inaccurate. He had crossed the estuary of the Thames, either without knowing it, or imagining that he was crossing the North Sea towards Holland. On resuming her course, at 10.42 p.m. she came in again over St Margaret's, going south-west, and finally went out to sea over Folkestone at 11.09 p.m. The crew saw clearly beneath them the three towns of Folkestone, Hythe and Sandgate as they passed over, but mistook the town lights of the Kent coast, that tend to runs north-east to south-west, for those of the Dutch coast, which follow a similar pattern. Nobody on board knew anything of the crossing of the Channel that followed.

L-49 reached the French coast near Cape Grisnez at about 11.40 p.m. and was driven by the north-west wind, against which she could make no headway, with only three motors running, in a steady south-easterly direction across north-eastern France. Soon after daybreak, two other Zeppelins were sighted, one 15 miles to starboard and another to port. The former was L-44, the latter L-50. Thinking that both of these Zeppelins knew their course, the commander of L-49 ordered them to be followed. At about 6.45 a.m. the leading Zeppelin (L-44, south of Lunéville) was seen to burst into flames. This was interpreted by the crew either as an accident or the result of Dutch AA fire.

The commander and his first officer stated categorically that, at that moment, they believed they were over either Westphalia or southern Holland. The second Zeppelin was, incidentally, observed to be steering an erratic course and this strengthened the commander's belief that Dutch AA guns were at work.

L-49 at once rose and, in order to avoid passing near the spot where L-44 had been set alight, headed north-west towards Nancy. The other Zeppelin, L-50, probably followed. It was at this juncture that the two engineers in the starboard car fainted from the effects of height and fatigue. They had been trying to fill the radiator with fresh water, as the greater part of its contents had boiled away. The motor seized soon afterwards and went out of action. The Zeppelin was then running on two motors only and could no longer struggle against the wind.

At about 8 a.m., when L-49 was near Neufchâteau, at a height of about 10,000ft, the commander decided to come down in order to fix his bearings and thaw his forward motor. Attacked by a number of aeroplanes, and feeling he had no speed and little capacity left for rising owing to loss of gas, the commander decided to land and save his crew. He saw a small town and a railway station

(Lamarche) and tried to head for them in order to fix his locality. The emergency landing flag was flown from the forward car and the Zeppelin came down, but as might be expected under such circumstances the landing was slightly rushed, with the result that L-49 came to earth, at 8.45 a.m. in the valley of the River Apance, 3 miles to the north of Bourbonne-les-Bains, on somewhat unfavourable ground.

As soon as the crew set foot on the ground the inhabitants, who had been watching the Zeppelin's manoeuvres, came running up in a state of wild excitement, many of them armed with old shot guns and farm implements. Three of the aviators who had been pursuing the Zeppelin crashed their aeroplanes in the adjoining field. The commander, now certain he had fallen in enemy territory, gave orders for the Zeppelin to be destroyed. An incendiary cartridge was placed in the pistol but misfired. Thereupon the crowd, which had assembled around the crew, so intimidated the exhausted Germans that they gave up all further attempts at firing the ship and surrendered to Lieutenant Lefèvre, one of the pilots who had attacked and followed the Zeppelin and crash-landed nearby.

L-50, commanded by Kapitänleutnant Röderich Schwonder, left Ahlhorn on the afternoon of 19 November, carrying a load of bombs, stated to have been 2,000kg in total. A machine gun fitted on the port side of her forward car was her entire armament. There were twenty crew on board and there would have been twenty-one, had the sail maker not been taken ill before the start.

Her orders were to cross the English coast at Flamborough Head, but a strong north wind rendered this impossible. Whilst over the North Sea, the forward engine of the after-car failed. The crew were certain that the English coast was crossed somewhere in Norfolk, probably due east of the Wash. Heavy, low lying clouds prevented the country from being seen at all clearly, but there were occasional gaps in the clouds and the helmsman had a fair notion as to whether they were over sea or land and they did occasionally catch glimpses of towns while over East Anglia. On passing over the coastline, from 8 p.m. onwards, a number of searchlights opened out on L-50 and the crew claimed they faced a good deal of gunfire. The Zeppelin managed to evade the searchlights and suffered no harm, although the impression of the crew was that she was by no means out of range of the guns.

Her bombs were dropped, the crew stated, in two instalments, roughly in the neighbourhood of two AA batteries, and the consensus of opinion was that these were located somewhat to the west of Norwich. She was flying at her maximum height of 19,000ft, and continued to keep roughly at this height for long after she left England.

As a matter of fact, no guns were in action against L-50 on the coast or inland in Norfolk, and her bombs were actually dropped between Narborough and Thetford. She had come in at Cley-next-the-Sea at 7.45 p.m. and proceeded south-west. At 8 p.m. she passed Fakenham and went on in the same south-west direction as far as Narborough where, at about 8.15 p.m. she turned south and, between 8.18 p.m. and 8.20 p.m., dropped five 50kg bombs at Barton Bendish, one 50kg at Beechamwell, and three 100kg and eight 50kg bombs (one of which did not explode) at Oxborough. No damage was caused.

The raider went off at high speed south-east and, at 8.25 p.m., dropped an incendiary bomb at Mundford, another at West Tofts and another, which failed to ignite, at West Wrentham. Then turning southward over Croxton Heath towards Thetford, one 300kg, five 50kg and two incendiary bombs fell at Croxton and two 50kg near Thetford, also causing no damage.

After having disposed of her bombs, L-50 abandoned the westerly course which she had in vain been trying to steer, and prepared to make for the east. Going off south-east, the raider passed north of Bury St Edmunds at 8.35 p.m., Haughley at 8.40 p.m., then north of Ipswich and out to sea at Hollesley Bay around 8.50 p.m.

The north wind, which was increasing in violence, drove her south, and the breakdown of a second engine in the forward car added to the difficulty of keeping the ship under proper control. Both the mechanics who were serving this engine had been suffering from acute height sickness, and it seems likely that the failure of the engine was due to their falling asleep. The navigators of L-50 believed she was well out to sea at about this time and some 60 miles east of the mouth of the River Thames.

L-50 drifted back over France and crossed to land at Dunkirk. The general impression of the crew was that they were over Holland or Westphalia and, when L-44 was seen falling in flames, some of the men concluded that she must have been shot down by the Dutch. Kapitänleutnant Schwonder now made off south-west away from the scene of the disaster.

At 10.45 a.m. the Zeppelin passed over Gray (between Langres and Besançon) going east, then north-east of Vesoul and south of Lamarche at 11.40 a.m. Suddenly, while still at a height of only about 1,500ft at about midday, the crew saw another Zeppelin (L-49) resting on the ground immediately beneath them, and Kapitänleutnant Schwonder decided to follow his consort's example and effect a landing in the same vicinity. He therefore ordered the emergency landing signals to be flown, and announced through the engine telegraphs that the ship was about to land. As L-50 came lower some rifles were fired at her, and a few aeroplanes bearing the tricolour cockade were noticed near the Zeppelin on the ground.

Schwonder realised he was over hostile territory and seems to have lost his composure, and gave immediate orders for the ship to turn west and climb again to her maximum height. She was stated to have reached close on 10,000ft when the order was countermanded, and the helmsmen were again told to land with all possible speed. According to French observers, L-50 then descended at an extreme angle of 20°–40°. As a result of this desperate manoeuvre, the Zeppelin grazed a small wood, and the forward car was torn off by trees on landing at Dommartin, near Montigny-le-Roi, a few miles west of Bourbonne-les-Bains. Sixteen men in all left the Zeppelin at this juncture, some jumping from the side gondola.

Released of the weight, the Zeppelin shot up in the air. It is thought that the two men who remained in the rear gondola were killed at the time, as this car seems to have been knocked out of shape by the concussion. Two further men were left on board, having been stationed in the gangway at the time. The derelict hull of L-50 then became the sport of the winds, and was chased by fighter aircraft. The Zeppelin drifted over Sisteron at about 4 p.m., to the consternation of the interned German officers who, for the second time in one day, saw one of their own Zeppelins at the mercy of the enemy. The Zeppelin was swaying from the stern, at one time horizontal then vertical, and was clearly helpless.

L-50 was finally seen drifting out to sea near Fréjus at about 5.30 p.m. and was pursued until nightfall by seaplanes from the station at St Raphaël. There is little doubt that she foundered in the Mediterranean during the night. No trace of L-50 or the four men left on board her was found on land, or on the surface of the water. The survivors of L-50 were deeply incensed with their commanding officer, and provided accounts of his bad leadership in which it was stated that the second in command tried to annul the order for the precipitate landing, by giving opposite instructions to the helmsman. It was denied by Schwonder's crew that a shortage of petrol rendered the landing necessary, and it was claimed that there was sufficient fuel on board to get the ship back to Germany. He is also blamed for not having given more notice of his intention to land, which would have enabled the parachutes to have been made ready and the four men left with the ship would probably not have been lost. Perhaps Schwonder had suffered, like his men, from navigating for many hours at a great height, and his skills and judgement were impaired.

L-52, under Kapitänleutnant Kurt Friemel, crossed the Lincolnshire coast near Mablethorpe at about 7.30 p.m., was south-west of Wainfleet at 7.45 p.m. and west of Spalding at 8.08 p.m. A direct south-westerly course was being followed. A 100kg bomb dropped at Gosberton was attributed to L-52, but it did no damage. At 9.45 p.m. she dropped her first confirmed bomb, of 300kg, at Kemsworth, near Dunstable, smashing some glass with its concussion.

Going on eastward, at 10.05 p.m. thirteen 50kg bombs were dropped in fields 3 miles south of Hereford, damaging five cottages seriously and five slightly. A man was slightly injured.

A 50kg bomb fell at Hoddeston, doing no damage and, at 10.20 p.m., thirteen incendiary bombs were dropped on Waltham Marshes, near Waltham Abbey, also to no effect. L-52 went on past Ashford, was heard at Lydd and then went out to sea north of Dungeness at 11.15 p.m.

L-53, commanded by Kapitänleutnant Eduard Prölss, came in over the Norfolk coast at Blakeney at 6.45 p.m. and went south-west, passing Walsingham at 6.55 p.m. and Fakenham at 7 p.m. She then turned south, and at 7.15 p.m. harmlessly dropped a 100kg bomb at West Bradenham. After passing Watton at 7.20 p.m. she turned south-east over Breckles, where one of her bomb doors slipped out of its guides and fell in a field.

Crossing into Suffolk, she passed Bury St Edmunds at 7.50 p.m. and dropped four incendiary bombs at Rivenhall, doing no damage. She passed Goldhanger at 8.15 p.m. and passed out to sea over Foulness at 8.25 p.m. Coming back overland at 8.40 p.m. at Reculver, she dropped two bombs in the sea, followed by three 50kg bombs on land, doing slight damage to an inn. Two 50kg bombs fell at Sarre and one at Chislet, slightly damaging crops, and she finally headed out to sea over Deal at 8.52 p.m.

L-54, under Kapitänleutnant von Buttlar, came in over the Norfolk coast at Haisborough at 7.55 p.m., going south-south-west. At 8.15 p.m. she was west of Great Yarmouth, heading west, and a 15 minutes later was at Southwold, turning north-west. Soon turning south again, the Zeppelin went back out to sea, coming back in at Aldeburgh at 8.45 p.m. She passed Hollesley at 8.55 p.m. and turned inland passing south of Ipswich.

At 9.05 p.m. she dropped her first bombs near Hadleigh. Nine 50kg bombs fell, one of which failed to explode, and no damage was done. The raider then turned south-east, and dropped one 300kg and two 50kg bombs at Wix, to no effect.

L-54 dropped its last bomb harmlessly on land at Little Clacton, and proceeded out to sea at 9.20 p.m., hugging the English coast as far north as Yarmouth. About 10 miles north of Yarmouth she was attacked at 11.30 p.m. by a RNAS aeroplane, piloted by Flight Lieutenant C.S. Nunn. The Zeppelin had descended to a height of about 5,000ft above the sea, and the aeroplane was at 8,800ft. Flight Lieutenant Nunn dived to attack the raider, but was unable to get into firing position owing to her superior speed. He pursued her out to sea for twenty minutes but could not overtake her. On his return the engine failed, and he crashed on landing at Burgh Castle Aerodrome, but escaped injury.

L-54 returned to Germany by way of the North Sea.

L-55, under Kapitänleutnant Hans Kurt Flemming, came in over the Lincolnshire coast at Anderby. At about 7.45 p.m. she was near Alford, at 7.55 p.m. near Horncastle bearing round to the southwards, and ten minutes later at Stickney, where she dropped a petrol tank.

After hovering in the Holbeach area for some minutes, she went away south-west, past March at 8.30 p.m. to Holme where, at 8.40 p.m. she dropped five 50kg bombs and one incendiary. The latter did not ignite. The bombs fell near the junction of the Ramsey branch of the Great Northern Railway main line, causing no harm. The Zeppelin then made directly south at high speed along the main line and, between 9.05 p.m. and 9.25 p.m., dropped sixteen HE bombs and one incendiary along a zigzag course roughly parallel with the line between Hitchin and Hatfield. The damage was not severe, and only one casualty was caused by the bombs, a man slightly injured by broken glass.

A 50kg bomb was dropped at Holwell Bury, in the parish of Lower Stondon, 3 miles north-west of Hitchin, but did no damage. An inn and a cottage were slightly damaged, and the GPO telegraph wires were broken by a 50kg bomb that fell 2 miles north-north-west of Hitchin. Midland Railway telegraph lines were affected by another 50kg bomb, 1 mile north of Hitchin, and cottage windows were broken by three 50kg bombs, one of which fell 100 yards south-east of the railway junction (Cambridge line) and two others in a field at Walsworth, east of the town.

Three bombs at Stevenage did no damage, a fourth demolished a farm building, where a man was injured, and the fifth and sixth slightly damaged some cottages. At Langley two 50kg bombs damaged farm buildings. At Codicote one 50kg and one incendiary bomb did no damage, while at Brocket Hall, 3 miles north-west of Hatfield, a few windows were broken by another 50kg bomb.

The Zeppelin then went directly south, losing touch with the railway and passing over Shenley Ridge at 9.30 p.m., going west. She appears to have passed unheard and unseen to the west of London, and was next reported south of Sevenoaks coming from the north-west and passing to the south-east at 9.58 p.m. At 10.16 p.m. she crossed the South Eastern Railway north of Battle, and probably went out to sea over Hastings at about 10.25 p.m., unobserved and no doubt without her commander and crew having any idea of their real position. She crossed the French coast at the mouth of the Somme about midnight, and crossed France safely back to her base in Germany.

The RNAS sent up a total of eight BE-2c aircraft from Frieston, Cranwell, Bacton, Burgh Castle, Yarmouth and Manstone. Only Flight Lieutenant Nunn from Burgh Castle managed to locate and actually engage one of the raiders. Other pilots saw no enemy. The RFC sent up sixty-five aeroplanes when enemy activity was spotted, but the Zeppelins themselves remained elusive to most, the exception being Second Lieutenant Pritchard, No. 39 Squadron, whose combat report stated:

Received orders at 22.17 to patrol between North Weald, Chingford, Enfield and Ware. I ascended at 22.20 and set off towards Chingford, climbing as I went. After patrolling for one hour fifteen minutes, I reached a height of 11,500ft and saw an airship travelling in a direction east-south-east from north London. It was at a height of approximately 15,000ft. I started off in pursuit and when in position, just south of the Mouth of the Medway, I opened fire. The airship was then in a position 150 yards in front and above me, My height was then 13,000ft and the time was 21.10. The height of the Zeppelin was still about 15,000ft.

The airship did not return my fire but ascended rapidly but could not do this on account of its height. Its speed was approximately 55 miles per hour. I followed up for another 35 minutes, climbing as I went but the airship gradually outdistanced me and got lost to view. I continued to chase with all speed and flying level until 1 a.m. when, owing to my petrol looking rather low, I turned and steered north-north-west for the coast. At 1.10 a.m. I had descended to within 600ft of the ground (aneroid reading) and fired my first parachute flare and saw that I was over the sea, so started north again until 1.15 a.m. when I dropped my last parachute flare. This failed to light. I then descended to within 50ft and lighted one of my wing-tip flares but the ground was impossible as a landing ground, so I again flew on for a couple of minutes when a searchlight picked me up from very close range and blinded me momentarily. As I was coming down I fired my last wing tip flare and saw a field and overshot, so I turned and tried to land in the field. My wing tip flare went out and I stalled from about 50ft and crashed into the field.

After extricating myself from the wreckage I turned out all lights, took the magazine off my gun and locked the trigger, secured my map and then found some assistance. I posted a guard on my machine and had the North Weald Aerodrome telephoned up and also the Eastbourne Hospital. An ambulance came for me and I was admitted to the Hospital.

A fatal accident also occurred among the RFC aeroplanes, when Second Lieutenant Hubert Philip Solomon (34), the pilot who had distinguished himself in pursuit of L-42 on 21/22 August near Hull, had his plane catch fire after taking off from Gainsborough and he was killed. He was buried with full RFC honours at Gainsborough General Cemetery, Lincolnshire.

FOUR

12–13 March 1918

On the afternoon of 12 March, five Zeppelins left their north German sheds to raid the Humber district, among them L-64, the newest vessel of the German fleet which had left the works at Friedrichshafen only the day before.

L-53, commanded by Kapitänleutnant Eduard Prölss, reached the Yorkshire coast between 9.45 and 10 p.m. but did not come over the land and soon went back to his base in northern Germany.

L-54, under the command of Kapitänleutnant von Buttlar, came only three quarters of the way over the North Sea and then dropped her bombs at sea, aiming at the 'V'-section of the Grimsby trawling fleet. The section, in the charge of Lieutenant Higman RNVR, was fishing at 54°2'N 1°47'E when, at about 6.30 p.m., three Zeppelins were seen, apparently close together and flying at a great height. Lieutenant Higman at once ordered the gun on his vessel to be cleared for action but, as the target at once rose and became indistinct in the haze and failing light, he did not fire. On sighting the fishing fleet, all three Zeppelins rose and after a short time separated, two carrying on to the westward while the third passed south-east.

Norfolk Constabulary.

PUBLIC NOTICE.

DAYLIGHT
HOSTILE AIR RAIDS

When hostile aircraft are within such a distance of

NORTH WALSHAM

as to render an attack possible, the public will be warned thereof by the following signals:—

Each Day of the week (including Sunday)
"Danger" Signal.

Police and Special Constables will patrol the streets on cycles exhibiting cards bearing the words, "POLICE NOTICE, TAKE COVER."

The Constables will blow whistles frequently to attract attention.

"All Clear" Signal.

Police and Special Constables will patrol the streets on cycles exhibiting cards bearing the words, "POLICE NOTICE, ALL CLEAR."

The Constables will blow whistles frequently to attract attention.

The signals will be given during the period of **half-an-hour before sunrise to half-an-hour after sunset.** Every effort will be made to give **timely** warning of the approach of hostile aircraft, but the public should remember that the **"Danger"** signal will be given when **Real Danger** is apprehended in which event they should **Seek Shelter** with all speed.

Persons still in the open should take what shelter is afforded by lying in ditches, hollows in the ground, &c.

It is recommended that, in the event of raids during school hours, children should be kept in the school until the **"All Clear"** signal is given. Likewise, workpeople at factories, etc., should remain on the premises in whatever shelter is afforded.

Horses and other animals should be stabled or secured in some manner and should not be abandoned in the streets.

J. H. MANDER, Captain,
Chief Constable.

County Police Station,
Castle Meadow, Norwich.
January, 1918.

One of the new series of hostile air raid alert posters issued to county towns across Norfolk, January 1918.

It was now dusk, and Lieutenant Higman hauled in his gear and removed the trawler section about 1½ miles north. He had been in his new position for about ten minutes when the flashes and explosions of a number of bombs were observed to the southward, approximately over the position which had been vacated. The detached Zeppelin, L-54, had dropped her bombs as her commander had supposed, on the fishing fleet. After dropping his bombs he seems to have wandered about for some time and then went back to base.

L-63 came straight in over the land at Hornsea at 8.30 p.m. and followed the railway directly to Hull. At 8.35 p.m. she passed Whitedale. She was fired on at 8.40 p.m., towards the sound of her engines, by the Marfleet gun, followed seven minutes later by Sutton. The Zeppelin did not reply, but swerved off westward and, when north-west of Hull at 8.55 p.m., turned south-east to bomb the city.

Owing to her height and the thick clouds, it was impossible for her commander to do more than guess approximately the position of the city, with the result that only six bombs actually fell within the municipal area and they did remarkably little damage. One 100kg and two 50kg bombs dropped in fields between Oak Road and the River Hull, north of the city, doing no damage, followed by two 50kg bombs further south, close to the crossing of the Hull & Barnsley Railway and the Hornsea branch of the North Eastern Railway. One of these, which fell in a field, did no damage; the other, which dropped close to the Hull & Barnsley Railway embankment, Montrose Street, badly damaged a house, destroyed a workman's cabin, damaged a signal box and made a hole in the embankment.

The Zeppelin turned east and dropped a sixth bomb, one of 100kg, on allotments at Southcoates Avenue, which damaged several houses. The only casualty was one woman, who died of shock.

At about 9 p.m. the raider went off north-east and dropped the rest of her bombs at Sutton and Swine. One 100kg, three 50kg HE bombs and two incendiaries fell at Sutton at about 9.10 p.m., followed by two 300kg and eight 50kg bombs at Swine. All remarkably fell in fields, did no damage and caused no casualties.

At 9.30 p.m. the Zeppelin went out to sea north of Tunstall, under fire from the guns at Hornsea. While over Hull and its neighbourhood, the Zeppelin was fired on by the guns at Harpings, Hessle, Paull, New Holland and Chase Hill between 8.56 and 9.10. p.m. The Chase Hill gun was never in range.

L-62 came in south of Flamborough Head at 9.15 p.m. and was fired on by the Flamborough guns from 9.17 p.m. until 9.23 p.m. She went south-west and then overland, and from 9.25 p.m. until 9.56 p.m. was seen circling north of Bridlington.

At 10.18 p.m. she was north-west of Driffield. She now turned south-east over the town but dropped no bombs, and then pursued her south-west course at very high speed, passing north-east of Pocklington at 10.30 p.m.

The guns at Walltoft, Hemingbrough, Brind Leys and Villa Farm opened fire on her by sound at 11.08 p.m., 11.10 p.m., 11.11 p.m. and 11.16 p.m. respectively. Immediately upon being fired at she turned sharply north-west, and at about 11.15 p.m. began dropping bombs immediately west of the village of Seaton Ross, circling round northwards during the bombing and then receding swiftly north-eastwards. In all, four 100kg, nine 50kg and ten incendiary bombs were dropped at Seaton Ross, followed by four 50kg bombs near the village of Melbourne, close by. All the bombs fell either in fields or woods; most of them did no damage, but four HE bombs that fell 75 yards west of the Black Horse Inn, Seaton Ross, did some damage to the inn and adjoining cottages. There were no casualties.

On her return journey to the coast, the raider was heard at 11.27 p.m. west of Beverley, and at 11.35 p.m. near Brandesburton. At 11.40 p.m. she passed out to sea at Barmston, under fire from the guns. It is possible that her commander intended to bomb Hull, and mistook the guns near Howden for those around Hull, but on the other hand it is quite possible that his objective was the naval air shed at Howden.

L-61 appears to have intended to attack Howden too, as her course was certainly directed on the same bearing as that of the previous ship. As her commander came into contact with no guns inland, he evidently came to the conclusion that he had missed his objective, be it Hull or Howden, and decided he was unlikely to find it so he went off without having wasted any of his bombs.

The Zeppelin was off Hornsea at 9.35 p.m. She then made three ineffectual attempts to come overland at 9.50 p.m., 9.54 p.m. and 10.05 p.m., being driven off on the first occasion by the Hornsea and Barmston guns and afterwards by the Hornsea guns alone. At 10.10 p.m., however, she got across the coast unheard by any guns and went south-west, where she was heard approaching Beverley at 10.23 p.m. At 10.30 p.m., however, when west of Beverley she passed suddenly northward and, at 10.40 p.m., was immediately south of Fimber. Here she passed south, and at 10.50 p.m. was near Burnby, south-east of Pocklington.

Apparently giving up the attempt to find any objective worthy of attack, she went off north-east, passing south-east of Malton at 11.02 p.m., south of Sherburn two minutes later and out to sea at Filey at 11.10 p.m.

Her speed throughout her course overland was high. This is apparently the first case in which one Zeppelin has made so long a journey overland and gone away without dropping any bombs. Kapitänleutnant Ehrlich must have known quite well that he was overland on account of the action of the coast guns, the position of which must have been well ascertained. He certainly wasted no time in looking for an objective, and evidently came to the conclusion that the search, on account of the weather conditions, was useless.

As a result of low clouds and mist, it was impossible for the Royal Flying Corps to take any effective action. Bad weather was reported, so it was almost impossible to send up aeroplanes or patrol. A total of only ten aeroplanes were sent up from Nos 33, 36 and 76 Squadrons. They saw nothing of the enemy. No 36 Squadron, in the north, experienced better weather, but the enemy Zeppelin never came near its patrol area.

13/14 March 1918

On the afternoon of 13 March, three Zeppelins headed across the sea for what was described in the German communiqué as a 'patrol attack' on the north-east coast of England.

L-42 patrolled a good way to the northward of L-52 and L-56, which at 6.37 p.m. were sighted near each other, 110 miles east of Whitby, by three of His Majesty's ships. On seeing the warships, they appear to have turned back and returned to their sheds. L-42, however, pursued her course and at 7.52 p.m. was reported 50 miles off the Northumberland coast.

L-42 crossed the coast north of Hartlepool without being heard, and getting inland, drifted in with the north-north-west wind over West Hartlepool with engines shut off, at a height of 17,000–18,000ft.

The first intimation of her presence was the dropping of bombs on the town at 9.20 p.m. Owing to the uncertainty of her position, an air raid warning had not been given, and consequently lights had not been put out. Those of the works in the district must have served as a very good guide to the Zeppelin, which easily located her objective and dropped her bombs with great accuracy.

The first four bombs fell near the workhouse, north-west of the town, followed by a fifth on Amberton Road, west of the railway. All were of 50kg weight. The fifth broke some glass; the others did no damage.

The Zeppelin had now started up her engines and was making for the docks. A sixth bomb of unknown weight fell in the Humber Dock on the east side of the railway, and then a seventh 50kg bomb exploded on the edge of No. 4 Timber Pond, doing slight damage to the bank, while two more bombs of unknown weight fell in the Central Dock.

The Zeppelin was now being engaged by the Hartlepool gun. The height of the target was given as 16,000ft, which was too low, as the shells, though good for line, were observed to burst below it. L-42 then swerved southward.

Two bombs of unknown weight dropped in the sea close to Dock Gate Cottages, breaking windows and splashing people on the shore with water. Another fell in the harbour, and the thirteenth in the mud by Messrs Furness, Withy & Company's graving dock. The fourteenth and fifteenth bombs, both

50kg, exploded in soft earth about 20 yards apart among the North-Eastern Railway Company's sidings and coal spouts, south of the Coal Dock; about a dozen railway trucks were damaged.

The raider then passed over the central ward of west Hartlepool, aiming her bombs apparently at the railway station. A 50kg bomb fell on an empty and neglected house in South Street, destroying it and wrecking the Normandy Hall public house next door. Another 50kg bomb fell on the pavement in Mainsford Terrace, blowing down a wall. Four 50kg bombs then fell together in Temperance Street, Frederick Street and Burbank Street. Four people were killed, three seriously and two slightly injured. The nineteenth bomb dropped in the road in Frederick Street and the twentieth on a house, which was destroyed; two others were wrecked, the remaining houses in the street being all more or less seriously damaged. Two people were killed, two badly injured and seven slightly injured. The twenty-first bomb, in Burbank Street, dropped in the roadway, seriously damaging two houses and affecting a number of others. One person was killed, one seriously injured and four others slightly injured.

No more bombs were dropped. The raider went off southward, parallel to the west of the railway and out to sea at Seaton Carew, under heavy fire from the AA defences.

All the bombs that fell on land were of 50kg weight, so if the Zeppelin had unloaded any of greater size, they were among those that fell in the docks. A flare, which fell on the sea bank at Carr House, opposite the Steel Works, was dropped by a British aeroplane.

The Royal Flying Corps sent up fifteen aeroplanes from 36 and 76 Squadrons. The pilots of 36 Squadron saw no enemy, but Lieutenant Wall of 76 Squadron, flying a BE-12b, was shelled by the Howden guns at about 11 p.m. The reason for this was that the gun commander at Hemingbrough was misled by the noise of the BE-12b which was new to him.

The pilots of 36 Squadron, in the area visited by the Zeppelin, had considerable trouble with their engines. Two crashed on landing, in one case the plane crashed at Pontop Pike, caught fire and the pilot was killed.

One aeroplane alone, flown by Second Lieutenant H.C. Morris with Second Lieutenant. H.D. Linford as observer, came into contact with the Zeppelin, and their combat reports stated:

Saw Zepp over Hartlepool heading due south, flying at not under 20,000ft. Time 9.35 p.m. I had only attained a height of 14,500ft when I first saw it, but kept climbing till I got to 17,300ft which was as high as I could get and then engaged it. My observer fired 280 rounds at Zepp and I fired 50 rounds, but it was out of reach and did not take any notice of us till I got to Redcar when it turned due east and went out to sea. I fired at Zepp with the idea of driving it off Middlesbrough and because I could not get any higher. We followed it 40 miles out to sea but it eventually got into the mist and evaded us. I then turned west and flew till I got to the coast and found the mist was too thick to attempt to get back to my own aerodrome (Seaton Carew) so I landed at Hylton.

After we had opened fire at the Zepp one searchlight got it for a second but did not hold it.

E.C. Morris, 2nd Lieutenant

9.35, sighted large Zeppelin flying at about 20,000ft over West Hartlepool, heading due south. 2nd Lieutenant Morris was then at 14,500ft and proceeded to climb until at a height of 17,300ft. I then opened fire at the enemy's airship, firing two double drums but without result. Firing was carried out in bursts of 10 or 15 rounds. 2nd Lieutenant Morris also fired about 50 rounds from his gun. As soon as he had fired, a searchlight from the ground found the airship, but was only able to hold it for an instant. This occurred over Redcar at 9.40 p.m. As soon as we had fired, the airship turned due east and we followed her out to sea. I fired at intervals until we lost her in the mist at 10.05 p.m. We were then about 35 or 40 miles out to sea. We then turned back but owing to the very heavy mist we were unable to locate Seaton Carew or Skinningrove. At 11.15 p.m. we landed at Hylton. It was impossible to see any part of the ground as low as 2,000ft.

R.D. Linford, 2nd Lieutenant
1/7th Bn. Royal Warwickshire Regiment
Seaton Carew
14th March 1918

Lieutenant Hill of 76 Squadron, in a BE-12b, went up from Helperby and patrolled for three hours between Hull, Middlesbrough and York at 17,000ft, and was commended in official reports as 'a very good performance.'

It was also noted in the Intelligence Report:

The ceiling of the Zeppelin in 1916 was only 12,000–13,000ft. Now, however, the efforts of the German constructors have enabled it to rise at least 20,000ft, the original heavy load which was assayed at the end of 1916 has been resumed, the danger from aeroplane attack having so greatly diminished at the great height at which the raiders now fly. The aeroplane danger, however, still exists for the airship and on this account it would appear that clear nights have been purposely avoided by the German airship commanders. The preference which they now show for very cloudy and misty weather greatly hampers the operations of our aeroplanes but on the other hand it makes it almost impossible for the airship commanders to know precisely where they are and so long as their operations are not directed against an enormous target such as London, Manchester of Glasgow, the chances are the majority of their bombs will fall harmlessly. Otherwise, landing ground flares, searchlights and guns now appear to be favourite targets

12/13 April 1918

Five naval Zeppelins left the north German sheds under the command of Fregattenkapitän Strasser on 12 April. Their chief objective would seem to have been the Midland manufacturing towns. In the communiqué issued after the raid the German naval airship command claimed to have bombed Birmingham, Nottingham, Leeds, Hull and Grimsby. In actual fact, of the places mentioned, only the environs of Birmingham and the river bank opposite Hull were bombed.

L-62, under the command of Hauptmann Kuno Manger, crossed the Norfolk coast at Overstrand at 9.30 p.m. and pursued a steady west-south-west course towards her target of Birmingham. Passing Bodham and Corpusty at 9.45 p.m., Foulsham at 9.50 p.m., and Litcham at 10 p.m., she was over Swaffham at 10.05 p.m. Turning west, the Zeppelin passed Downham Market at 10.15 p.m. and then veered off to the north-west, dropping her first bombs 200 yards west of Middle Drove Station at 10.20 p.m. The two 100kg bombs both exploded, breaking some windows in the properties near where they fell. A searchlight located about 1 mile from where the bomb dropped was probably the objective.

L-62 then made its way to Tydd St Mary where, at 10.25 p.m., three 100kg bombs were dropped in a cornfield 1,000 yards east of the aerodrome, and five incendiaries 600 yards further on, doing no damage whatever.

An aviator who endeavoured to engage the Zeppelin described her as 'sitting' over the aerodrome at about 18,000ft while she dropped her bombs, and she went off before he could get near her, turning sharply southward to elude pursuit and then apparently circling about for a time before resuming her westward course. At 10.35 p.m. the raider was near Wisbech, and at about 10.50 p.m., north of Murrow. The aviator appears to have lost her as soon as she doubled south, as he was under the impression he was still pursuing her in a north-westerly direction, until he finally crashed owing to lack of petrol at Buckminster.

The Zeppelin actually went west-south-west after passing Murrow, and dropped a 50kg bomb, aimed at a searchlight 1 mile east of Nassington, at 11.05 p.m. breaking a window. Five minutes later she was at Oundle, and then passed between Glendon and Corby.

At 11.35 p.m. she was heard from Lutterworth, and at 11.42 p.m. the Wyken and Radford guns near Coventry opened fire upon her, continuing in action for five minutes; although she was only momentarily visible, the lights failing to pick her up properly. In answer to this fire, she turned south-west and, at 11.45 p.m., dropped thirteen bombs south-east of Coventry on a length of about 1 mile, running from south-east to north-west parallel with the Willenhall–Coventry Road, between Baginton Sewage Farm and Whitley Abbey Park. Of these bombs, two were of 300kg weight, two of 100kg and nine incendiaries – all fell in fields. At Baginton a bullock, a heifer and a lamb were killed, at Whitley Abbey windows were broken by the concussion of the 300kg bomb that fell in the park. There were no casualties.

From Coventry the Zeppelin passed on north of Kenilworth, and between 11.50 p.m. and 11.55 p.m. dropped six 100kg HE bombs, in groups of two, between Packwood and Monkspath. No damage was done, except some window glass at Hockley Heath. There were no casualties.

L-62 then approached Birmingham directly at a speed of 60mph. At 11.57 p.m. the gun at Lodge Hill opened fire with anti-Zeppelin (AZ) shells, followed a minute later by guns at South Yardley and Brandwood End. The Zeppelin was now between Yardley and Hallgreen; she suddenly turned and dropped two 300kg bombs roughly at midnight, one at Hallgreen and the other at Shirley while receding south-east, and went off over Solihull in a north-easterly direction. The first bomb caused slight damage to a farm and cottages, and the second broke some glass in shops and houses, there were no casualties.

At approximately 12.10 a.m. the Zeppelin was near Water Orton, and five minutes later passed over Baxterley. When over Atherstone a minute or two afterwards, she had turned again south-west and, at 12.24 a.m., was reported between Longford and Coventry, going east.

The Zeppelin then circled north of Coventry, under the fire of the Wyken gun which again opened at 12.30 a.m., followed at 12.34 a.m. by Radford. They had ceased fire by 12.38 a.m., the Zeppelin having gone off north-east towards Hinckley.

She then turned suddenly southwards, and at 12.45 a.m. passed over Rugby, going east. Two or three British aeroplanes were searching for the raider in their neighbourhood, and one of them engaged her, the pilot receiving a wound to his head. The presence of the aircraft no doubt accounted for the erratic movements of L-62.

At 12.50 a.m. she was east of Lutterworth and at 1 a.m. just north-east of Daventry pursuing a steady course eastward. At 1.15 a.m. she passed between Northampton and Brixworth, at 1.20 a.m. between Burton Latimer and Wellingborough, and at 1.30 a.m. between Thrapston and Raunds. Veering south-east at 1.45 a.m., she passed over Huntingdon and five minutes later over St Ives.

At 2 a.m. she was near Littleport, and 15 minutes later near Lakenheath and Brandon. The raider was spotted north of Thetford at 2.25 a.m., Attleborough at 2.35 a.m. and Forncett at 2.45 a.m. From 2.46 a.m. to 2.49 a.m. one of the

Pulham 18-pdr anti-aircraft guns at 'The Beeches' was in action against her, firing seventeen AZ rounds.

At 2.50 a.m. she passed Long Stratton, and five minutes later was west of Loddon. Here, she appeared to stop her engines in order to ascertain her whereabouts and then went past Loddon at about 3 a.m. There were several British aeroplanes in the neighbourhood. Owing to the heavy mist prevailing at that time she was generally invisible, and the sound of her engines appears to have been confused with that of the aircraft pursing her.

Seen from Pakefield, Suffolk, at 3.10 a.m., she descended to 11,000ft to escape the headwind and approached Yarmouth from the west. Yarmouth was cloaked in fog so the 3in 20 cwt AA gun at Nelson's Monument could only open fire towards her sound at 3.20 a.m. The firing was suspended soon afterwards, on account of the presence of British aircraft in pursuit of the Zeppelin. The firing, however, had driven L-62 south-westward, and she passed over Herringfleet at 3.25 a.m., then Gorleston at 3.34 a.m. before going north-east and out to sea.

The guns later fired towards the sound of targets at 3.55 a.m. and again at 4.15 a.m., when there was no hostile aircraft overland anywhere in England, thus illustrating the confusion and danger that could arise in cloudy and misty weather if British aeroplanes passed over the areas defended by AA guns.

L-63 was first heard by Skegness, passing to the south at 10.05 p.m. and steering west until 10.25 p.m. when south of Coningsby. Three minutes later, when reported by Billinghay, she was steering north-west and at 10.29 p.m. the Brauncewell gun came into action against her and continued to fire at her until 10.34 p.m.

The first bomb thrown by the Zeppelin was a 100kg HE that harmlessly dropped at 10.30 p.m. on a field at Blankney Farm, a few miles north-east of Cranwell Aerodrome, the flares of which may have been the intended target.

The raider now steered due north, and at 10.35 p.m. eighteen bombs were dropped 1 mile east of Metheringham. Two of these were of 300kg weight, fifteen were of 50kg and the last was an incendiary bomb. No casualties were caused, and the only damage was to windows in the vicinity. After dropping the bombs the raider made off to the southward, and was not observed again until half an hour later when she reappeared near Spalding, going east.

At about 11.10 p.m. she dropped one incendiary bomb at Fleet and five at Little Sutton, to no effect. She then headed northward over the Wash to Skegness, which she passed going north at 11.45 p.m. The raider appears to have lingered around the mouth of the Wash for half an hour and then went south-east to Hunstanton, which she passed going east at 12.30 a.m. Hugging the coast until 1.10 a.m., she was spotted off Cromer, after which she went out to sea.

L-64 came in at Saltfleet at approximately 9.45 p.m. and was seen at Louth steering south-west at 10 p.m. At 10.02 p.m. an incendiary bomb was dropped at Biscathorpe, but did no damage. At 10.15 p.m. the raider circled to the north of Lincoln, during which the gun at Burton Road opened on her at 10.28 p.m. She passed on south-west, and retaliated by dropping two 100kg bombs at Skellingthorpe, followed by twelve 50kg HE bombs at Doddington, one of which did not explode. At Skellingthorpe the railway, an engine shed and telegraph wires were damaged. Three men were wounded by anti-aircraft shell fragments, and some windows were broken at Doddington. There were lights showing at Skellingthorpe and Doddington, there having been no warning given that a raid was in progress, because the buzzer at Lincoln which usually gave the warning was not heard. The raider was followed by gunfire until 10.45 p.m., when the sound of her engines had died away to the south-west.

Shortly after, between 10.40 and 10.45 p.m. she doubled back, and dropped four more HE bombs: a 50kg HE at Waddington, and two 50kg and one 300kg at Mere, south of Lincoln. The 300kg bomb failed to explode, and the others caused no damage. The flares of the aerodrome at Waddington were lit, and almost certainly the target of these bombs.

At 10.54 p.m. the Brauncewell gun came into action again against the Zeppelin, and fired upon her until 11.07 p.m. when she was to the north of Swineshead. At 11.35 p.m. Spilsby reported her to the southward, travelling east and at midnight she was over the Wash, in the neighbourhood of Wainfleet. Turning north overland again at 12.05 a.m., near Burgh-le-Marsh, twenty minutes later she was off Skegness. She then went north along the coast, finally circling in the neighbourhood of Mablethorpe until 1.05 a.m. when she went out to sea.

L-60, commanded by Korvettenkapitän Flemming, was only overland about an hour. She came in south of Spurn Head at 9.20 p.m. and went up the estuary under the fire of the Cleefield, Scartho, Immingham and Chase Hill guns. She probably struck inland at East Halton, where her first and last bombs were dropped. Four 100kg bombs, three of which did not explode, eight 50kg and one incendiary were thrown, killing two sheep and slightly damaging a signal box.

Pursuing an easterly course, a number of bombs were dropped in rapid succession roughly between 9.40 p.m. and 9.55 p.m. at Thornton Abbey and Thornton Curtis, on the outskirts of Barton-on-Humber, and at Horkstow and at Saxby. Three 100kg, eight 50kg and ten incendiary bombs fell, but beyond damaging some telegraph wires and glass there was no further damage.

After throwing her last bombs at Saxby, the raider turned north, then north-east, passing around Hull and coming under fire from Harpings and Sutton guns at 10.14 p.m. and 10.16 p.m. respectively. The Sutton gun reported the Zeppelin departing east-north-east, but she did not immediately go out to sea as she was heard at Tunstall to the north-west, flying east. The German communiqué stated that the commander of the raider claimed he had bombed Grimsby.

L-61, under Kapitänleutnant Herbert Ehrlich, appeared to have Leeds as its target. Crossing the coast near Withernsea at 9.30 p.m., at 9.35 p.m. fire was opened on her by the Paull and Marfleet guns, to the north of the Humber estuary, and these may have caused the raider to divert south-west, as at 9.50 p.m. she passed over Brigg.

Passing over the Peak District, she reached Bold, between Warrington and St Helens, Lancashire, where at about 11.10 p.m. she dropped two 50kg bombs, one falling in the main road and one in a field close to Clock Face Colliery. There were no casualties, but there was damage to telegraph wires and a water main.

Proceeding north, the raider dropped bombs at Ince on the southern outskirts of Wigan at about 11.30 p.m. As no warning had been received, the blast furnaces of the Wigan Coal & Iron Company had not been damped down and were in full blaze. Three 100kg bombs, one 50kg, which did not explode, and four incendiaries, one of which did not ignite, were dropped on Ince. Only one man was injured, but considerable damage was done to cottage property. A signal cabin was also damaged, two railway wagons were wrecked and a portion of the permanent way of the Lancashire & Yorkshire Railway main line was destroyed.

Proceeding over Wigan, fifteen more bombs were dropped on the borough, of which seven were 100kg, six 50kg and two incendiary bombs.

Five people were killed:

Bamlett, John Robinsom (42), a discharged soldier, killed in the street.

Fordham, Jane Ann (67), buried in the kitchen of her wrecked house.

Kershaw Henry Wright (8) was going to take shelter in a large cupboard with his mother, who was carrying his younger brother, when a bomb fell outside their front door.

Middleton Joseph (11)

Readman Ellen (4)

A further 14 people were injured. Three small houses were completely demolished, and considerable damage was done to others.

Steering north-east, the raider was over New Springs, Aspull at about 11.38 p.m. and there dropped four 300kg bombs. They fell in a field, but considerable damage was done by the concussion to cottages in the vicinity and a fire broke out at the Crown Brewery, resulting in damage estimated at £1,630. Four people were injured. From Aspull, L-61 went off due east, dropping her remaining bombs, both incendiaries, at Little Hulton about 11.40 p.m. and at Outwood, Radcliffe, a few minutes later. Both fell harmlessly in fields. Passing over Middleton the raider dropped a flare, and at 11.59 p.m. the sound of her engines was heard north-west of Oldham.

Resuming her easterly course, the raider passed over Pontefract at 12.40 a.m. and, at 12.51 a.m., came under fire from the Selby and Howden defences. She turned south-east, and was next heard north-east of Goole at about 1 a.m. and over Staddlethorpe at 1.09 a.m.

At 1.25 a.m. the Hull guns came into action against her. The raider hovered over the town for half an hour, and it was not until 2.03 a.m. that the guns on the east side of the city and south of the estuary opened on her. She then turned south-east, and at 2.38 a.m. was fired on by the Spurn and Kilnsea guns and was last heard from Withernsea, out at sea at 2.50 a.m., where HMS *Albion* also fired several rounds at her.

The consensus of opinion from guns, pilots and observers, was that the Zeppelins had navigated at 16,000–20,000ft during this raid.

5/6 August 1918

Götterdämmerung

On the afternoon of 5 August 1918, five airships were despatched to carry out a raid on the Midlands. Three of them, **L-53** (under the command of Kapitänleutnant Eduard Prölss), **L-65**, (under the command of Kapitänleutnant Walter Dose) and **L-70**, (under the command of Kapitänleutnant Johann von Lossnitzer) which was also carrying Zeppelin chief Peter Strasser, approached the coast together.

At 8.10 p.m. the three Zeppelins were sighted by the *Leman Trail* lightship, bearing east about 8 miles, steering west-north-west and flying low. Messages were relayed across the sea until the east coast defences were alerted and the daylight warning was sounded. The *Eastern Daily Press* reported: 'There was no sign of alarm; just the anxiety to get cover in the event of anything happening. The police and special constables took up their positions and all arrangements in case of emergency were made.'

At 9.25 p.m. the leading pair, L-70 and L-65, were seen from the *Haisborough* lightship bearing east-north-east about 12 miles, at a very high altitude. Both were plainly visible in a clear patch of sky. They then passed behind the clouds and were not spotted again until half an hour later when L-53, which had fallen behind, passed north of the *Leman Trail* lightship at 9.25 p.m., and went directly over the *Haisborough* lightship heading south-west. The three airships were thus disposed in a 'V' formation, and the leading pair appeared due north of Cromer at 9.50 p.m., dropping flares. Ten minutes later they were north of Wells and L-70 dropped three bombs that fell in the sea very close by the schooner *Amethyst*.

In the meantime, the telephone of the Great Yarmouth RNAS station had rung with the warning of the approach of the Zeppelins. The duty officer,

Captain Robert Leckie, commander of 228 Squadron, the 'Boat Flight', acted immediately. Informing the station commander, Lieutenant Colonel Nicholl, orderlies were sent post-haste across the town to call in officers and men. These included Temporary Major Egbert Cadbury, commander of 212 Squadron, who had been at Wellington Pier where he had been in the audience of a charity concert at which his wife had been due to sing (they were newly married and it had been a wartime romance – she was Mary, the eldest daughter of the Reverend Forbes Phillips, rector of Gorleston.)

At the cinema, the film clattered to a stop and the glass lantern slide projected on the screen stating: 'All officers wanted at Air Station immediately.' Rushing back on foot, bicycle or bundling onto the station's Ford tender, the aircrew and pilots knew the drill and soon, thirteen aircraft – DH-4s, DH-9s and Sopwith Camels – were in the air from Great Yarmouth, Burgh Castle and Covehithe, with twenty more soon joining them in the air over the county from airfields farther inland.

Leckie turned over the station to its permanent commander, and jumped in the observer-gunner seat behind Cadbury, in a DH-4 as its 375hp Rolls-Royce Eagle engine warmed up. Cadbury and Leckie were seasoned in Zeppelin combat. Both had already shot down airships and had been recognised for their gallantry.

Once over the sea, in an attempt to increase his rate of climb, Cadbury ditched the two 100lb bombs used against surface craft that had been fitted to the DH-4's rack. This appeared to make little or no difference; Cadbury put it down to the barometric pressure.

At 9 p.m. Strasser sent a Morse code message to his command: 'To all airships, attack according to plan from Karl 727. Wind at 5,000 meters [16,250ft] west-south-west three doms [13.5 mph]. Leader of Airships.'

At 9.25 p.m. the three Zeppelins were sighted by the pursuing aircraft 'in line abreast with their noses pointing landward.' The other two airships, L-56 and L-63, which were near Yarmouth at the time, were apparently not observed.

At 9.45 p.m. Cadbury emerged from cloud to see the shadowy shapes of the Zeppelins altering course north-westward at 17,000ft, some 2,000ft above the DH-4. The little aircraft's climb was painfully slow, and ten minutes later Cadbury was 400ft below the Zeppelins when he pointed to the airship in the centre, indicating it to Leckie as their target. Once they were within range, at approximately 10.10 p.m., Cadbury concentrated on keeping the plane steady head on, and slightly to port, as Leckie took aim with the twin Lewis guns. He rattled a steady concentration of Pomeroy explosive bullets into the bow of L-70, blowing a great hole in the fabric, and a fire was started which quickly ran along the whole length of the airship.

The burning Zeppelin was seen from the *Leman Trail* lightship, about 40 miles away, and was described as 'a large red flame', the glare of which was seen on land as far away as Reedham (25 miles away). West of Blakeney Bell buoy, and having been narrowly missed by the bombs the Zeppelin had dropped earlier, the *Amethyst* had another lucky escape. When the Zeppelin was coming down, several more bombs fell near the ship, followed soon after by the flaming superstructure of L-70, which landed only 300 yards from the schooner!

On seeing the fate of their companion, the other two airships immediately altered course east and made off at high speed. At about 10.25 p.m., L-53 dropped a large number of bombs in the sea.

Major Cadbury's engine now failed temporarily, but he managed to get it going again and closed with L-65. He again attacked bow-on, and Captain Leckie opened fire within 500ft of it. Fire immediately broke out in the midships gondola, owing probably to the ignition of a deposit of oil and grit accumulated from the exhaust of the forward engine on the underside of the gondola. There is little doubt that this Zeppelin would have also been destroyed had Captain Leckie's gun not jammed with a double-feed which, at the critical moment and in the darkness and perishing cold conditions, could not be cleared; a situation that was not helped since, in the rush to leave Yarmouth, Leckie had forgotten his gauntlets.

The fire in the gondola was extinguished by the crew of the Zeppelin and, though Major Cadbury maintained contact with L-65 for about five minutes, the gun could not be got into working order again and the action had to be broken off. The machine gun fire from the Zeppelin during the attack had been heavy but fortunately it had also been inaccurate.

Another aeroplane attempted to attack L-65 after Cadbury had broken off the fight, but it was too low, being 2,000–2,500ft beneath the target. L-65 and L-63 were fortunate enough to effect their escape into the now dark and rainy night, and eventually returned safely to their sheds.

The remaining two airships, L-56 (under Kapitänleutnant Walter Zaeschmar) and L-63 (under Kapitänleutnant Michael von Freudenreich) had proceeded from Great Yarmouth up the coast, being seen from Caister (over which one of them passed), the *Newarp* lightship and Winterton between 9.10 p.m. and 9.36 p.m. and off Haisborough at 9.42 p.m. They appear to have dropped flares between Mundesley and Overstrand at 10.05 p.m.

The catastrophe to L-70 occurred at 10.25 p.m., as Kapitänleutnant Freudenreich was to record:

I was nearing the coast when we suddenly saw an outbreak of flame on L-70, amidships or a little aft. Then the whole ship was on fire. One could see flames all over her. It looked like a huge sun. Then she stood up erect and went down like a burning shaft. The whole thing lasted thirty, maybe forty-five seconds.

Major Egbert Cadbury DSC DFC (left) and Captain Robert Leckie DSO DSC DFC, photographed at RNAS Great Yarmouth a few hours after they shot down Zeppelin L-70.

Fearing that the same fate would befall him and his crew, L-63 made off north-north-east, making a considerable detour in order to ensure her safe return.

Meanwhile, L-56 went south-east back along the coast and, over an hour later at 11.45 p.m., appeared off Great Yarmouth. The GHQ Intelligence Section notes state:

> Apparently her commander had no mind to return to Germany without having attempted to do some damage and the danger from aeroplanes which had attacked L-70 and L-65 having now presumably been evaded, he essayed an attack upon Lowestoft. Three bombs were dropped in the sea and then the airship seems to have crossed the coast at 11.56 p.m. flew from north-west to south-east over Lowestoft (evidently without her commander knowing where he was, as she dropped no bombs) and out to sea again and between midnight and 12.15 a.m. dropped fourteen or fifteen bombs in the water. No damage was done to any craft.

Cadbury and Leckie were a long way from their base at Yarmouth but, sighting the flares of Sedgeford, made a safe landing at 11.05 p.m. It was only when he jumped down from the cockpit that Cadbury realised why the DH-4 had climbed so sluggishly. The two bombs that he thought he had released into the sea were still in place (and primed), but by some miracle were still intact despite a rough landing. Just one more jolt and the DH-4 could well have been blown to pieces! Later, both Cadbury and Leckie were recognised for their action, with the award of the Distinguished Flying Cross (DFC).

The RAF (the RFC and RNAS had combined) had sent up thirty-three aeroplanes but, owing to bad weather, the inland machines were not able to see anything. Lieutenant Benitz of 33 Squadron was killed when his plane crashed, and a Sopwith Camel flown by Lieutenant G.F. Hodson, and a DH-9 crewed by Captain B.G. Jardine and Lieutenant E.R. Munday, did not return.

Aftermath

The German official communiqué was particularly imaginative in its account of the action:

> In the night 5th–6th August the so often successful leader of our airship attacked, Fregattenkapitän Strasser, with one of our airship squadrons, again damaged severely the east coast of Middle England with effectual bomb attacks, especially on Boston, Norwich and the fortifications at the mouth of the Humber. He probably met a hero's death in the raid with the brave crew of his flagship. All the other airships that took part in the attack returned without loss

or damage in spite of strong opposition. Besides their experienced fallen leader the airship commanders Korvettenkapitän J.R. Prölss, Kapitänleutnants Walther Zaeschmar, von Freudenreich and Dose took part with their brave crews in the success.

A search of the sea soon revealed that neither Strasser nor any of his twenty-two crew had survived. The British military authorities did not wait for the sea to give up its dead. Within two days, a trawler in naval service had located and buoyed the wreckage of L-70. Over the following three weeks, much of the wreck was recovered by divers and wire drags. From this watery grave came code books and all manner of airship intelligence, even the barograph that recorded that she had just reached 5,500m when she fell.

The body of Peter Strasser and a number of crew were also found; others were washed up on the nearby coastline. All were buried at sea, and with them the future of the Zeppelin in combat.

APPENDIX 1

Aircraft Raids 1914–18*

As an aid to researchers and to avoid confusion between what were Zeppelin or aeroplane raids, what follows is a list of aeroplane raids conducted 1914–18, originally published in *The German Air Raids on Great Britain 1914–1918* by Captain Joseph Morris (London, 1925).

Note: Those attacks shown as taking place at night occurred between dusk and dawn. Where the raids continued beyond midnight the double dates are given.

Date	Day or Night	Locality	Number of bombs dropped	Casualties
1914				
24 December	Day	Dover	1	0
25 December	Day	Thames up to Erith	2 (on Cliffe, Kent)	0
1915				
21 February	Night	Essex, Braintree, Coggeshall, Colchester	4	0
16 April	Day	Kent, Faversham, Sittingbourne, Deal	10	0

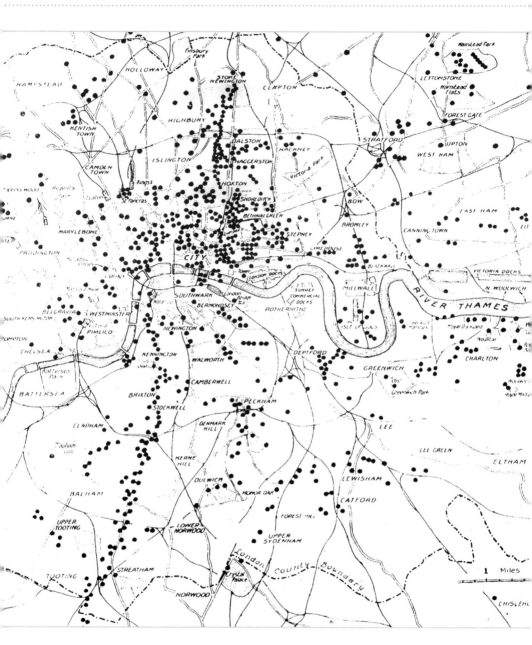

Map of the Zeppelin and aeroplane bombs dropped on London during the First World War, produced in 1919.

Date	Day or Night	Locality	Number of bombs dropped	Casualties
3 July	Day	East Suffolk	No bombs dropped on land	0
13 September	Day	Margate	10	2 killed, 6 injured
1916				
9 January	Day	Dover	No bombs dropped	0
23 January	Night	Dover	9	1 killed, 6 injured
23 January	Day	Dover, Folkestone	5	0
24 January	Day	Dover, Folkestone	No bombs dropped	0
9 February	Day	Broadstairs, Ramsgate	13	3 injured
20 February	Day	Walmer, Lowestoft	25	1 killed, 1 injured
1 March	Night	Broadstairs, Margate	7	1 killed
19 March	Day	Dover, Deal, Margate, Ramsgate	48	14 killed, 26 injured
23 April	Day	Dover	No bombs dropped	0
24 April	Day	Dover	No bombs dropped	0
3 May	Day	Deal	9	4 injured
20 May	Night	Kent, Dover	59	1 killed, 2 injured
9 July	Day	North Foreland	No bombs dropped	0
9–10 July	Night	Dover	7	0
12 August	Day	Dover	4	7 injured
22 September	Day	Dover	7	0
22 October	Day	Sheerness	4	
23 October	Day	Margate	3	2 injured
28 November	Day	London	6	10 injured

Date	Day or Night	Locality	Number of bombs dropped	Casualties
1917				
14 February	Day	Deal	No bombs dropped	0
1 March	Day	Broadstairs	9	6 injured
16 March	Night	Margate, Westgate	21	0
17 March	Day	Dover	5	0
5 April	Night	Kent	8	0
6–7 May	Night	London	5	1 killed, 2 injured
25 May	Day	Kent, Folkstone	159	95 killed, 192 injured
5 June	Day	Essex, Kent	64	13 killed, 34 injured
13 June	Day	Kent, Margate, Essex, London	126	162 killed, 432 injured
4 July	Day	Harwich District	42	17 killed, 30 injured
7 July	Day	Margate, London	75	57 killed, 193 injured
22 July	Day	Harwich District	55	13 killed, 26 injured
12 August	Day	Essex, Southend, Kent, Margate	37	32 killed, 46 injured
22 August	Day	Kent, Margate, Ramsgate, Dover	50	12 killed, 25 injured
2 September	Night	Dover	14	1 killed, 6 injured
3–4 September	Night	Kent, Chatham, Margate	46	132 killed, 96 injured
4–5 September	Night	East Suffolk, Essex, Kent, Dover, Margate, London	90	19 killed, 71 injured
24 September	Night	Kent, Dover, East Suffolk, Essex, London	118	21 killed, 71 injured
25 September	Night	Kent, London	60	9 killed, 23 injured

Date	Day or Night	Locality	Number of bombs dropped	Casualties
28 September	Night	Kent, Essex, East Suffolk	45	0
29–30 September	Night	Kent, Essex, London	55	14 killed, 87 injured
30 September	Night	Kent, Margate, Dover, Rochester, Essex, Southend, London	92	14 killed, 38 injured
1 October	Night	Kent, Essex, London	73	11 killed, 41 injured
29 October	Night	Essex	8	0
31 October	Night	Kent, Dover	16	0
31 October–1 November	Night	Kent, Thanet, Essex, London	278	10 killed, 22 injured
6 December	Night	Kent, Thanet, Essex, London	421	8 killed, 28 injured

Crater left by a bomb dropped from an aircraft onto Coggeshall, 21 February 1915.

Date	Day or Night	Locality	Number of bombs dropped	Casualties
18 December	Night	Kent, Thanet, Essex, London	142	14 killed, 85 injured
22 December	Night	Westgate, Broadstairs, Ramsgate	No bombs dropped on land	0
1918				
28–29 January	Night	Kent, Margate, Ramsgate, Sheerness, Essex, London	69	10 killed, 10 injured
16 February	Night	Kent, London	29	12 killed, 6 injured
17–18 February	Night	London	16	21 killed, 32 injured
7–8 March	Night	Essex, Bedfordshire, Hertfordshire, Kent, London	29	23 killed, 39 injured
19–20 May	Night	Essex, Southend, Kent, Margate, Dover Rochester, London	155	49 killed, 177 injured
17 June	Day	Kent Coast	No bombs dropped	0
18 July	Day	Kent Coast	No bombs dropped	0
20 July	Day	Kent Coast	No bombs dropped	0

APPENDIX 2

German Airship and Zeppelin Aircrew now buried at the Friedhof – the German Military Cemetery at Cannock Chase, Staffordshire.

SL-11

(Sixteen crew) brought down at Cuffley on 3 September 1916, East Terrace, Grave 1.

Obermaschinist Jakob Baumann

Obermaschinist Hans Geitel

Goltz, Vizefeldwebel Rudolf

Feldwebelleutnant Karl Paul Hassenmuller

Gefreiter Bernhard Jeziorski

Untermaschinist Fritz Jourdan

Untermaschinist Karl Kachele

Obersteuermann Fritz Kopischke

Obermaschinist Friedrich Modinger

Obermaschinist Reinhold Porath

Obersteuermann Rudolf Sendzik

Unteroffizier Heinrich Schlichting

Hauptmann Wilhelm Schramm

Unteroffizier Anton Tristram

Oberleutnant Wilhelm Vohdin

Untermaschinist Hans Winkler

WILHELM SCHRAMM HAUPTMANN

JAKOB BAUMANN OBERMASCHINIST
HANS GEITEL LEUTNANT
RUDOLF GOLTZ VIZEFELDWEBEL
KARL HASSENMÜLLER FELDWEBEL-LEUTNANT
BERNHARD JEZIORSKI GEFREITER
FRITZ JOURDAN UNTERMASCHINIST
KARL KACHELE UNTERMASCHINIST
FRITZ KOPISCHKE OBERSTEUERMANN
FRIEDRICH MÖDINGER OBERMASCHINIST
REINHOLD PORATH OBERMASCHINIST
RUDOLF SENDZICK OBERSTEUERMANN
HEINRICH SCHLICHTING UNTEROFFIZIER
ANTON TRISTRAM UNTEROFFIZIER
WILHELM VOHDIN OBERLEUTNANT
HANS WINKLER UNTERMASCHINIST

SL 11 + 3 · 9 · 1916

WERNER PETERSON OBERLEUTNANT ZUR SEE

ADOLF BLEY OBERSIGNALMAAT
ALBIN BOCKSCH OBERMASCHINISTENMAAT
KARL BORTSCHELLER FUNKENTELEGRAFIEOBERMAAT
WILHELM BROCKHAUS OBERHEIZER
KARL BRODRUCK LEUTNANT ZUR SEE
PAUL DORFMULLER MASCHINISTENMAAT
RICHARD FANKHANEL OBERMASCHINISTENMAAT
GEORG HAGEDORN OBERMASCHINISTENMAAT
FRIEDRICH HEIDER OBERBOOTSMANNSMAAT
ROBERT KLISCH FUNKENTELEGRAFIEOBERGAST
HERMANN MAEGDLFRAU OBERMASCHINIST
BERNHARD MOHR OBERSEGELMACHERSGAST
AUGUST MULLER MATROSE
FRIEDRICH PASCHE BOOTSMANNSMAAT
KARL PAUST OBERMASCHINISTENMAAT
EWALD PICARD OBERSIGNALMAAT
WALTER PRUSS MASCHINISTENMAAT
PAUL SCHIERING OBERMATROSE
BERNHARD SCHREIBMULLER STEUERMANN
KARL VOLKER OBERMASCHINISTENMAAT
ALFRED ZÖPEL OBERBOOTSMANNSMAAT

L·32 + 24 · 9 · 1916

HEINRICH MATHY KAPITÄNLEUTNANT

EUGEN BOUDANGE MASCHINISTENMAAT
ARTHUR BUDWITZ BOOTSMANNSMAAT
KARL DORNBUSCH OBERMATROSE
NIKOLAUS HEMMERLING MASCHINISTENMAAT
KARL HIORT OBERMASCHINISTENMAAT
ERNST KAISER SEGELMACHERSMAAT
ERNST KLEE FUNKENTELEGRAFIEOBERGAST
SIEGFRIED KORBER STEUERMANN
GUSTAV KUNISCH SIGNALMAAT
KARL MENSING MASCHINISTENMAAT
FRIEDRICH PETERS OBERSTEUERMANNSMAAT
HEINRICH PHILLIPP OBERMATROSE
FRIEDRICH ROHR MASCHINISTENMAAT
HUBERT STENDER MASCHINISTENMAAT
JOSEPH WEGENER MASCHINIST
JOCHEN WERNER LEUTNANT ZUR SEE
HEINRICH WITTHOFT BOOTSMANNSMAAT
VIKTOR WOELLERT OBERMASCHINISTENMAAT

L·31 + 2 · 10 · 1916

Grave markers and burial plots of the Zeppelin aircrew buried in the German Military Cemetery, Cannock Chase, Staffordshire.

The crew of SL-11were given a military funeral at Potter's Bar Cemetery on Mutton Lane. Men of the Royal Flying Corps bore the coffins, the service was conducted by a vicar and a military chaplain who conducted an Anglican form of burial service, and buglers from the Grenadier Guards sounded the 'Last Post'.

L-32

(Twenty-two crew) brought down at Snail's Hall Farm, Great Burstead, 24 September 1916, East Terrace, Grave 2.

Obersignalmaat Adolf Bley
Obermaschinistenmaat Albin Ernst Bocksch
Funkentelegraphieobermaat, Karl Bortscheller
Oberheizer Wilhelm Otto Brockhaus
Leutnant Zur See Karl Friedrich Brodrück
Maschinistenmaat Paul Dorfmüller
Obermaschinistenmaat Richard Hermann Fankhanel
Obermaschinistenmaat Georg Hagedorn
Oberbootsmannsmaat Friedrich Heider
Funkentelegraphieobergast Robert Klisch
Obermaschinist Hermann Franz Oswald Maegdlfrau
Obersegelmachersgast Bernhard Johannes Mohr
Matrose August Müller
Bootsmannsmaat Friedrich Paul Pache
Obermaschinistenmaat Karl Hermann Paust
Oberleutnant Zur See Werner Chustau von Peterson,
Obersignalmaat Ewald Johann Heinrich Picard
Maschinistenmaat Walter Prüss
Obermatrose Paul Rudolf Schiering
Steuermann Berhard Johann Schreibmüller
Sailor Karl Johann Petin Volker
Oberbootsmannsmaat Alfred Robert Johannes Zöpel

The crew of L-32 were buried in the nearby churchyard of St Mary Magdalene, Great Burstead, three days after the crash. Officers and men of the Royal Flying Corps bore the coffins, a vicar and an army chaplain conducted the graveside service. The bodies were transferred to the German Military Cemetery at Cannock Chase in November 1962, where they were buried in caskets in one grave.

L-31

Brought down at Potters Bar, 2 October 1916.

Kapitänleutnant Heinrich Ferdinard Fredrich Mathy
Maschinistenmaat Eugen Boudange
Bootsmannsmaat Arthur Budwith
Obermatrose Karl Dornbusch
Maschinistenmaat Nikolaus Hemmerling
Obermaschinistenmaat Karl Hiort
Segelmachersmaat Ernst Kaiser
Funkentelegraphieobergast Ernst Klee
Steuermann Siegfried Korber
Signalmaat Gustav Kunish
Maschinistenmaat Karl Friedrich Wilhelm Mensing
Obersteuermannsmaat Friedrich Joahnn Peters
Obermatrose Heinrich Phillipp
Maschinistenmaat Friedrich Karl Christian Rohr
Maschinistenmaat Hubort Karl Ernst Asmin Stender
Maschinist Joseph Friedrich Wegener
Leutnant Zur See Jochen Julius Otto Hubertus Werner
Bootsmannsmaat Heinrich Witthöft
Obermaschinistenmaat Viktor Karl Friedrich Wilhelm Woellert

The bodies of the German aircrew from L-31 were buried at Mutton Lane Cemetery, alongside their comrades from the SL-11 Cuffley crash on 5 October 1916. Each crew were buried in caskets in one grave at the German Military Cemetery in September 1962.

L-34

Brought down at the mouth of the Tees, 28 November 1916.

Signalmaat Julius Wilhelm August Petitjean. Grave Reference: Block 14, Grave 397.
Maschinistenmaat Alfred Rueger. Grave Reference: Block 14, Grave 398.

Initially none of the bodies of the twenty-two crew were recovered from the submerged wreck, but the following January five bodies were washed up and buried in Redcar Cemetery, Yorkshire. Only two of them – Petitjean and Rueger, – were identified. They were transferred to the German Military Cemetery in June 1962. Around the same time, two further unidentified comrades were exhumed from Holy Trinity Churchyard at Seaton Carew, and laid to rest in Graves 402 and 403 in the same row.

L-48

Brought down at Theberton, Suffolk, 17 June 1917.

Kapitänleutnant Franz Eichler
Korvettenkapitän Viktor Schutze
Obermaschinistenmaat Heinrich Ahrens
Maat Wilhelm Betz
Obersignalmaat Walter Dippmann
Obermaschinistenmaat Wilhelm Glückel
Obermaschinistenmaat Paul Hannemann
Signalmaat Heinrich Herbst
Bootsmannsmaat Franz Konig
Funkentelegrafiemaat Wilhelm Meyer
Obermaschinistenmaat Karl Milich
Obermaschinistenmaat Michael Neunzig
Obermatrose Karl Ploger
Obermatrose Paul Suchlich
Obermaschinistenmaat Hermann Van Stockum
Steuermann Paul Westphal

Sixteen crew died, fourteen of them were buried three days later in St Peter's Churchyard extension, Theberton. Another two were buried a further two days later, after their bodies had been recovered from the wreckage. They were granted military honours, the funerals were conducted by the rector, a military chaplain and a Roman Catholic priest. Upon the coffin of Korvettenkapitän Schütze was a wreath bearing the inscription 'To a brave enemy, from R.F.C. Officers', a gesture criticised by local people who were fed up with the Zeppelin menace. (Indeed, one story tells of how local men refused to dig the grave and munitions girls from Garrett's Works at nearby Leiston were called in to do the job.)

The bodies of the crew of L-48 were transferred to the German Military Cemetery at Cannock Chase in January 1962. Although the bodies are gone, the site of the graves in Theberton is still marked and retains the epitaph from Romans XIV–IV: 'who art thou that judgest another man's servant?'

L-70

Brought down in the sea off Wells-next-the-Sea, Norfolk, on 5 August 1918.

All twenty-two crew were lost, including Zeppelin chief Peter Strasser. Personnel from Admiralty Intelligence spent three weeks searching the wreck. Any bodies that were recovered were searched, weighted down and sunk again. Some of the bodies were washed up on a beach in Lincolnshire, but they were ordered to be taken out by boat and committed to the deep, as local people refused them burial in their parish.

Later still, in October 1918, a body was washed ashore in north Norfolk and was buried in All Saints Churchyard, Weybourne. The body was claimed to be that of Leutnant Kurt Krüger, but the official records simply show him as an unknown officer. In January 1963, the remains were removed to the German Military Cemetery at Cannock Chase and interred in Block 16, Row 9, Grave 116.

APPENDIX 3

'I was London's First Zepp Raider' by Major Erich Linnarz

Major Linnarz, who was commander of Zeppelin LZ.38, had been four times over England before, on May 31, 1915, he succeeded in reaching London. This was London's first air raid.

It was not until January 1915 that the Kaiser at last sanctioned the bombing of England, and not until four months later that he was prevailed upon by his advisers to give his consent to attacking London. The proud ship LZ.38, the latest product of Count Zeppelin's works at Friedrichschafen on Lake Constance, which I commanded, was one of those detailed for the job.

On the morning of May 31 orders in cipher were brought from Berlin to me at Brussels to raid London. Preparations for the flight were carried out all that day. Engines were tested, ballast tanks examined, the radio apparatus thoroughly overhauled, and the huge deflated envelope closely inspected for flaws. Presently there was the hiss of gas and slowly the monster took a more rigid shape. Then the bomb-racks were loaded. One hundred and nineteen bombs there were in all – eighty nine incendiary, thirty high explosive ones. A ton and a half of death.

As the perspiring soldiers wheeled the infernal things on trucks before placing them in position, the setting sun sank behind the shed and stained the sky a deeper and more ominous red. My crew, clad in their leather jackets and fur helmets, were standing in groups on the landing ground. A siren sounded shrilly and they moved to the shed, entered the gondola and took up their posts. Gently guided by ropes the ship slid smoothly forward. The sounding of a second siren indicated that the ship was clear of its shed.

'Hands off, ease the guides,' I shouted. The men at the ropes let go.

Great Eddies of dust swept through the air as the final test to the mammoth propellers was given. An officer approached and told me all was ready, I stepped in, gave a signal, and mysteriously the ship soared upwards. We were on our way to London.

From my cabin, with its softly lit dials – everyone with a story to tell – its maps, its charts, and compass, I could hear the rhythmic throb of the engines; feel the languorous swing of the gondola as we rode smoothly through space. Over invaded Belgium we flew. Here it was that one of my crew at the helm reported that he had sighted what he thought to be a hostile airship approaching. For safety I altered course and steered in the direction of Ostend.

Often raiding Zeppelins, on their way out from Belgium to England, encountered enemy craft endeavouring to intercept their passage. But, as it afterwards turned out, this one was only Captain Lehmann, who was killed in an airship crash in America last year, on one of the other Zeppelins detailed to raid England. He had left Namur earlier, also with London as his aim, but over the Channel he had broken a propeller, which had pierced his gas-bag and forced him to return to his base.

He was on his way back when we saw him. None of the other ships reached London that night, but discharged their bombs on East Coast towns. On, on we sped. It was a beautiful night – a night of star spangled skies and gentle breezes, a night hard to reconcile with a purpose as grim as ours. And then the glimmer of water showed below and we knew we were over the sea. Tiny red specks winked at us. They were patrol boats keeping their ceaseless watch in the Channel, and we were looking down their funnels into the glowing heart of their stoke-hold furnaces. England!

We crossed the black ridge of the coast. Immediately from below anti-aircraft guns spat viciously. We could hear the shells screaming past us. We increased our altitude and our speed. Across the Thames estuary we raced, wheeling inland at Shoeburyness, over Southend, which I had raided the week before – and then, following the gleaming river, we made straight for the capital. Twenty minutes later we were over London. There below us its great expanse lay spread. I knew it all so well. I had spent several months there five years before. There seemed to have been little effort to dim the city. There were the old familiar landmarks – St. Paul's, the Houses of Parliament, and Buckingham Palace, dreaming in the light of the moon which had now risen.

I glanced at the clock. It was ten minutes to eleven. The quivering altimeter showed that our height was 10,000 feet. The air was keen, and we buttoned our jackets as we prepared to deal the first blow against the heart of your great and powerful nation.

Inside the gondola it was pitch dark save for the glowing pointers of the dials. The sliding shutters of the electric lamps with which each one of the crew was provided were drawn. There was tension as I leaned out of one of the gondola portholes and surveyed the lacework of lighted streets and squares. An icy wind lashed my face.

I mounted the bombing platform. My finger hovered on the button that electronically operated the bombing apparatus. Then I pressed it. We waited. Minutes seemed to pass before, above the humming song of the engines, there arose a shattering roar.

Was it fancy that there also leaped from far below the faint cries of tortured souls?

I pressed again. A cascade of orange sparks shot upwards, and a billow of incandescent smoke drifted slowly away to reveal a red gash of raging fire on the face of the wounded city.

One by one, every thirty seconds, the bombs moaned and burst. Flames sprang up like serpents goaded to attack. Taking one of the biggest fires, I was able by it to estimate my speed and my drift. Beside me my second in command carefully watched the result of every bomb and made rapid calculations at the navigation chart.

Suddenly from the depths great swords of light stabbed the sky. One caught the gleam of the aluminium of our gondola, passed it, retraced, caught it again, and then held us in its beam. Instantly the others chased across the sky, and we found ourselves moving through an endless sea of dazzling light. Inside the gondola it was brighter than sunlight. Every detail of the car was thrown in sharp relief. The crew at their posts looked like a set of actors grouped in the limelight without their make-up. And so began a game of hide and seek in the sky. The helmsman and I tried every way of eluding the searchlights, practising every trick of navigation.

Then came the bark of the batteries. Shells shrieked past us, above us, below us. There were glowing tracer shells which we had never seen before, but had heard all about — slim projectiles that tore a hole in the ship's fabric and then burst into flame. It was this thought that sent us home quickly. We had been over London for an hour. Soon we left the thrusting searchlights behind. We could see ahead of us the sea, through which the moon had laid a silver path to guide us home. As we crossed the black ridge of the shore we were met with a further attack from the anti-aircraft guns at Burnham and Southminster. I think our gondola light, now alight and casting a feeble glow over the cabin, perhaps had betrayed us. I put it out. Shell after shell whizzed past, some of them the dreaded incendiary type. Some burst dangerously near. On, on we flew, and at last we were out of range and the firing died down.

Now a new menace threatened us — aeroplanes. We went in dread of these since your pilots had orders that if they failed to reach us with the machine-gun fire they were to climb above us and ram our gas-bags with their machines.

Evidently the supreme sacrifice meant nothing to these brave men. One by one they came from the airfields that had been established round the coast to intercept returning raiders. My look-out thought he spotted one flying towards us. Higher we rose out of reach. The British aeroplanes were faster than we were, but they couldn't reach our height limit.

Presently in the fading moonlight, we could see the waves beating against the Belgian coastline far below. We were feeling cold and hungry, exhausted and spent from the high-pitched hours of that night – rather like the remorseful reveller returning in the hour before the dawn. It was almost dawn. The first vague light was edging the horizon as we flew over invaded Belgium. We had been away ten hours. The first attack on London had been accomplished. Our bomb rack was empty. Behind us we could faintly make out the red glow of fire on the sky's rim. It was ravaged London. And as we sank to the earth and the gondola bumped across the landing-ground at Brussels-Evere, the sun, rising in front of the Zeppelin sheds, smeared the sky with crimson streaks as though fingers dipped in blood had been drawn across the horizon.

(First published in *The Great War: I Was There* Part II, 1939)

SELECT BIBLIOGRAPHY

Books

Andrew, Christopher, *Secret Service: The Making of the British Intelligence Community* (London, 1985)

Baker, Brian, *The Zeppelin Graves on Cannock Chase* (Cannock Chase, 2002)

Buttlar, Horst von, *Zeppelins over England* (London, 1931)

Campbell, Erroll A., *A History of Sheringham and Beeston Regis* (Sheringham, 1970)

Castle H.G., *Fire over England* (London, 1982)

Corbett, Sir Julian and Henry Newbolt, *History of the Great War Naval Operations* (5 vols.) (London, 1921–31)

Davy, Terry, *Dereham in the Great War* (Dereham, 1990)

Dudley, Ernest, *Monsters of the Purple Twilight* (London, 1960)

Hood, Harold, *Illustrated Memorial of the East Coast Raids by the German Navy and Airships* (Middlesbrough, 1915)

Gamble, C.F. Snowden, *The Story of a North Sea Air Station* (London 1928)

Gliddon, Gerald (ed.), *Norfolk & Suffolk in the Great War* (Norwich, 1988)

Gore, L.L., *The History of Hunstanton* (Bognor Regis, 1983)

Hallam, Squadron Leader T.D., *The Spider Web* (Edinburgh, 1919)

Ingleby, Holcombe, *The Zeppelin Raid in West Norfolk* (London, 1915)

Jones, H.A., *The War in the Air*, vols III and V, (London, 1922–35)

Lehmann, Ernst A. and Howard Mingos, *The Zeppelins* (New York, 1927)

McCaffery, Dan, *Air Aces: The Lives and Times of Twelve Canadian Fighter Pilots* (Toronto, Canada, 1990)

Marben, Rolf (ed.), *Zeppelin Adventures* (London, 1931)

Mee, Arthur, *The King's England: Norfolk* (London, 1940)

Morris, Captain Joseph, *The German Air Raids on Great Britain* (London, 1925)

Neumann, George Paul (ed.), *Die Deutshen Luftstreitkräft in Weltkreig* (Berlin, 1920)

Poolman, Kenneth, *Zeppelins over England* (London, 1960)

Rawlinson, A., The *Defence of London 1915–18* (London, 1923)

Rimmell, Raymond, *L-Zeppelin! A Battle for Air Supremacy in World War I* (1984)

Robinson, Douglas H., *The Zeppelin in Combat* (Atglen, Pennsylvania 1994)

Wyatt, R.J., *Death from the Skies* (Norwich, 1990)

Newspapers & Journals

Aeroplane Monthly

B.P. – the works magazine of Boulton & Paul Ltd, Norwich

Daily Express

Daily Mirror

Dereham & Fakenham Times

East Anglian Daily Times

Eastern Daily Press

Illustrated War News

Lynn News

Norfolk Chronicle

Norfolk Fair

Norwich Mercury

Observer

Punch Magazine

Sphere

The Great War: I Was There

The Times

War Illustrated

Wing

Yarmouth Mercury

Sources at the National Archives

AIR 1/2123/207/73/2 to AIR 1/2123/207/73/28, Intelligence Section reports on enemy airship raids on Britain (January 1915–August 1918).
AIR 1/552/16/15/38, Police reports, correspondence and press cuttings re. hostile air raids in England 19 January–24 June 1915.
MEPO 2/1652, Inquest and funeral arrangements for the bodies of the crew of SL-11, Cuffley.

Internet Sources

Commonwealth War Graves Commission: www.cwgc.org/debt_of_honour.asp
The Aerodrome: Aces and Aircraft of World War One: www.theaerodrome.com

ACKNOWLEDGEMENTS

The National Archives; Imperial War Museum, London and Duxford; RAF Museum, Hendon; The Commonwealth War Graves Commission; Kath Griffiths, Norfolk Library and Information Service Heritage Library; Dr John Alban, Susan Maddock and Freda Wilkins-Jones at The Norfolk Record Office; Great Yarmouth Library (Local Studies); Alan Leventhall, King's Lynn Library (Local Studies); BBC Radio Norfolk; The Fleet Air Arm Museum, Yeovilton; Sue Tod, Felixstowe Museum; The Association of Friends of Cannock Chase; The Long Shop Museum, Leiston, Suffolk, Woodbridge Museum; Dr Paul Davies and Andrew Fakes of Great Yarmouth Local History & Archaeological Society; Reverend Barry K. Furness, Smallburgh Benefice; Geoffrey Dixon; Scouting archivist Claire Woodforde, Norfolk Family History Society; Kevin Asplin; Helen Tovey, *Family Tree Magazine*; Dr Stephen Cherry, Stewart P. Evans, Mike Covell, Amanda Hartmann Taylor and my loving family.

INDEX

Visit our website and discover thousands of other History Press books.

www.thehistorypress.co.uk

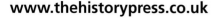

The History Press